SAVING PORT MORESBY

Fighting at the end of the Kokoda Track

Copyright © David W. Cameron

First published 2022

This book is copyright. Apart from any fair dealing for the purposes of private study, research, criticism or review as permitted under the Copyright Act, no part may be reproduced, stored in a retrieval system or transmitted in any form or by any means, electronic, mechanical, photocopying, recording or otherwise, without written permission.

All inquiries should be made to the publishers.

Big Sky Publishing Pty Ltd
PO Box 303, Newport, NSW 2106, Australia
Phone: 1300 364 611
Fax: (61 2) 9918 2396
Email: info@bigskypublishing.com.au
Web: www.bigskypublishing.com.au

Cover design and typesetting: Think Productions

Printed and bound in Australia by Griffin Press

 A catalogue record for this book is available from the National Library of Australia

SAVING PORT MORESBY
Fighting at the end of the Kokoda Track

www.bigskypublishing.com.au

DAVID W. CAMERON

OTHER BOOKS BY DAVIDS CAMERON

The Battle for Isurava: Fighting on the Kokoda Track in the Heart of the Owen Stanleys

The Battles for Kokoda Plateau: Three Weeks of Hell Defending the Gateway to the Owen Stanleys

Convict Hell: Macquarie Harbour

Convict era Port Arthur — 'Misery of the deepest dye'

Australians on the Western Front 1918, Vol 1: Resisting the Great German Offensive

Australians on the Western Front 1918, Vol 2: Spearheading the Great British Offensive

The Charge: The Australian Light Horse Victory at Beersheba

The Battle of Long Tan: Australia's Four Hours of Hell in Vietnam

Our Friend the Enemy: Anzac from both sides of the wire

The Battle for Lone Pine: Four Days of Hell in the Heart of Gallipoli,

Shadows of Anzac: An Intimate History of Gallipoli

Gallipoli – The Final Battles and the Evacuation of Anzac

"Sorry lads the order is to go" The August Offensive at Anzac Gallipoli, 1915

25 April 1915: The Day the Anzac Legend was Born

Hominid Adaptations and Extinctions

Bones, Stones & Molecules: 'Out of Africa' and Human Origins (with Colin Groves)

MAP LEGEND

Military Symbols

III	Brigade/Regiment
II	Battalion
I	Company
•••	Platoon
⊠	Infantry Unit
⊓	Headquarters
•	Artillery
⊓⊓	Combat Engineers
↕	Medium/Heavy Machine-gun
↕	Mortars
⊠ (in oval)	Defensive Position
I/144 ⊠ 3 JAP	Japanese 3rd Company, 1st Battalion, 144th Regiment
39 ⊠ B AUS	Australian B Company, 39th Battalion
39 ⊠ 20 AUS	Australian 20 Platoon, 39th Battalion
39 ⊠ B(-) AUS	Australian B Company, 39th Battalion (minus elements)
39 ⊠ B(+) AUS	Australian B Company, 39th Battalion (with additional supporting elements)

Natural Features

	Jungle
	Waterfall
	River
	Track
	Wetland
	Garden
	Thick Scrub

This book is dedicated to Emma, Anita,
Lloyd and little Naomi – love you

CONTENTS

Prologue 1

PART 1: THE LOSS OF MYOLA 9

1 '. . . they need to be blooded' 10

2 'We wanted the soldiers to be alive' 15

3 'You don't even see the Jap who gets you!' 24

4 '… the position WAS occupied, but by Japanese!' 39

5 '… introduced themselves as Chester Wilmot and Osmar White' 55

6 'I was bloody lucky that time' 63

7 '… the Jap always goes for the high ground' 71

PART 2: THE BATTLE FOR MISSION RIDGE 79

8 '… continuous stream of lights' 80

9 'They should be very proud of their performance' 89

10 'The aircraft thundered up the valley' 98

11 'At about midday another Jap was discovered' 110

12 '… he was struck in the centre of his throat by a bullet' 118

13 'An attempt is to be made to force our way through' ... 127

14 'earned … the bloody name of Butcher's Hill' 139

15	'… give the bastards a full mag'	158
16	'… hour and a half later the Japs were sending mortar-bombs among the houses'	170
17	'There's your Bren back, Slim'	177
18	'We were all in tatters, half-starved, muddy and bloodstained'	190
19	'Nauro was a repetition of Menari'	196
20	'Captured Lieut. Colonel Key and four others'	205

PART 3: THE BATTLES FOR IORIBAIWA RIDGE 217

21	'One enemy bomber bombed us this morning'	218
22	'Take as many Japs with us as we can'	228
23	'… prepared for an attack on Moresby'	234
24	'The men slither down a clay slope, and then, miracle of miracles …'	248
25	'… the "woodpecker", which slowly spits out leaden death'	262
26	'A party of about 70/80 Japs went past'	276
27	'I consider I have just arrived in time'	289
28	'Spirited Jap attack again concentrated on D Coy'	307
29	'Ken. You'll die there if necessary. Do you understand'	319
30	'Everyone know his job…'	328
31	'The Sea! Look! It's the sea of Port Moresby!'	339

32	'Returned empty-handed'	348
33	'Good morning, Honner, you've just arrived from Australia, have you?'	355
34	'… to see an old warrior of the spear age place down his weapons and shouldered a Bren gun'	364
35	'Our 25-pounders opened fire on Jap positions'	375
36	'I never told them about the weevils'	385
37	'General turned on a good brew'	394
38	'If we don't go forward, we will only starve to death'	403
39	'The order to retreat had crushed the spirit of the troops'	417
40	'Two days were spent rafting down the river amid beautiful scenery'	431
41	'Did I tell you I'm trying to collect 3 decorations of the highest?'	443
42	'The posn on Ioribaiwa Ridge now consolidated'	451
43	'Left a good deal of stuff here'	466
44	'… the way lay open for our troops to advance'	472

Epilogue 482
Acknowledgements 490
Bibliography 492
Endnotes 502
Index 530
About the Author 544

PROLOGUE

Japanese Major General Horii Tomitarô, commanding the *South Seas Force*, was originally tasked with conducting a reconnaissance in force to determine whether it was feasible to capture Port Moresby from the northern coast using the Kokoda Track. His initial force to conduct this operation was led by Lieutenant Colonel Tsukamoto Hatsuo, commanding the *1st Battalion, 144th Regiment* (*I/144th Regiment*).

Facing these Japanese were the militiamen of the 'B' Company, Australian 39th Battalion along with members of the 1st Papuan Infantry Battalion (1st PIB) and the Royal Papuan Constabulary (RPC) who were eventually pushed off Kokoda Plateau, three weeks after the Japanese landings at Gona and Buna on 21 July 1942. During the fighting for the plateau, the 39th Battalion lost its commanding officer (CO), Lieutenant Colonel Bill Owen; just days before, the father figure of the battalion, Captain Samuel 'Sam' Templeton, had been killed at Oivi. A week later, on 8 August, with the arrival of the rest of the 39th Battalion, and its new temporary CO, the then Major Allan Cameron, an attempt was made to recapture Kokoda and its airstrip, with the men of 'A' Company succeeding in occupying the plateau. After two days of fighting, these men were forced to fall back to Deniki where the rest of the battalion had fallen back after failing to push through to support them.[i]

Two days after retaking Kokoda Plateau, Tsukamoto's troops had climbed the northern slopes of the Owen Stanley Range, using the

---

[i] For a detailed account of the Japanese invasion, the fighting in the lowland fetid swamps just north of the Owen Stanleys, and the tragedy that befell those trapped behind the Japanese lines during the first month of the invasion – military and civilian – see Cameron, *The Battles for Kokoda Plateau: three weeks of hell defending the gateway to the Owen Stanleys*.

Kokoda Track, and after two days of fighting, captured Deniki on 14 August. Within 24 hours, the Australians were digging-in at Isurava about ten kilometres south of Deniki, as the crow flies. For two weeks, Tsukamoto and his men remained at Deniki, building up their supplies and awaiting the arrival of the rest of their regiment.

Meanwhile, Horii, who was commanding the main body of the *South Seas Force*, landed near Buna on 19 August – his mission had now changed from conducting a reconnaissance in force to conducting combat operations to cross the Owen Stanleys to capture Port Moresby, using the Kokoda Track. His force was built around the two fresh battalions of the *144th Regiment*, consisting of Major Horie Tadashi and his men of the *II/144th Regiment* and Lieutenant Colonel Kuwada Gen'ichirō and his men of the *III/144th Regiment* (a Japanese regiment is essentially the same as an Australian brigade, each with three battalions numbering approximately 500 men per battalion).

At once, Horie and Kuwada hurried forward to the front where they joined Lieutenant Colonel Tsukamoto and his command on 24 August, who had spearheaded the invasion weeks earlier. Soon, the Japanese *41st Regiment* and its three battalions commanded by Colonel Yazawa Kiyoshi would land in Papua and rush forward to support the advance. Unlike the Australians, both Japanese regiments were well supported with medium to heavy machine guns, mortars, and mountain artillery.

Horii's plan was for Tsukamoto and Kuwada's battalions to take Isurava, supported with combat engineers, dismounted cavalry and mortars and mountain guns. Tsukamoto's battalion would attack, advancing down Kokoda Track, while Kuwada's battalion would attempt a western-flanking manoeuvre along the high ground just west of Isurava. Further east, along Misima Ridge, Horie's battalion also conducted a flanking movement, but his objective was to capture Alola, placing him behind the Australians dug-in at Isurava.

General Horii finally launched his attack on 26 August. Just two days later (28 August), acknowledging the difficulties at Guadalcanal

PROLOGUE

and the disaster at Milne Bay, the Imperial Headquarters in Tokyo while ordering Horii to continue his advance, ordered him not to attack Port Moresby, rather he was to take up a defensive position just north and await orders for a renewal of the offensive. Horii would not receive these orders until mid-September, and he would select Ioribaiwa Ridge as his base for future operations against the township.

* * *

As planned, Tsukamoto and Kuwada attacked the Australians militiamen at Isurava, who were now led by their new CO, Lieutenant Colonel Ralph Honner, who had arrived at Isurava a week before. Meanwhile, Horie was developing his flanking manoeuvre to the east, pushing back the ill-equipped and poorly trained militiamen of the 53rd Battalion who had arrived just days earlier to defend Alola and the eastern and western tracks that converged at the village. To lose Alola would mean Honner and his men at Isurava would be completely cut off and surrounded.

Luckily for Honner and his men, the veterans of the Australian 21st Brigade, led by Brigadier Arnold Potts, were now advancing along the track to reinforce Honner. These men of the 2nd Australian Imperial Force (2nd AIF) had fought with great distinction in Syria, and had arrived back in Australia months before. The battle-hardened veterans of the 2/14th and 2/16th battalions were now being pushed up the Kokoda Track. The first men of the 2/14th Battalion arrived at Isurava on the afternoon of the day that Horii launched his attack. The men of the 2/16th Battalion arrived soon after at Alola, taking over its defence from the militiamen of the 53rd Battalion who were soon after ordered back to Port Moresby, as they were considered more of a hindrance, and Potts needed to relieve his supply situation.

Potts and Honner's men suffered heavily after a week of constant combat and were forced to conduct a fighting withdraw down the track – a major problem was the lack of supplies. The Australians were fighting well forward of their main supply base, while the Japanese at this point were still close to their supplies built-up at Kokoda. It was now that Potts finally got the welcomed news that his remaining unit, the 2/27th Battalion, was moving up the track to reinforce his depleted battalions. Many of his men from the 2/14th and 2/16th battalions, along with some from the 39th Battalion were missing from the fighting in and around Isurava and Alola, lost in the jungle and cloud-covered ridges and valleys of the Owen Stanleys, blindly pushing south to hopefully contact the brigade. Many of these men would never be seen again, including the commander of the 2/14th Battalion, Lieutenant Colonel Arthur Key.

With the arrival of the 2/27th Battalion, Honner and his men were finally relieved and ordered down the line to regroup and re-equip, having fought the Japanese alone for over a month, and in doing so, suffered around 50 per cent casualties. Traditionally, losses approaching ten per cent are considered a 'blood bath'.

At this point, Horii was determined to outflank, surround, and finally destroy the Australian force thought to be taking up a position around Templeton's Crossing and Myola. Horii was now reinforced with the arrival in the forward area of two battalions of the *41st Regiment*: Major Koiwai Mitsuo and his men of the *II/41st Regiment*; and Major Koboyahi Asao and his men of the *III/41st Regiment*. These men had landed at Buna on 21 August and immediately committed. Horii now had five battalions numbering around 2500 men, strongly reinforced with additional combat engineers, artillery, and service troops of various kinds. Potts' force was now essentially a composite battalion of around 700 battle-weary men from the 2/14th and 2/16th battalions, having suffered over 30 per cent casualties combined, and unlike the Japanese, they had no heavy machine guns or artillery and only a few mortars.

* * *

Potts' men momentarily took up a position just north of Templeton's Crossing. For the first time in a week, they were able to brew themselves

PROLOGUE

a mug of tea and get a hot meal. Potts realised he could not hold any position for long unless he was heavily reinforced and until the Japanese were burdened with an extended supply line, slowing their advance and sapping their strength.

On 1 September, he informed his commanding officer, Major General Arthur 'Tubby' Allen, a 48-year-old accountant from Hurstville in New South Wales, who was now commanding the Australian 7th Infantry Division based in Port Moresby, that he intended to establish a defensive position halfway between Myola and Efogi.

Next morning – 2 September – the 2/16th Battalion was positioned an hour's march north of Templeton's Crossing while the 2/14th Battalion were still further north trying to hold back the Japanese advance. The men of the 2/16th Battalion had been strengthened that morning with the return of some of their lost mates who had been cut off over the last few days. Captain Phillip Rhoden, second-in-command (2i/c) of the 2/14th Battalion and now commanding the unit in Key's absence, led the survivors of his battalion through the 2/16th Battalion's lines at about 10 am.

About an hour later, the Japanese were attacking the men of 'A' Company, 2/16th Battalion, while other Japanese troops moved around their western flank along the high ground. On being told his position was being flanked, the commanding officer of the 2/16th Battalion, Lieutenant Colonel Albert Caro, had no option but to fall back towards Templeton's Crossing sooner than planned.[ii]

To make matters worse, that night (2 September), a Japanese convoy disembarked a further 1000 fresh troops and supplies near Buna, which included Colonel Yazawa's *I/41st Regiment*. The *41st Regiment* was based on four companies per battalion, unlike the *144th Regiment*, which had just three companies per battalion. The men of the *I/41st Regiment* were rushed inland to help reinforce their comrades. Horii

---

[ii] For a detailed account of the fighting around Deniki, Isurava, Alola, and Eora Creek, see Cameron, *The Battle for Isurava: fighting on the Kokoda Track in the heart of the Owen Stanleys*.

would soon have a two brigade-group to conduct combat operations against the out-numbered Australians, which could barely muster a brigade, even with the soon to arrive 2/27th Battalion.

The Japanese were now approaching Templeton's Crossing. The race was on for Potts to establish a defensive position somewhere near Efogi, located in the very heart of the Owen Stanleys; luckily for him, the 2/27th Battalion was fast approaching that very position, which was soon to be known as 'Brigade Hill', or better known to those who would fight there as 'Butcher's Corner'.

PROLOGUE

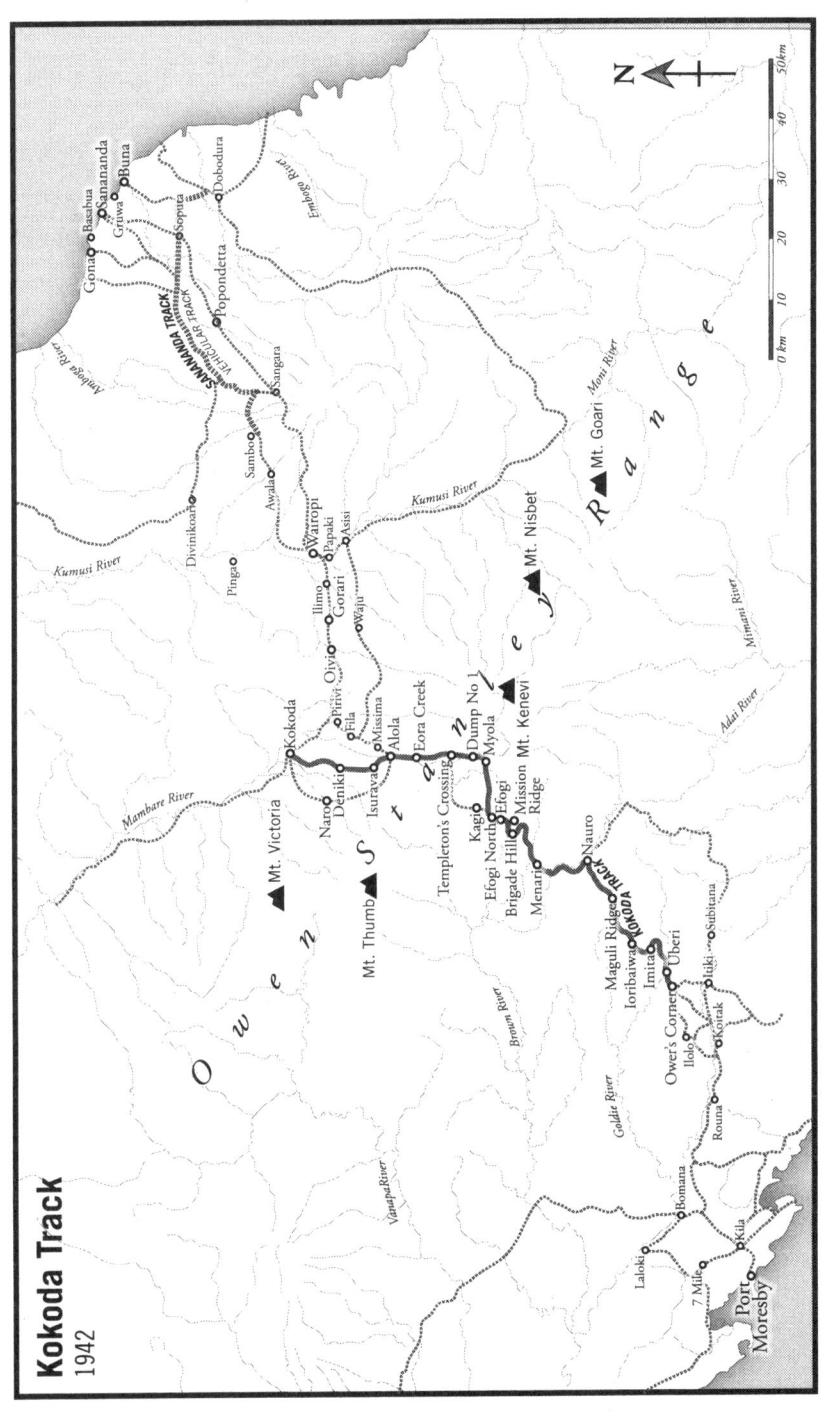

SAVING PORT MORESBY

PART 1
THE LOSS OF MYOLA

1
'... THEY NEED TO BE BLOODED'

In late August 1942, Australian Lieutenant General Thomas Blamey – a 58-year-old professional soldier from Victoria, who was commanding the Allied Land Forces, South West Pacific Area (SWPA) – and his superior, the 62-year-old American general, Douglas MacArthur, who was the Supreme Commander of Allied Forces in the SWPA, were finally coming to the realisation that something had to be done to support the Australians fighting along the Kokoda Track. For over a month, both men let the Japanese dictate offensive operations, barely even reacting to the Japanese. However, panic now had seemingly set in at their respective headquarters in Australia.[1] On 30 August, MacArthur radioed alarmingly to the Joint Chiefs of Staff in Washington:

> Unless the strategic situation is constantly reviewed in the light of current enemy potentialities in the Pacific and unless moves are made to meet the changing conditions, a disastrous outcome is bound to result shortly; it is no longer a question here of preparing a projected offensive; without additional naval forces, either British or American, and unless steps are taken to match the heavy air and ground forces the enemy is assembling, I predict the development shortly of a situation similar to those that have successfully overwhelmed our forces in the Pacific since the beginning of the war.[2]

Australian Lieutenant General Sydney Rowell, a 48-year-old professional soldier from South Yarra in Victoria – who commanded all Australian

CHAPTER 1

troops in New Guinea, known as the New Guinea Force (NGF) – was concerned about the condition of the 21st Brigade now trying to hold back the Japanese on the Kokoda track. Even so, he was far from convinced that this Japanese force alone could capture Port Moresby. By the time they fought their way to the outskirts of the township – if they could get that far – Rowell was sure the Japanese would be strangled by the long supply line back to the northern side of the Owen Stanleys. Indeed, the problem of supply for the Japanese increased the further south they advanced.

The opposite was true for Brigadier Arnold Potts, a 46-year-old farmer from Kojonup in Western Australia, and his men of the 21st Brigade, as they got closer to Port Moresby. Rowell later stated: 'At no time did I consider that the capture of Moresby by the enemy from the north was possible'.[3] That said, if this Japanese force was the vanguard of a much larger force that was to descend upon Port Moresby via the track, or a supporting Japanese invasion force along the south coast – all bets were off. It was also known that with each passing day, the Japanese were landing more troops at the Buna and Gona beachheads.

Blamey and MacArthur remained ignorant of the facts on the ground, with MacArthur making disparaging remarks about the failure of Potts' men of Maroubra Force[iii] to halt the Japanese advance; many of whom now lay dead or dying on the northern lowlands or in the heart of the Owen Stanleys, while the survivors continued to conduct a fighting withdraw back towards Port Moresby. MacArthur complained about the Australians while he stayed safely and comfortably ensconced in Australia, having made no serious attempt to support Lieutenant General Rowell and NGF.[4]

Even with the definitive Australian victory against the Japanese invasion at Milne Bay, located at the easternmost tip of Papua in late August, MacArthur and his headquarters staff remained extremely nervous and jittery. In a personal letter to Rowell on 1 September,

---

[iii] Maroubra Force was the name given to the men fighting the Japanese along the Kokoda Track.

SAVING PORT MORESBY

Major General George Vasey, a 47-year-old professional soldier from Victoria and Deputy Chief of the General Staff (DCGS) wrote: 'GHQ is like a b----y barometer in a cyclone – up and down every two minutes ... they need to be blooded'.[5]

Rowell judged he could not rely on the militia forces that formed the bulk of the strength available to him for the defence of Port Moresby and for operations in the mountains – even though the courage of the militiamen of the 39th Battalion, 30th Brigade, indicated otherwise. He was now trying to get additional 2nd AIF forces sent to him – none of his men of the militia 14th Brigade had been committed to combat, being based in Port Moresby since arriving almost six months before – most having been used as wharf labourers, unloading supplies.[6] On 29 August, Rowell was informed that the 2nd AIF battalions of the 25th Brigade were going to Milne Bay, and he urgently signalled Blamey:

> Allen has now had good chance to see 14 Infantry Brigade and is by no means impressed with their efficiency and general standard of training. This is no reflection on their courage, but units contain a large number of young men not yet properly developed or trained. His view with which I concur is that 25 Infantry Brigade is required here if this place is to be regarded as reasonably secure from major seaborne attack …. Your decision will naturally depend on the outcome of the operations now in progress at Milne Bay. But bearing in mind the difficulties of quick reinforcement I submit that it is advisable to spare no sacrifice to collect ships to bring these seasoned troops in now.[7]

Blamey initially rejected the request, but Rowell emphasised the need for the brigade to be sent to Port Moresby: 'After the experience of the 53rd Battalion I can have NO repeat NO confidence that any A.M.F. unit will stand'.[8] Rowell's assessment along with the

CHAPTER 1

Australian victory at Milne Bay bore fruit, and Rowell was informed on 3 September that the destination of the 25th Brigade had changed to Port Moresby. That said, the availability of troop transports remained a serious problem.[9]

Blamey had apparently awoken to the dangers to Port Moresby; alarmed about developments along the Kokoda Track and the likelihood of another Japanese invasion along its southern coast to support the attack against the township. He now informed Rowell that both the 2nd AIF 16th and 17th brigades, which had just arrived from Ceylon (present day Sri Lanka), were now available. Rowell replied that he would like the 16th Brigade sent forward in addition to maintaining the two-militia 14th and 30th brigades as well as 2nd AIF 25th Brigade. No sooner had he submitted his request, Blamey confused the issue by proposing that the 2nd AIF 16th Brigade would relieve one of the militia brigades.[10]

It was now that American Major General Richard Sutherland, MacArthur's Chief of Staff, arrived at Port Moresby. It was quickly decided that the US 128th Regiment, 32nd Division, would be sent to Port Moresby. This regiment, however, was directed by MacArthur to find a new track across Papua to flank the Japanese and strike them in the rear. MacArthur was obsessed with flanking manoeuvres against the enemy, even when terrain, vegetation and just as critically logistics dictated that such tactics were futile. In Papua, such flippant decision-making could all too easily see his own force strangled by logistical bottlenecks. It has been said that military amateurs focus on strategy and tactics, while professionals concentrate on logistics – nowhere was this truer than in the jungles and highlands of the Owen Stanleys of Papua New Guinea.[11]

While these high-level discussions were going on, Rowell had to halt the Japanese advance, but he assessed that the imminent danger was not that Port Moresby would be captured, but rather, Japanese infiltration parties would cause havoc with Allen's lines of supply and morale. Enemy parties could work their way down to the Moresby area, using the Brown, Goldie and Laloki rivers. This threat became increasingly likely as the main Japanese force got closer to the Ioribaiwa and Imita ridges just north of Port Moresby. The township, however, would be in danger if the Japanese landed a supporting force along the southern coast of Papua like the one just repulsed at Milne Bay.[12]

Major General Allen, head of the Australian 7th Division, had been unable to exclusively focus his attentions on operations in the Owen Stanleys, as he was also responsible for the defence of Port Moresby, not just from the Japanese advancing down the Kokoda Track, but also from a possible seaborne invasion. Lieutenant General Rowell's task as head of NGF had become even more critical, with Japanese operations against Milne Bay. Realising the urgency playing itself out along the track, Rowell was now able to free Allen from the responsibility of defending Port Moresby with the Australian victory at Milne Bay. Finally, Allen could give his undivided attention to the forward operations along the Kokoda Track.[13]

2
'WE WANTED THE SOLDIERS TO BE ALIVE'

In the heart of the cloud-covered mountains and valleys of the Owen Stanleys, Maroubra Force was preparing to fall back towards Myola. Captain Burt Kienzle – a 37-year-old planter and gold miner who had established his home at Yodda Valley in the mid-1930s, which was not far from Kokoda – had observed two large dry lakes when flying from Yodda to Port Moresby before the war. With war and the desperate supply situation, he had cut a track from what became known as Templeton's Crossing on the main track to the dry lakes. He named the position Myola, an Australian Indigenous word meaning 'dawn of day'.

These lakes now represented the main supply depot for Maroubra Force because the Allied 'biscuit bombers' could readily drop supplies in the clearly defined area, and they could quickly be recovered by waiting ground troops. Within days, Kienzle and his workforce had cut another track directly linking Myola to Efogi, bypassing the need to go through Kagi. This part of the Kokoda Track from Kagi to Templeton's Crossing became known as the 'pre-war track' and was now seldom used. It would, however, enable flanking manoeuvres by the enemy to bypass Myola.[1]

In December 1941, Bert Kienzle had been in Sydney with his wife, Meryl, and their two children, Carl and Katherine, on an overdue vacation. It was there that they heard of the Japanese bombing of Pearl Harbour. Bert left his family safely in Sydney and returned to Kokoda and Yodda. On returning, he was attached to the Australian military government just established in Papua, which was the Australian New Guinea Administrative Unit (ANGAU).

In June 1942, he became responsible for organising Papuan carrying parties for Maroubra Force along the Kokoda Track through the Owen Stanleys. These carrying parties were crucial in getting supplies to the frontline and getting the gravely wounded out. Even with the superhuman effort of the carriers, however, they alone could not hope to keep the troops supplied; air drops by biscuit bombers at Myola were a critical part of the Australian logistical hub on the track.[2]

American historian Eric Bergerud wrote: 'As the advance continued the Japanese learned some bitter lessons. Supply was a terrible problem for everyone concerned. The Australians had a [position] at Myola suitable for airdrops, but the Japanese lacked even that limited capability. Vehicles were useless. Australians later found the Kokoda Trail littered with the bicycles the Japanese had made famous in Malaya. Nor were there any 'Churchill supplies', enemy supplies captured as they fled. Unlike the supply depots the British used in Malaya, the Australians had very little to abandon. Furthermore, the Australians fell back but were never routed. This meant that Japanese supplies had to be carried forward'.[3]

* * *

Brigadier Arnold Potts knew his current position was hopeless for a concerted defence. During the early morning, Captain Clifford Thompson – a 27-year-old accountant from Caulfield in Victoria, who had been a member of the brigade staff – now rejoined the 2/14th Battalion[iv] as its adjutant. With the disappearance of the battalion CO, Lieutenant Colonel Arthur Key, a 36-year-old assistant sales manager from Armadale in Victoria, and many of its officers during the fighting around Isurava, Thompson was desperately needed.[4] Captain Phillip Rhoden, a 28-year-old solicitor from Essendon in

---

iv The 2/ indicates the 2/14th was a battalion of the 2nd AIF (serving in the Second World War) differing it from its parent unit the original 14th Battalion of the 1st AIF (serving in the First World War) – militia battalions such as the 3rd and 39th battalions were not part of the 2nd AIF and were not allocated this distinction.

CHAPTER 2

Victoria, was now in temporary command of the 2/14th Battalion. Thompson recorded in his dairy a first-hand account of Potts' strategy as of 2 September:

> Most of us thought that the Brigade would make a stand in the Templeton's Crossing area. This was soon found to be of no real value because the enemy could go up to Kagi [forcing a new path to the pre-war track] and so cut the whole Brigade off. I remember hearing Brigadier Potts say that he could easily order with confidence the Brigade to stand astride the track and hold till the last, but that would achieve nothing, as the track would still be open to Port Moresby, either by the enemy going round our perimeter or up through Kagi. Brigadier Potts, in his own confidant way, decided to go back in long bounds to Myola until he could gain contact with Moresby Headquarters and get a decision on future action. At no time did Brigadier Potts have any misgivings about what he would do. He could always turn round to fight the enemy, but he considered that his main role was the defence of the track (and, incidentally, Port Moresby) until such time as new troops could be employed to drive the enemy back.[5]

Thirty-year-old Lieutenant Colonel Geoffrey Cooper, from a prominent beer brewing family in Adelaide, and now commander of the 2/27th Battalion, years later succinctly recalled the situation that Potts faced: 'There are no good positions in the jungle. The situation was a bit like being a halfback on a football field when darkness comes down. You're trying to guard your goal, but you can't see where the other team is running with the ball. Well, you have to drop back towards the goal …. Arnold realised well before HQ that his job was to keep his force between the enemy and Port Moresby and to whittle them away as he went …. It was a classic Fabian tactic – when Fabius Maximus fought a war of gradual attrition against the superior forces of Hannibal and depleted them until they could be defeated'.[6]

SAVING PORT MORESBY

Radio communications between Maroubra Force and Port Moresby were also complicated at this point, as the large AWA (Australian Wireless Association) radio, which was operated by 21-year-old privates Clive Turnbull from Junee and 21-year-old Bernard 'Spud' Murphy from Wentworth Falls, both in New South Wales, was hit by Japanese fire. As members with the New Guinea Air Warning Wireless Company (NGAWWC), both men were instrumental in keeping communications open with Port Moresby and had done outstanding service since landing at the Kokoda airstrip on 24 July with Lieutenant Colonel Bill Owen, a 37-year-old bank clerk from Kew in Victoria, who was killed four days later during the first battle for Kokoda.[7]

The historian of the NGAWWC wrote: 'The Japanese opened fire on the wireless lean-to for the last time. The Teleradio was hit. The vibrator was ruined, and the set therefore became useless. Time was taken to remove the crystals and the transmitter was dropped down a 'thunder box' [latrine] before the spotters took off. Murphy and Turnbull returned to the Convalescent Camp at Koitaki. Communications were taken over by the 1st Australian Corps'.[8] Both men were awarded the Military Medal for their actions during the Kokoda Campaign. All communications with Port Moresby now depended on a single telephone cable strung from the front line back to Moresby.

* * *

Falling back with the exhausted Australians of the 21st Brigade along the track was Damien Parer, a 30-year-old combat cameraman from Melbourne. With him initially were war correspondents Osmar White, a 33-year-old journalist with Melbourne's *The Herald and Weekly Times* and Chester Wilmot, a 31-year-old news broadcaster with the Australian Broadcasting Commission (ABC), also from Melbourne. They had just days before arrived at the Australian frontlines to cover the fighting but were now falling back with the troops heading south. While Parer remained with the frontline troops to film the fighting

CHAPTER 2

withdrawal, which would later become the source material for his ground-breaking film *Kokoda Front Line*, White and Wilmot left for Port Moresby at the request of Potts, to help emphasise the need for reinforcements and supplies. They had also heard of the fighting that had broken out at Milne Bay and hoped to cover the fighting there.[9]

In mid-September, White concluded to his readership in Australia his experience as they trekked back towards Port Moresby passing through the wounded: 'I am ashamed of my aching legs, the weakness in my stomach, because the wounded are walking – more bravely by far, more cheerfully than I. Surely no war was ever fought under worse conditions then these: Surely no war has ever demanded more of a man in fortitude .... There can be no more fortitude than I saw on that track. Fortitude is not a thing to pity or be sad for. I wish, by some magic words, I could make all the mothers, the sweethearts, the wives – even the shirkers and place seekers and cowards at home – feel what I feel so passionately now. That what men suffer in their minds and bodies for a good cause is not suffering to be pitied or want for. It is suffering for which to be grateful and proud, with a calm heart'.[10]

* * *

During 2 September, the remaining unit from Potts' 21st Brigade, the 2/27th Battalion, was finally approaching Menari. Lieutenant Colonel Geoffrey Cooper, and his men had only been released to Potts just days before. Given the logistical difficulties in crossing the track, the advance was made in two stages, with the Battalion CO going forward with 'B' and 'A' companies along with select HQ personnel ('A' Company was one day's march behind 'B' Company). The second group was led by the 2i/c of the battalion, 33-year-old Major Denis White, from Kigwigil in New South Wales, with 'C' and 'D' companies and the remaining battalion HQ personnel; they were a few days march behind Cooper's force.[11]

The battalion war diary records for 2 September that 'B' Company was followed by the headquarters personnel, who moved out at out

8 am from Nauro, with most arriving at Menari by 4 pm, where a hot meal was provided and all men slept under shelter. At this stage, Cooper's force had lost three men (two unable to go any further while another suffered from dysentery). At Menari, Cooper met the divisional liaison officer and staff captain of the 30th Militia Brigade returning to Port Moresby.

On 1 September, White and his men had left the Port Moresby area, and the next day, they left Uberi. His advance party left at around 7.30 am, with the main body following an hour later. The bulk of his two companies arrived at Ioribaiwa by 4 pm, although some stragglers did not arrive until after 10 pm. All men were served a hot meal on arrival and remained under cover for the night.[12]

* * *

Regimental Medical Officer (RMO) Major Rupert Magarey, a 28-year-old physician from Adelaide, with the 2/6th Field Ambulance, was the senior medical officer in the forward area. Typically, an Australian Field Ambulance unit consisted of 12 medical officers (MO) and 225 other ranks (ORs), but they rarely operated at full strength during the war, and this was certainly the case for the track.[13] Magarey had already seen the worst of what the Owen Stanleys and the Kokoda Track had to offer: 'It was necessary to be quite ruthless .... Every man who could possibly walk had to, and over and over again men arriving at medical posts could be given only short rests and then had to be pushed on again. The fortitude and cheerfulness shown by the majority of these men was beyond praise and the feats of endurance performed by some of the wounded, particularly those with wounds of the lower limbs, were almost incredible'.[14]

The medical staff responsible for the treatment of the sick and wounded had begun evacuating Myola on 1 September after it became clear that no airstrip would be constructed to fly the wounded and sick back to Port Moresby. They were now establishing their aid post at Menari, south of Efogi. The three other medical officers charged

CHAPTER 2

with caring for these men had already been in the thick of the fighting, during the battles for Isurava, Alola, and Eora Creek.[15]

In charge at Menari was Major John Oldham, a 30-year-old physician from Sydney with the 2/6th Field Ambulance; however, at this point, he was moving south with the stretcher-cases from Myola. Assisting him at Menari were Captain Douglas Wallman, a 46-year-old physician from New Hindmarsh in South Australia, and Captain William McLaren, a 28-year-old medical practitioner from Chatswood in New South Wales – both were with the militia 14th Field Ambulance. Around 60 stretcher-cases and over 200 sick and walking wounded were received at their medical aid post on the first day alone. Within days, additional staff from the militia 14th Field Ambulance would arrive to assist, while another 24 members of the 2nd AIF 2/6th Field Ambulance, who had seen action in the Middle East, would also soon arrive.[16]

Among these new arrivals was Major Donald Brummitt, a 29-year-old physician from Dulwich in South Australia, who had just come forward with a second detachment of the 2/6th Field Ambulance to set up an Advanced Dressing Station at Efogi. Brummitt was to stage patients from there to the main aid post. A series of small medical posts had already been established at Templeton's Crossing, Myola, Kagi, Efogi, Menari, Nauro, Ioribaiwa, Uberi, and finally Owers' Corner.

The sick and wounded were now leapfrogging all the way back to the 2/9th Australian General Hospital at Port Moresby. Soon after establishing the surgery station at Menari, it was decided to move the medical post and its patients further south to Nauro, where all surgery could be better accomplished, closer to the base of supply.[17]

Crucial to the care of the sick and wounded were the Indigenous carriers from Papua and Rabaul. Papuan carrier Havala Laula, from Kagi, recalled years later helping to carry the stretcher-cases: 'When Australians were wounded, we took them to Owers' Corner .... We took off their bandages and rubbed their wounds with bush medicines. And then we wrapped leaves around their wounds. That made them feel better – they felt all right …. When a wounded soldier died, we'd bury him on his stretcher'.[18] His friend and fellow carrier, Lubini Helia from Alola, also recalled years later: 'We'd light their cigarettes and give them anything they wanted …. It was very, very hard [journey]. We wanted the soldiers to be alive; we didn't want them to be dead. So even though it's steep down and up we went as fast as we could'.[19]

Colonel Frank Norris, a 51-year-old physician from Melbourne, and chief medical officer of the Australian 7th Division, said of the stretcher-bearers on the Kokoda Track:

> With improvised stretchers – one or two blankets lashed with native string to two long poles spread by stout traverse bars – as many as eight or ten native bearers would carry day after day. To watch them descend steep slippery spurs into a mountain stream, along the bed and up the steep ascent, was an object lesson in stretcher bearing. They carry stretchers over seemingly impassable barriers, with the patient reasonably comfortable. The care which they show to the patient is magnificent. Every need which they can fulfil is tended. If night finds the stretcher still on the track, they will find a level spot and build a shelter over the patient. They will make him as comfortable as possible, fetch him water and feed him if food is available – regardless of their own needs. They sleep four each side of the stretcher and if the patient moves or requires any attention during the night, this is given instantly.[20]

* * *

Further north, the men of the 2/14th and 2/16th battalions were leapfrogging their way back towards Myola as ordered. From around noon to 5 pm, 2 September, the men of the 2/16th Battalion had fought a delaying action against the Japanese of the *II/41st Regiment* about one kilometre north of Templeton's Crossing. As darkness set in, they were able to use an eastern subsidiary path from the main

CHAPTER 2

Kokoda Track to skirt around Templeton's Crossing (then held by the men of the 2/14th Battalion) to rejoin the main track about halfway between the crossing and Myola; just behind was the steep grade up to the highest point of the track at close to 2100 metres above sea-level (about one-quarter the height of Mount Everest). By the time they reached the track junction during the late afternoon of 3 September, the men of the 2/14th Battalion had been forced to vacate Templeton's Crossing, taking up a holding position just south of where Eora Creek crosses the main track, also south of the track juncture. Losing Templeton's Crossing meant the Japanese could now easily flank and bypass Myola by using the old pre-war track. As the men of the 2/16th Battalion settled into this position, the men of the 2/14th Battalion set off for Myola at around 4 pm. The Japanese were not far behind.[21]

3

'YOU DON'T EVEN SEE THE JAP WHO GETS YOU!'

On 3 September, Brigadier Arnold Potts and his staff had by 12.30 pm arrived at Myola from Templeton's Crossing. At Myola, Potts received word from Major General Arthur Allen, which had a mixture of good and bad news: 'Expect to have Ken here for dinner approximately 6 September …. Supply situation demands vital necessity for holding Myola as dropping area. No other area suitable to our planes and we cannot allow enemy to use Myola. Therefore, we must establish firm base forward of Myola as soon as possible. Is it possible to revert to offensive action now?'[1]

Potts know that this news implied that Brigadier Kenneth Eather, a 41-year-old dental mechanic from Bankstown in New South Wales, and his men of the Australian 25th Brigade were due to arrive in Port Moresby. The question about offensive operations, however, merely indicated the completed ignorance by his superiors in Port Moresby of the situation that he and his men were facing. Brigade Major Hugh Challen, a 36-year-old manufacturer from Melbourne, exploded: 'If only they could see for themselves'.[2] Potts took heed of Challen's suggestion and requested that Allen or his General Staff Officer 1, Colonel Charles Spry, a 32-year-old regular soldier from Brisbane, come forward to see the situation for themselves, but before Allen could respond, Maroubra Force would be retiring south from Myola. Potts now requested that Challen move forward to assess the situation just south of the track junction where the men of the 2/14th Battalion were located.

CHAPTER 3

* * *

Lieutenant Colonel Albert Caro, a 37-year-old accountant from Nedlands in Western Australia, and his battle-weary men of the 2/16th Battalion were still struggling over the rough country skirting to the east of Templeton's Crossing using the eastern subsidiary track. The battalion the night before had rested along the narrow path before setting off again at around 5.30 am. Sergeant John Scott, a 28-year-old farmer from Katanning in Western Australia, with the intelligence section, 'C' Company, was scouting the way forward. Coming up behind the intelligence section, the men of the battalion were exhausted as they continued to push through the thick vegetation, forced to drink water oozing from the moss-covered trees. This subsidiary track led in a south-easterly direction down a spur. The track had probably been made by troops of their battalion who had been cut off during the fighting around Abuari days before. It was now that Lieutenant Stan Bisset, a 30-year-old clerk from Surrey Hills in Victoria, intelligence officer with the 2/14th Battalion, and some of his men now came across the rear elements of the 2/16th Battalion strung out along the narrow track.[3]

Just days before, Lieutenant Bisset had been with his brother 'Butch' as he took his final breath at the Isurava Rest House area after having been mortally wounded during the battle for Isurava. Within 24 hours of burying his brother, the young lieutenant and some of his men were swept into Eora Creek Valley as the Japanese broke through the Australian perimeter. With Bisset at the time was Warrant Officer (WO) Leslie Tipton, a 31-year-old truck driver from Melbourne, and several of their men. As the young lieutenant cautiously approached the track in front of his men, he was happily surprised not only to see Australians on the track, but a good friend of his from the 2/16th Battalion – 30-year-old Lieutenant Allan Haddy, from Laverton in Western Australia – who on seeing his mate replied: 'G'day Stan, they were wondering where you had got to'.[4] The young lieutenant recalled with great relief on finally reaching the Australian lines:

I could see a battle at Eora Creek, but we couldn't get there, because of sheer drops and precipice and everything else on the other side. And I had to make a detour around and used the stars one night to check my position … we had to get back to Alola [but] …. we found that Alola was occupied by the Japs, so we then had to go back to Eora Creek …. I had to make a detour a little bit further to the east and ultimately, we got onto a ridge …. We'd been about four days [out] and wet every night and we only had one emergency ration per man for the five days and it was raining pretty [heavily] every afternoon and we were wet and cold, and we decided it was raining so heavily that we'd try and make a bit of a lean-to. So, we cut a tree and made some bamboo supports and picked up pandanus leaves and put them over the top, then the eleven … or thirteen of us … lay virtually on top of one another on the ground to keep our body warmth. To try and keep it, get some sleep for the few hours and I know that will live in my memory because the smell from the wounds of the three-walking wounded was pretty high and pretty terrible, but it was something that did live in your memory ….

Bill Lynn, my sergeant, he had a number of spare dressings …. I delegated him to look after the wounded and he treated them very well. Once a day he [put] new dressings on their wounds … after the fifth day I was … getting worried and we came onto a track on this high ridge and I could see by the track that recently somebody had been over it and I couldn't immediately tell whether it was our troops or Japanese troops but it was heading in the right direction and suddenly I was about two or three hundred metres ahead of the rest of them and Les Tipton … said, 'Stan you're going too quick. They can't keep up'. I said, 'Les, I've gotta push them along because if we don't get in there quickly, well, the Japs will be in front of us'. So fortunately, in

CHAPTER 3

about another … half an hour, I came face to face with the rear guard of the 2/16th Battalion and it was Alan Haddy … he was acting as the rearguard there and we faced each other with rifles … I said, 'Thank God for you, Alan!' I said, 'Think you could drop off a few tins of bully beef or something. My blokes are pretty starving'.[5]

By 3.30 pm, Scott and his small party had reached the main Kokoda Track. Here they were instructed by the Brigade Major Challen that when the battalion arrived, it was to take over the defence of the position that was now occupied by the men of the 2/14th Battalion just south of the track juncture.[6] It was not until around 4 pm that the bulk of the men of the 2/16th Battalion emerged from the jungle onto the main Kokoda Track. On arriving, these men took over the position from Captain Phillip Rhoden and his men of the 2/14th Battalion.

It was now that another party of lost men from the 2/14th Battalion came in. Lieutenant George Pearce, a 33-year-old orchardist from Shepparton in Victoria, came in with his section from 16 Platoon. Pearce and these men had last been seen trying to take out a Japanese machinegun position during the fighting for Isurava. Most believed that they had been massacred as they had ventured out beyond the Australian frontlines, and no word had been heard from them since. During the battle for Isurava, Sergeant William Irwin, a 29-year-old labourer from Seddon in Victoria, was with him. Pearce had sent Irwin back to try to contact their CO at Isurava, but that was the last they had seen of him. At this point, Irwin and a few others who had joined him were still lost in the jungle heading south.[7]

Lieutenant Colonel Caro placed 'A' Company to the left of the track and 'B' Company on the right with 'C' and 'D' companies to the rear in support, collectively all forming a tight perimeter defence around the Battalion Headquarters. With this, Rhoden and his men, which now included Bisset and his men, moved south down the track, climbing the steep slope to the high ground to the rear that then descended to Myola.[8] Soon after, Lieutenant Colonel Caro and Brigade Major Challen got word that they were both to proceed to Myola, as a brigade conference would commence on their arrival.

Once there, Caro was also to reconnoitre for defensive positions in the area. Caro took with him Major John 'Ben' Hearman, a 32-year-old farmer from Donnybrook in Western Australia. Caro placed the capable Captain Frank Sublet – a 22-year-old public servant from Perth, commanding 'B' Company, 2/16th Battalion, who had commanded the fighting east of Alola just days before – in command of the battalion in his absence. He was ordered to hold the position until 6 am, 5 September.[9]

About 30 minutes after Caro, Hearman and Challen headed down the track, advanced elements of Major Koiwai's force encountered several Australians who were washing their dixies in the creek just north of the track junction. The official Japanese history of the campaign records: 'The pursuit party went up the mountain road that ran along the river in the valley. As they approached the crossing point for one of the small creeks during the evening, they came across five or six Australians washing their mess kits while whistling a tune. They were only 50–60 metres apart …. After a rapid machinegun burst from the pursuit party all the Australians fled into the jungle close to the nearest bank of the stream. Return fire then came from that area. This area was the second-highest point in the range, where a pass ran under the highest peak'.[10]

* * *

In darkness, the men of the 2/14th Battalion were now pushing their way towards Myola, Captain Clifford Thompson recorded in his diary: 'The track was particularly muddy. On this occasion, in getting down some hillsides, I dislocated my shoulder twice within a few hours. [Medical Officer] Don Duffy put it back, Roy Watson holding me while Don Duffy demonstrated his medical skill. Troops held onto each other in single file. Someone in front would have a torch, but

CHAPTER 3

the remainder stumbled, swore and scrambled on'.[11] Indeed, at one point, 23-year-old Private Geoffrey Scammell[v], from Boyup Brook in Western Australia, who was absent-minded at times, let go of the bayonet scabbard of the man in front and wondered off into the scrub, taking with him all of those behind him, including Captain John Thurgood, a 28-year-old salesman from Malvern in Victoria and the rest of 'D' Company. The mistake was quickly realised, much to the chagrin of Scammell, but it took some yelling in the darkness to put things right. Soon a halt was called for, but they would renew their march in the early morning hours.[12]

* * *

Meanwhile, Japanese Kokoda and Isurava veteran 2nd Lieutenant Noda Hidetaka with *No. 8 Company, III/144th Regiment* recorded in his diary of 3 September: 'The place which we thought yesterday was IORA [*sic* Eora] was found to be wrong. We rested at IORA today. We received rations for two days'.[13] Noda and his men had days before been part of the western flank attack against Isurava, using Naro Ridge.

* * *

Earlier that morning, Lieutenant Colonel Geoffrey Cooper with 'B' Company, 2/27th Battalion came up to reinforce Potts. During the early morning hours, they had left Menari for Efogi, while a few days march behind was Major White and the men of 'C' and 'D' companies who left Ioribaiwa for Nauro. The diarist of the battalion recorded that Cooper and 'B' Company moved out from Menari at 7 am for Efogi, arriving there at about 4 pm. As they pushed north, they encountered many wounded from 2/14th and 2/16th battalions, while around 250 relatively fit men of 53rd Battalion were also on the track heading south. It was reported, 'morale of wounded, but more particularly in

---

[v] The battalion history records this individual as being David Scammell, but the author could only find one individual with this surname serving in the 2/14th Battalion: Geoffrey Scammell.

case of 53 Bn men, was very low'.[14] Pushing north from Nauro was 'A' Company, who did not stop at Menari but pushed further north to join their CO and 'B' Company at Efogi, with most of these men arriving on dusk. Rain fell heavily from 4 pm, and due to the large number of stretcher-cases being brought in from Myola, a hot meal could not be provided for Cooper and his men. The battalion war diarist recorded: 'Hot tea, however, was served to all men. Shelter for the night 3/4 Sep was taken under huts. Receiving information from 21 Bde that Bde at MYOLA. 2/27 Bn Force to move to KAGI and relieve 39 Bn on 4 Sep 42'.[15]

Corporal John Burns, a 23-year-old baker from Waikerie in South Australia, would later write the history of the 2/27th Battalion; he recorded the journey through the mountains: 'The weather in the Owen Stanley Ranges at this time followed a set pattern; each morning clear and hot, clouding over at midday with heavy, incessant rain from about 1400 hours until nightfall. By night, the troops were hot and perspiring through their physical efforts and drenched by rain. Each night they sought rest and sleep by lying down on the wet ground covered by a half blanket or groundsheet. Those who slept did so as if in a Turkish bath for the first part, but inevitably woke in the early hours of the morning shivering as if in midwinter at Woodside, and strangely the troops did not catch cold'.[16]

Sergeant Clement Makings, a 32-year-old market gardener from Adelaide with the 2/6th Field Ambulance, was at Major Brummitt's medical aid post at Efogi. Makings recalled the state of the wounded and sick, and the arrival of the leading elements of the 2/27th Battalion:

> We were up early and after feeding [the patients] shocking food (plain hard rice the only village where our cooks didn't cook), dressed patients using same wet clothes which we took off them. Had all patients away before 9.am.
>
> Was a wonderful sight to see all the native carriers and stretcher-bearers lined up for their rations, was said there were 990 of

CHAPTER 3

them there. It was amazing to see them carrying the wounded over the track. The track either side of Efogi is very steep and dangerous and slippery, a tree crossing the river and used as a bridge, was quite a feat for Blondin [famous tightrope walker] to cross. To see the track, one would say it was impossible to carry a stretcher over it, yet these natives did it. No praise is too rich for them.

Later in the afternoon the 2/27th Battalion began to arrive and kept coming until dark, [it] was quite a mess up because patients filled the huts, and no provision was made anywhere along the track for troops and others they had to sleep out in the rain. None of the troops carried blankets, only a groundsheet, some just a gas cape.

I cannot imagine wounded ever having to bear what the wounded in the Ranges had to. Some of the stretcher-cases were 6–8 and 10 days on their stretchers, anyone who could possibly walk, arm wound, many with bullets through their legs had to walk. Often a week or more before they reached the road [end of the track].

I have spoken to many soldiers some who have been in last war and most of this, through Greece, the desert etc. and they all say the New Guinea show was the worst.

The infantry had shocking conditions. Firstly, the physical strain, the shocking food, quite often they went for days on bully beef alone, not even biscuits, no tea, no fires (when in contact with the Japs, often sleeping standing up against a tree with just a gas cape for covering), wet almost the whole time, especially feet. Many didn't take their boots off for weeks and when we took them off, we'd often have to cut their socks off.[17]

A few days behind Cooper's force, White and his men had left Ioribaiwa at 8 am, with most arriving at Nauro by 5 pm. They had encountered the lead elements of the 53rd Battalion retiring to Port

Moresby. At this point, the battalion war diary also records: 'stragglers had rejoined main body. Hot evening meal was served and blanket per man provided, and all slept under cover. Rain from 1500 hrs to 1700 hrs. Large numbers of 53 Bn personnel passed through on journey to MORESBY'.[18]

Meanwhile, the 180 or so militiamen of the 39th Battalion were at Kagi and were targeted by Allied aircraft; the battalion war diary records: 'Four Allied planes dive-bombed KAGI. No casualties'.[19] Papuan carrier Havala Laula, from Kagi, recalled years later: 'It was a very sad time for us, when the war came to Kagi. They [Japanese] destroyed the villagers, ruined the gardens and killed all our livestock. But the Australian troops treated us well, they gave us food and supplies'.[20]

With the battalion here was 21-year-old Private Jack 'Laurie' Howson, from Clayton in Victoria, who like most of the men there had been involved in the thick of fighting for over a month.[21] While the diary records no casualties, Private Howson was actually wounded by one of these bombs: 'At Kagi some Yankee Airacobras were sent to strafe and divebomb the Japs to slow them down a bit. But instead of coming down to a couple of thousand feet they let go their bombs at about twelve thousand, so that half of them fell close to us and I copped the blast from one of them. We stayed at Kagi for a couple of days, and I started to get a piercing shrieking in the ears. I realised then that the bomb-blast must have done some damage to my headpiece'.[22]

Sergeant Major George Cops, a 22-year-old grocer from Beeac in Victoria, with the 'E' Company, 39th Battalion, was out on a patrol and recorded in his diary sometime that night: 'Move out early on a standing patrol, it's a good position so should be able to delay the enemy for some time. We spent a darned good night last night under a roof, what a luxury! Nothing happened during the day for a change. Hear that our forward troops are still holding on but have been forced to withdraw again – the situation is still grim'.[23]

CHAPTER 3

It was also now that signalman 20-year-old Kenneth Phelan, from East Malvern in Victoria, with the Battalion Headquarters Company, volunteered – he recalled he should have known better: 'When we stopped at this place at Kagi. Volunteers were called from the 39th Battalion to man this village in case the Japanese should come through, and I volunteered to go "never volunteer" and I became, then, a rifleman, not a signaller anymore … and I had to do my shift at guard, and naturally we had a guard on each part of the track that came into this village 24 hours a day because we were expecting the Japs any moment. There is no one in front of us …. "Who is in front of you, mate" "No one", and the sergeant, who is still living, and I say to him every time we have a reunion, "Remember that time you put me on that guard post at Kagi?" I was the furthest down the track. Every time I went on duty he put me furthest down the track, and he said to me each time when he left me, he said, "When you get killed, I know they are coming"'.[24]

* * *

At this point, while the bulk of the men of the troubled 53rd Militia Battalion were on their way back to Port Moresby, Captain James Bryce, a 22-year-old university student from Melbourne, with his men of 'C' Company, had been ordered to remain at Myola to assist in guarding the supplies.[25] Waiting for the bulk of the men as they reached Efogi was their new commanding officer, Lieutenant Colonel Allan Cameron, a 33-year-old lawyer from Elwood in Victoria. A month earlier, as a major, Cameron had taken over temporary command of the 39th Battalion with the death of Lieutenant Colonel William Owen during the first battle for Kokoda Plateau. On taking command of the battalion, Cameron organised the controversial attack that resulted in the second Battle for Kokoda (8 to 10 August). After being relieved by Lieutenant Colonel Ralph Honner, a 38-year-old lawyer from Nedlands in Western Australia, Cameron was soon back at Port Moresby and promoted to taking command of the 53rd

Battalion with the death of its CO, Lieutenant Colonel Kenneth Ward, a 39-year-old public servant from Chester Hill in New South Wales who was killed near Abuari waterfall the week before during the fighting around Isurava.[26]

Cameron remained to command Bryce and his men then at Myola as well as another two parties under Major Lee Fargher, a 43-year-old insurance inspector from Goulburn in New South Wales, and Captain Joseph Gilmore, a 29-year-old draftsman from Cairns in Queensland. The battalion historian recorded the situation: 'Under Order CO Lt Col Cameron. Maj. [Gerry] Spring moves back with Maj Hawkins and balance of troops .... Orders received from Maj. Fargher, 21 Bde to hold 40 fit men at IFOGI [sic] to collect stores dropped by plane .... Troops are in bad condition. Feet, due to being in wet boots, are rubbing and in poor condition. Socks had to be cut off before they could be removed .... Adjt kept Capt. Gilmore ... at IFOGI to take charge of 40 ORs kept at this station ... 213 all ranks moved on from IFOGI [back] to NAURO'.[27]

* * *

Just days before, Captain 'Doc' Geoffrey Vernon, who was fast becoming a living legend of the Kokoda Campaign, was again on the move. The week before, he had left Myola for Port Moresby to organise the transport of medical supplies to the aid post he had established at the dry lakes. He was now again heading north towards the scene of the fighting.[28]

Doc Vernon was a Great War veteran, who had served with the Australian Light Horse during the Gallipoli and Middle East campaigns. After returning to Australia, he moved to New Guinea to work as a physician and planter. When the Japanese entered the war, the 60-year-old lowered his age by ten years and joined ANGAU to look after the health of the Papuan carriers along the Kokoda Track. He had arrived at Deniki just before the first battle for Kokoda in late July, when the commander of the 39th Militia Battalion, Lieutenant Colonel Owen,

CHAPTER 3

was killed during the fighting on the plateau. Vernon had been ordered to stay at Deniki, as the wounded would be brought to him there; however, he ignored Owen and went down to Kokoda Station.

While the Japanese attacked during the early morning hours of the next day, Vernon, along with 2nd AIF veteran of the Middle East, Corporal Jack Wilkinson – a 35-year-old mining operator at Misima in Papua, now with the 39th Battalion – treated the wounded, including the mortally wounded battalion CO. The old Gallipoli veteran was among the last to leave Kokoda as it fell to the Japanese, concerned to make sure that all had got out. Since then, he had been going up and down the track, establishing aid posts, and looking after the health of the Australian troops and Papuan carriers, including the sick and wounded.[29]

Doc Vernon was now approaching the southern slopes of Ioribaiwa on his way back to Myola, having spent the previous night at Uberi. He recorded in his official report to ANGAU for 3 September: 'Walked on to Ioribaiwa. I had some reinforcements and relief orderlies for the line with me'.[30]

* * *

Since the fighting began around the Isurava Rest House, Captain Sydney 'Ben' Buckler, a 23-year-old professional soldier from Coffs Harbour in New South Wales, commanding 'A' Company, 2/14th Battalion was leading a lost party of men south through Eora Creek. This was their fourth day struggling through the thick jungle trying to reach the Australian lines. With him was Lieutenant Maurice 'Mokka' Treacy, a 27-year-old shop assistant from Mildura in Victoria, who was 2i/c of 'B' Company. He and his men of 11 Platoon had been in the thick of the fighting, helping to hold the northern perimeter at Isurava. Treacy had stayed behind to attend to the wounded and construct stretchers; it was noted by his men that the young officer continued to bandage his wounded men even as he received 'attentions of an enemy sniper'.[31]

Lieutenant Treacy recorded in his official report that by 5 am, he had placed covering parties in position; only then did the difficult task of getting the wounded across the swiftly flowing stream begin. At this point of the crossing was a log bridge, and it was impossible to cross elsewhere with stretchers. This crossing, however, was within sight of Alola, and the possibility of enemy parties moving along the track 'was not remote'. To the relief of all, the last stretcher was safely across before 6 am, and the men climbed up the track that led to Abuari village. He concluded in his report for 3 September:

> At 1430 hrs after a particularly stiff climb, the party reached the bivouac area chosen for the night. This area was directly opposite ALOLA due EAST. From here the enemy at ALOLA was kept under observation and the following points observed.
>
> (1) The enemy was using animal transport.
>
> (2) Native gardens where being ransacked for food, all growth being pulled up and the gardens evidently destroyed – natives were working with the JAP.
>
> (3) A party of 93 enemy infantry troops seen to move SOUTH along the track from the ISURAVA area.
>
> (4) Some huts in ALOLA village had collapsed; this may have been done purposely for camouflage or could have been caused by the [Allied aerial] bombing the previous day.[32]

* * *

Meanwhile, west of Captain Buckler was Lieutenant Robert McIlroy, a 31-year-old clerk from Geelong in Victoria, 'C' Company, 2/14th Battalion, who had two days before been sent out to lead a 20-man patrol to cover the pre-war track that led to Kagi. They had been forward of Templeton's Crossing and were tasked with cutting their way through the jungle to the pre-war track. However, their local guide disappeared within hours, and they had become hopelessly lost

CHAPTER 3

in the jungle. The night before, he and his men had been forced to sleep on any available ledges they could find on the ridges west of the main Kokoda Track. He recorded in his diary for 3 September that he and his men got up early in their rain-soaked uniforms, having spent a miserable sleepless night, not knowing they were also now west of the pre-war track. The young lieutenant recorded in his report: 'Continued our climb up the slope and during the morning crossed the ridge and started on the way down a [creek] re-entrant on the other side. It later proved that this ridge was the top of the main range. After following the general direction of this re-entrant all day, we slept alongside the creek'.[33]

* * *

On that day, George Johnston, an Australian correspondent based in Port Moresby, reported the return of the wounded, which were almost certainly the Victorians of the 39th Battalion:

> Our wounded are beginning to come back from the Owen Stanleys – thin, bearded, gaunt men with hollow cheeks and the marks of strain and pain around their eyes. Their uniforms are ripped and covered with dried mud. Their slouch hats are pulpy and shapeless. Stretcher-cases are jolting back on crude log litters on the shoulders of brawny Papuan natives. The others have walked all the way along the terrible track that links Moresby with Kokoda. They don't try to conceal what they've been through.
>
> 'Up there we've been fighting on the worst battlefield in the world,' one man said. 'Before I was hit, I spent 46 hours without shutting an eye. There are mists creeping over the trees all day, and sometimes you can't see your hand in front of your face under the cover of the jungle. Most of our chaps haven't seen a Jap! You don't even see the Jap who gets you! It's like fighting the invisible man. Those Japs are tough, hard fighters and their

camouflage is perfect. They can move through scrub or tall grass without making a sound and without showing a sign except – if your eyes are good – an occasional stirring in the vegetation ....'

Another Victorian (he had celebrated his 19th birthday in the Owen Stanleys) had been out with our scooting patrols. 'The Jap patrols travel much lighter than we do, and they don't stick to the tracks,' he explained. 'It's nerve-raking work. You mightn't see a thing four hours and then suddenly two patrols meet on a narrow twisting track through the jungle trees. Most of the fighting is done at a range of a few yards [metres], and the man who is quickest on the drop is the man who comes home. We saw very few live Japs, but plenty of dead ones. The Japs certainly know all there is about jungle fighting, but, oh boy, wait till we get 'em in the open!'[34]

4
'... THE POSITION WAS OCCUPIED, BUT BY JAPANESE!'

During the early morning of 4 September, Brigadier Arnold Potts had sent Brigade Major Hugh Challen, 37-year-old Captain Bert Kienzle and a small party south to conduct an extensive reconnaissance towards Efogi and a potential fallback position south of the village. It was the only area that could be defended if the Japanese should flank Myola. This position was defined by a broad spur soon to become known as 'Mission Ridge', due to a Seventh Day Adventists mission hut erected there. Within hours, Kienzle and his Papuans of ANGAU were blazing a new track to bypass Efogi, using a ridge that looped south of Myola that connected with the main track at Menari.[1]

On returning to the brigade sometime before noon, it was confirmed the proposed fallback position just forward of Myola was not suitable for defence. The brigade's *General Narrative of Events* records that this position proved impossible for defence, as the high ground formed a basin and command of these features would be essential; however, with the number of troops available, this would not be possible. Even so, it was decided that an attempt would be made for a time to hold Myola with 2/14th Battalion in close defence, while the 2/16th Battalion would move forward to conduct strong patrols along the tracks to Eora Creek crossing, while other parties would patrol the pre-war track from Kagi to Templeton's Crossing, as well as the high ground to the west. The Brigade Headquarters would remain at Myola. Meanwhile, the track being cleared by Kienzle from

Myola along the high eastern feature to Menari bypassing Efogi would reduce travelling time between the village and Myola by five hours.[2]

* * *

By dawn, Captain Phillip Rhoden and most of the 2/14th Battalion had reached Myola and were resting. They had a hot meal, washed and were given clean clothes to replace their stinking, tattered uniforms and took advantage of the broad open area to expose their puffed waterlogged feet to the sun; some could not take off their socks, they had to be peeled or cut away. Corporal George Clark, a 27-year-old window-dresser from Hampton in Victoria, was now the unit chiropodist who pared off rotten tissue and socks. Among these men was Sergeant Robert Thompson, a 25-year-old farmer from Moorooduc in Victoria, with the headquarters' Signal Platoon. He had been in the fighting at Isurava, and he recalled reaching Myola: 'We moved back again. I think we got to Myola and for the first time in ten days we got supplies and I was able to take my boots and socks off and about three layers of skin [came off] as I pulled my socks off. And to have a proper shower or to tip a bucket over me and get into clean clothes, and there they decided that we should have greens, but we didn't have any greens. They had green dye and we all dyed our trousers because these things stand out and we managed to get a good meal'.[3]

The battalion diarist also recorded: 'By 0700 hrs D and C coys had arrived at MYOLA and had a hot meal as did the remainder of the unit. A clothing and equipment … reissue was made to all troops. The camp resembled more a market than anything else, with troops enjoying the full advantage of a stream in which they had their first wash since Aug 25th., clean dry clothes, and feet wet and pulpy from constant damp conditions over a period of very many days were treated by the unit chiropodist. CO, Adjt and IO made recce to large dry lake area, and other offrs and NCOs made recce in the immediate vicinity of MYOLA, all with a view to the close defence of MYOLA'.[4]

CHAPTER 4

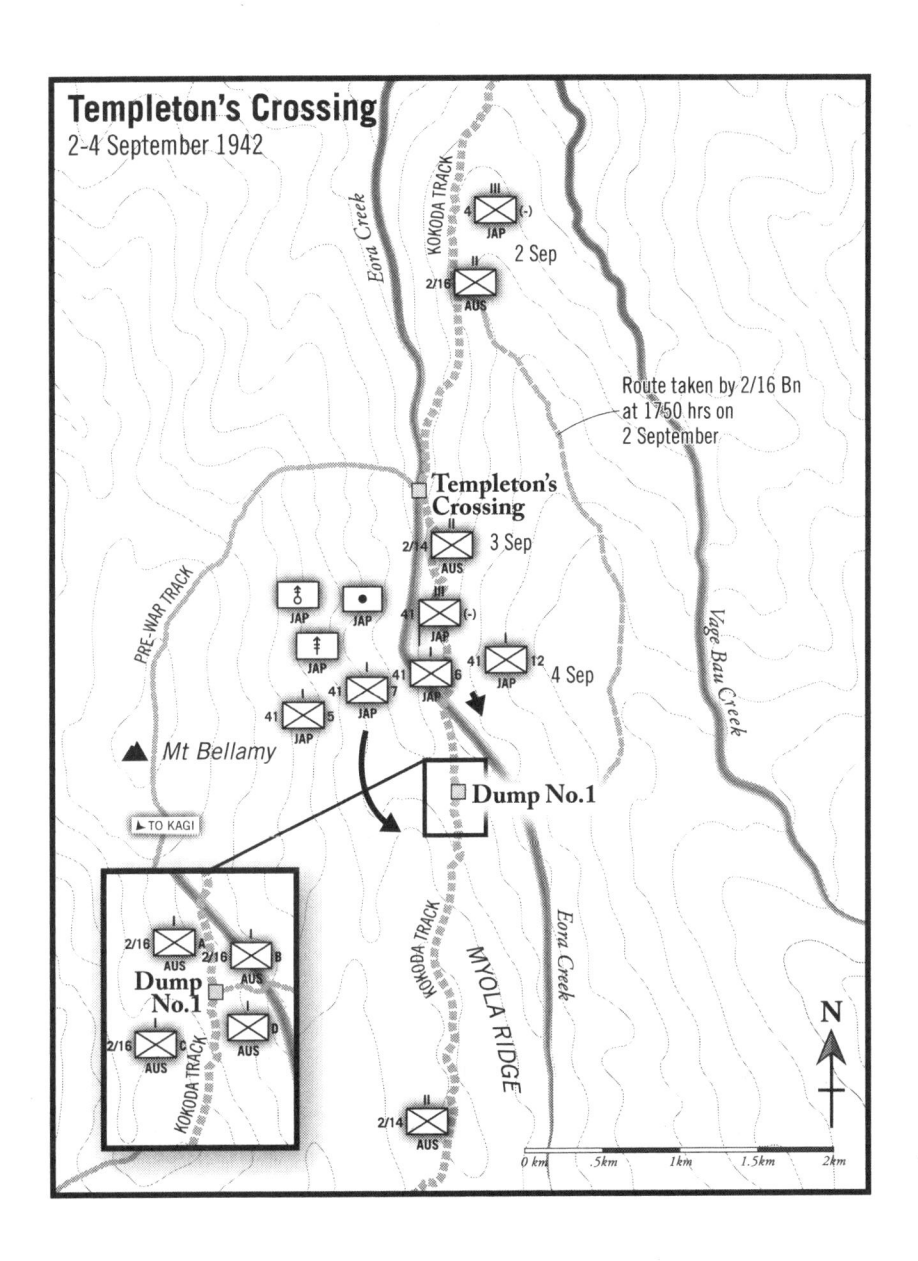

Further south, Lieutenant Colonel Geoffrey Cooper and his men of the 2/27th Battalion were advancing along the track heading for Myola. The war diary of the 2/27th Battalion records: 'A' and 'B' companies along with Cooper and his headquarters' personnel moved out of Efogi by at 7 am., for Kagi. About two hours later, the lead elements had pushed just south of the village where they found 26-year-old Lieutenant John Bennie from Wayville in South Australia, 2/27th Battalion liaison officer with 21st Brigade Headquarters who was then supervising store dumps about one kilometre at the rear of the Kagi. By 2 pm the bulk of Cooper's force had arrived at the village after descending into the deep and steep riverine gorge which separated Efogi from Kagi, with just a few stragglers coming in at dusk. Here Cooper found Lieutenant Colonel Ralph Honner and his militiamen of the 39th Battalion. Honner was a veteran of the 2nd AIF having been a captain with the 2/11th Battalion, being awarded the Military Cross (MC) for his actions against elite German paratroopers during the Greek Campaign. On Crete, he was part of the rearguard that helped the bulk of the Anzac and British force to escape off the island. He and several others evaded the Germans for months before commandeering a small boat and sailing to Egypt. On returning to Australia, he was promoted to lieutenant colonel and took command of the gallant militiamen of the 39th Battalion, arriving to take command a week before the Battle of Isurava.[5]

* * *

The Japanese of the *144th Regiment* had been involved in the fighting since the day of the invasion and had fought the Australians at Kokoda, Deniki and Isurava. For the last few days, they had finally been placed in reserve, as the men of Colonel Yazawa Kiyoshi, commanding the *41st Regiment*, led the advance. That day, 2nd Lieutenant Noda Hidetaka with *No. 8 Company, III/144th Regiment* recorded in his diary: 'At 1030 hrs we handed over the colour guard to No. 7 Company and rejoined the 3rd Battalion'.[6] That morning,

CHAPTER 4

however, Horii ordered the advance by the men of the *41st Regiment* to come to a halt, the veterans of the *II/144th* and *III/144th Regiment* would again lead the advance. Horii now issued Nankai Shitai Operational Order A-112:

> Post of Mount LEHUNTE
> 4 Sep, 0900 hrs.

The SHITAI's advance has encountered renewed resistance from the enemy, but on the whole is progressing favourably. Since last evening the pursuit TAI met with enemy resistance near IORA [*sic* Eora], but it is planned to defeat this morning. 1 Bn of the YAZAWA Regt and the L of C Hospital landed at GIRUWA, on 1 Sep with the 5th wave, and the L of C Hospital [No. 67 Eastern New Guinea L of C Hospital] is expected to be set up in the vicinity of KOKODA.

The SHITAI will alter the disposition of the pursuit TAI and will then continue the pursuit towards the western end of the range.

The YAZAWA pursuit TAI will defeat the enemy in the sector and prepare to be relieved.

Col KUSUNOSE commanding 144th (-1 Bn) becomes the new pursuit TAI. They will leapfrog and relieve the YAZAWA, Pursuit TAI, and carry on the pursuit towards the heights on the west [south[vi]] of WAMI [Nauro].

The other TAI under command of Col YAZAWA, will follow the Pursuit TAI ... 1st Bn of YAZAWA BUTAI will immediately upon landing, follow the main strength.

The KOBAYASHI BUTAI of the Ind Engr Regt will continue

---

vi   While it is recognised that much of the Kokoda Track lies off a south-west bearing, for simplicity's sake, it is simplified here as a north-south bearing i.e., north coast versus south coast.

in its present assignment.

Owing to the occurrence of casualties, a Fd Hosp will be established at IORA or in an area west of IORA and will take care of wounded.

HORII Tomitarô
SHITAI Comdr

Method of Issuing:

To be communicated verbally to the comdrs of 41st and 144th and all Pursuit TAI. Thereafter, order receivers will be assembled and will receive this order verbally and take it down in writing.[7]

Horii was clearly dissatisfied with the performance of the *41st Regiment* over the last few days, as he considered their advance too slow, covering just ten kilometres in five days, averaging two-kilometre per day, and its senior officers had missed opportunities to inflict serious casualties against the Australians. When Colonel Kusunose Masao and the *144th Regiment* resumed the advance, they covered 40-kilometres in eight days, averaging five-kilometres per day. Indeed, the regimental commander, Colonel Kusunose, had to be carried forward on a stretcher, as he was suffering from malaria and had a high fever. However, he refused to be taken to the rear because he was determined to remain in command of his regiment.[8]

Meanwhile, Japanese Major Koiwai Mitsuo, commanding the *II/41st Regiment*, was still leading the Japanese advance. Two days before, he had received a serious dressing-down by his commanding officer for not crushing the Australian resistance and pursuing the enemy force after occupying the vacated Australian position a few kilometres north of Templeton's Crossing. Koiwai was now determined to show more aggression in continuing the advance, not knowing that he and his men were about to be relieved of this 'honour'. His scouts had reported the presence of a significant Australian force where Eora Creek crosses the Kokoda Track, about two kilometres south of Templeton's Crossing.

CHAPTER 4

He relieved the men of *No. 8 Company, II/41st Regiment* because they had spent the last few days trying to flank the Australians through the jungle. These exhausted men would be relieved by *No. 7 Company, II/41st Regiment*, who would lead a night-flanking assault against the Australian blocking party.[9]

Koiwai would focus his attack against the Australians and use a frontal assault along the axis of the track as a feint, using *No. 6 Company, II/41st Regiment*, and *No. 12 Company, III/41st Regiment*, while the main thrust would be against the Australian left flank, by the men of *No. 7* and *No. 5 companies, II/41st Regiment*. He recalled: 'I was looking forward to the arrival of the *7th Company* because of the quality of the command. Lieutenant Nakao, the company commander, had fought in China, and his warrant officer, Kaneshige, was also a superb fighter with lots of experience'.[10] In addition, his regimental commander had given him permission to use the gunners of *2nd Mountain Gun Company*. The Japanese officer recalled his plan for a night attack and the importance of psychological warfare against the enemy:

> Though called a company they had only one gun and I was told that they were allowed to fire no more than ten rounds .... Even ten rounds would be enough. My goal was not to destroy the enemy position, but to scare them with the mountain gun. There can be no better tactics than to crush the enemy morale .... I wanted to fire it right before the night attack of 7 Company .... I assigned it a position and a target after closely observing the enemy with binoculars. It is usually considered wasteful [of ammunition] to fire an artillery gun in a night attack [but] human psychology in the front line was ... beyond tactics. My plan was that, after firing shells at the position ... and terrifying the enemy out of their senses, we would charge with bayonets at them with the result they would be in fear of our attack in the darkness of the jungle.[11]

In the end, Koiwai would attack without artillery, as his commanding officer was concerned about the Australians digging-in – it was commonly assessed by the Japanese: 'If the Australians were given time to dig-in, it was almost impossible to drive them out'.[12] As he was waiting for the gun to make an appearance, the major recalled that his regimental commander made an appearance and became increasingly agitated: '"What are you waiting for?" Yazawa demanded. "Attack now!" Koiwai replied, "I'm waiting for artillery support before giving the order to attack." Yazawa snorted back in reply, "There won't be any artillery support"'.[13]

* * *

Meanwhile, the 2/16th Battalion diarist recoded after the initial confrontation with advance Japanese elements at the creek, that it had been a 'fairly quiet night – very wet, rained all night. B Coy claim four enemy casualties'. The diarist also recorded that just after 9 am, a patrol from the Brigade Headquarters Company moved back along the track towards Myola to confirm that no Japanese infiltration parties had got behind their position. At around 9.20 am, Lieutenant George Hicks, a 26-year-old tailor from Angaston in South Australia, leading 14 Platoon, 'C' Company moved out to conduct a patrol in the low ground on the left, forward of the creek, to give warning of any Japanese flanking movements in that area. At around 11 am, another party of lost men from the 2/14th Battalion came through their right flank, having been cut off at Isurava the week before.[14]

It was not until 2 pm that Lieutenant Hicks and his men returned from their patrol, reporting a large Japanese force was moving around their western flank in strength. These were the Japanese troops of *No. 7* and No. *5 companies, II/41st Regiment*. Hicks informed Sublet that they had found a Japanese telephone cable that appeared to have been laid just hours earlier, judging by footprints and the disturbance of undergrowth – he cut a section of the cable. More worryingly, he had observed several hundred Japanese forming up for a frontal attack not

CHAPTER 4

far from the creek crossing. These were the men of *No. 6 Company, II/41st Regiment*, and *No. 12 Company, III/41st Regiment*, and behind them in reserve were the men of *No. 8 Company, II/41st Regiment*.[15]

The war dairy records: 'C Coy patrol returned with disturbing information that the enemy had apparently moved around our left flank. Estimated one [Japanese] coy taking a signal wire with them'.[16] With this information, Sublet assessed that the flanking Japanese were attempting to capture the high ground about 500 metres behind his position to cut him off from the rest of the brigade. The brigade's *General Narrative of Events* also records: 'During the afternoon, the situation in front became suddenly serious. The enemy had crossed the river northwest of the 2/16 Bn position and a patrol 2/16 Bn came … with the news that there were approx. 200 enemy opposite the Bn on the far side of the river and a force estimated at 100 had crossed the river and were moving round the Bn left flank. The enemy had laid a telephone line across the river along the track [which] they cut as they advanced'.[17]

* * *

At Myola, Potts was informed by Sublet at around 3 pm of the developing Japanese attack and that he planned to retire to the high ground before being cut off. The men of the 2/14th Battalion were now ordered to prepare for an advance back up the track to help extricate Captain Sublet and his men.[18] The men of 'A' Company moved forward with Major Hearman, who was returning to the 2/16th Battalion. At 1.30 pm, 'B' and 'C' companies, supported by the Headquarters Company of the 2/14th Battalion also moved further up the track to take up a blocking position for the men of the 2/16th Battalion to pass through, while 'D' Company remained at Myola to defend the position.[19]

Meanwhile at Kagi, Lieutenant Colonel Cooper was placing his men of 'A' and 'B' companies, 2/27th Battalion in the defence positions prepared by Honner and his militiamen of the 39th

Battalion. Honner's battalion, due to casualties and sickness over the last month of combat, was now reduced to just 180 men. Arrangements were made for the militiamen to have a complete night's rest; in the morning, they would head back to Port Moresby. At 5 pm, 26-year-old Lieutenant Reginald Applebee, from Gawler in South Australia, leading 10 Platoon, 2/27th Battalion, left Kagi to relieve a patrol of the 39th Battalion that was then two hours out along the track forward of Myola.[20] Fifteen minutes later, Brigadier Potts rang from Myola, informing Cooper and Honner that the 2/16 Battalion was hard-pressed by enemy attacks but were holding the line midway between Templeton's Crossing and Myola.

Elements of the 2/14th Battalion and the Brigade Headquarters were at Myola in defensive positions. Potts ordered the 39th Battalion to now remain at Kagi, while a company of the 2/27th Battalion was to move to Templeton's Crossing to attack the Japanese from the rear, using the old pre-war track to relieve pressure on 2/16th Battalion. Another company from the 2/27th Battalion, using the main Kokoda Track, was to move immediately to Myola, reaching there that night or early the next morning.[21]

At 5.30 pm, 'A' Company, 2/27th Battalion, in accordance with Potts' orders, left Kagi for Myola, using a cross-country track for the first three kilometres before reaching the main Kokoda Track; while 'B' Company was ordered to prepare to move to Templeton's Crossing and along the way pick up Lieutenant Applebee and 10 Platoon at the forward post. They were then to contact the Corps Guard Platoon (part of the Headquarters Company) who were patrolling on the track junction between Myola and Templeton's Crossing.[22]

Captain Charles Sims, a 26-year-old electrical fitter from Wallaro in South Australia, leading 'A' Company, 2/27th Battalion, recalled the journey: 'Brigade Commander ordered Coy to leave Kagi and make Myola by night fall, as a result Coy left Kagi at 1730 hours for Myola taking cross country track for the first one and a half miles to the main Myola track. The going was very treacherous owing to the

CHAPTER 4

rain …. Coy reached main track and continued on towards Myola travelling until 2100 hours that night. They lay either side of that track in the undergrowth and spent a most uncomfortable night in the rain. It was found out afterwards that native police had been sent out to order our return to Kagi'.[23]

* * *

Back at the position held by the men of the 2/16th Battalion, and after conferring with his company commanders, Sublet was preparing to fall back to the high ground. Captain Bret Langridge, a 23-year-old bank clerk from Inglewood in Western Australia, volunteered to act as the battalion rearguard to hold back the enemy as the rest withdrew to the new defensive position. Before this retirement could be made, however, the Japanese attacked. Unknown to Hicks, the Japanese of *No. 7 Company* with elements of *No. 5 Company, II/41st Regiment* had followed him back to the sector held by 'A' Company, by using the same tactics they had used so effectively the month before during the second battle for Kokoda Plateau.[24]

It was now that Lieutenant Grear McGee, a 29-year-old professional soldier from Launceston in Tasmania, commanding 'A' Company on the left in the forward position, was under attack, as recorded in the battalion war diary: 'A Coy was attacked strongly. A further five casualties, three killed, but the Japanese attack was repulsed without further casualties averted only by the quick action and tenacious fighting of Private Brian Maloney[vii] who used his Bren gun with devastating effect'.[25] Indeed, it was here that Japanese Warrant Officer Utsumi reported Australian resistance as 'strenuous' and that he had suffered considerable casualties.[26] Lieutenant Araki, with *No. 5 Company, II/41st Regiment*, also recalled that while dug-in, Australians would sometimes stand up in their fighting pits fully exposed to enemy fire in order to better throw their grenades.[27]

---

vii Private Brian Maloney was a 27-year-old miner from Kalgoorlie in Western Australia.

Captain Sublet ordered the Battalion Headquarters Company to stand-to, as they were to reinforce 'A' Company. Sublet now attempted to withdraw his force back to the high ground, while facing a determined 300-strong Japanese force. It was critical then that his men gain the heights before the Japanese cut them off. McGee and his men of 'A' Company fought to hold back the Japanese advance, giving the remainder of the battalion a chance to withdraw. It was now that Captain Langridge and his men of 'D' Company moved in to act as the battalion rearguard, enabling the men of 'A' Company to retire through them.[28]

With 'D' Company now acting as the rearguard, the battalion began its withdraw south. As Sublet's two rear units, 'C' and 'D' companies fell back; however, they were ambushed, but McGee came to their rescue, helping them push through the flanking Japanese of *No. 7* and *No. 5 companies, II/41st Regiment*. Assisting McGee and his men of 'A' Company was Lieutenant Henry 'Bluey' Lambert, a 29-year-old farmer from Burracoppin in Western Australia, and his men of 18 Platoon, 'D' Company. The battalion diarist recorded: 'C & D coys ambushed near top of hill. Fortunately, 18 Pl & A Coy had been holding an extricating position on top of hill which enabled our tps to get through with few casualties. Enemy had apparently not had time to get all their force into position – very fortunate, for us'.[29] The brigade's *General Narrative of Events* concludes: 'The Bn were in a bad position, on a fwd slope with only one platoon on the high ground in their rear. Capt. SUBLET, left in charge of the Bn ... withdraw the Bn to the high ground in the rear. This was accomplished just in time, five minutes later the Japs renewed their attack'.[30]

Indeed, the Japanese quickly reorganised and lunged forward, this time with a force estimated at around 400 men. Major Kiowa had committed his four companies to assault the Australian position, leaving *No. 8 Company, II/41st Regiment* in reserve. Sergeant George Morris, a 31-year-old farmer from Harvey in Western Australia, and his men of 11 Platoon were ordered to take up his position on the

CHAPTER 4

high ground; however, he was 'put out' on discovering others were occupying it. Sublet recalled almost comically years later, which would not have been funny at the time: 'When I arrived back at the previously selected holding position Sgt Morris very irately told me that I had mistakenly allocated his company to a position that was already occupied. However, a quick inspection disclosed that the position (a key one) WAS occupied, but by Japanese! The dynamic Morris reacted like a flash and waded in with his men, evicting the Japs in short order. Morris' determined and full-blooded action no doubt saved the battalion from a difficult situation'.[31]

A week before, Sergeant Morris had supported Private Alexander Thornton, a 28-year-old labourer from East Ryde in New South Wales, when he stormed several Japanese positions during the attack northeast of Abuari on 30 August. Thornton, who would die from wounds suffered in these attacks, was recommended for the Victoria Cross (VC); however, he was instead awarded the Distinguished Conduct Medal (DCM) – the highest honour below the VC; Sergeant George Morris was awarded the Military Medal (MM) for his actions that day. Morris and his men would figure prominently in the fighting to breakthrough to the brigade during the fighting at 'Butcher's Corner' on Mission Ridge.[32]

Darkness was setting in, and it was now that Major Hearman appeared, and by 7.30 pm, he was leading the bulk of the men back towards Myola. However, Lieutenant Bert Madigan, a 32-year-old sign-writer from West Perth, and his men of 12 Platoon, assisted by the ever-dependable Sergeant Morris and his men of 11 Platoon, remained behind as the rearguard. These men stood their ground while advancing Japanese elements attempted to push south down the track. These Australians enabled the rest of the battalion to continue unmolested, slipping and falling most of the way back to Myola in the churned-up mud of the track. They struggled to get their wounded back, each man held the clothing or bayonet scabbard of the man in front so as not to lose his way; they had to stop and wait for daylight because progress on the track here without light was impossible.³³

Meanwhile, two warrant officers with ANGAU, Ronald Preece, a 24-year-old plantation manager from the Rigo District in Papua, and 30-year-old John Davies, from Sydney, were leading a party of stretcher-bearers that had earlier evacuated the wounded from Templeton's Crossing. They were now a few kilometres south of Sublet's position, but now they came to his assistance.³⁴ An un-named soldier with the 2/16th Battalion later recorded his experience of retreating towards Myola and coming across the wounded men that were being cared for by Preece and Davies:

> We set out ... [with] the disturbing thunder of the Japanese mortars. As those who have traversed it will know, the trail twists and turns, ascends or descends here around a log or wicked splintered stump. Sometimes the trail slopes outward at an awkward angle and sometimes it is bolstered up by logs hewn from the jungle. Little creeks cross it very frequently and there is always mud – and roots. In this area there were countless thousands of these roots exposed by the rough usage of blucher boots. Hour after hour we struggled over these slippery, treacherous things, stumbling, lurching, sliding, cursing. If a root was on the slightest angle and you were inexperienced enough to put your foot on it, over you went. We were only going down. Up this slope was coming the strangest medley of men I ever saw. Wounded men of the 2/14th and 2/16th battalions – sick and exhausted men .... Here and there a party of [Indigenous carriers] battled with a stretcher-case.³⁵

* * *

Further south, Lance Corporal William Cousens, a 26-year-old jackeroo from Jarrahdale in Western Australia, with 'B' Company, 2/16th Battalion and several of his mates were about to finally stumble

CHAPTER 4

into Myola after having been lost in the jungle during the withdrawal from the fighting around Abuari on 30 August. They reached the main track just south of the dry lakes, with Cousens recording many years later:

> At the same time, we sighted a native carrier coming down the track to our left, carrying a tin ration on his head. He had not spotted us, so we stood quietly concealed until he came abreast of us when we all pounced upon him with the demand of 'Which way Japanese man?' and heaved sighs of relief when he pointed off to his right. Then to another question of 'Where you come?' he replied with what we took to be Myola. He had been very shaken by the ordeal, but rapidly regained his composure when he saw we were not Jap troops and took off down the track rather rapidly when we let him go, leaving his tin of jam behind him. We had nothing to open this tin with and when we smashed it open with rocks, the contents scattered on mud and foliage, to offer us very little nourishment at all.
>
> What to do now? Being half dead with hunger and deficient in arms, we decided to try to reach Myola to get some food and more weapons to rejoin our troops, who were falling back in any case.
>
> Myola proved to be only a short distance up the track from where we were, and after making our plight known to the commander of that post, he gave us food and rearmed my two mates, while stating that he had no .45 ammunition available for my Tommy gun, also that we were to remain in the camp to defend it from the many Japs who were around the place. We knew that that was a bit farfetched, so after scrounging around the camp for a while, I was able to get hold of a drum magazine from a Tommy gun conveniently leaning up against a tent before we silently melted into the jungle to come out on to the track again about

200 yards down from the camp, in time to mingle with our mates again, who were withdrawing back to extend the enemy lines of communication and to find a suitable spot to re-engage him again ….

It was almost dark by now, so it was decided that we camp here for the night, with no more movement possible in this inky black darkness of the night. Me and 'Nugget' Goode [William Goode] were assigned the task of guarding a bridge over a small creek just out of Myola which could be used by Japs, as well as our own chaps, still coming back from down the trail, with the plan being for Nugget to do the challenging behind a big tree on the right front of the bridge and me to do the shooting right in front of the bridge if it happened to be the enemy. But this plan just didn't materialise, as Nugget must have gone to sleep, leaving me to do all the challenging with little chance of surviving had there been any attempt by the Japs to cross the bridge that night, lucky for me they didn't.[36]

* * *

Meanwhile, the Japanese now occupied Sublet's former position just south of the track junction near where Eora Creek crossed the Kokoda Track. The Japanese official history only provides a brief account focusing on the flank attack to take the high ground behind Sublet's front-line position: 'The pursuit party swung around to the west of an Australian camp. Skirting along the ridgeline, the party suddenly attacked a group of Australians in the evening as they gathered in a saddle of the mountains. A torrential downpour started just as night fell. The two forces held their positions while enduring the cold. And waited for the rain to stop. Sometime during the night, the Australians withdrew'.[37]

5

'... INTRODUCED THEMSELVES AS CHESTER WILMOT AND OSMAR WHITE'

Hours earlier, further south, Lieutenant Norman Cairns, a 26-year-old newsagent from Collingwood, with the 2/14th Battalion, had arrived at Kagi from Myola and briefed lieutenant colonels Geoffrey Cooper and Ralph Honner on the tactical situation. About an hour later, new orders arrived from Brigadier Arnold Potts, cancelling his previous orders; the situation was obviously very fluid. Cooper was to now withdraw 'A' Company, his 10 Platoon, and the Corps Guard Platoon back to Kagi as soon as possible. Papuan police constables at Kagi were sent out to contact 'A' Company but were unable to reach them. Sims and his men of 'A' Company were eventually met by brigade personnel evacuating Myola the following day. Lieutenant Applebee and his men of 10 Platoon, 2/27th Battalion were connected by phone and received orders to send a runner forward to the Corps Guard Platoon, instructing them to return to Kagi at first light the next morning. Applebee was also issued with the same orders.[1]

By 6.30 pm, Potts rang again from Myola, informing Cooper and Honner that they were heavily pressed by a superior Japanese force and that the brigade would withdraw to the high ground south of Efogi. Cooper's two companies of the 2/27th Battalion, along with Honner and his men of the 39th Battalion were also to move back from Kagi to the high ground immediately south of Efogi, to cover the withdrawal of the 2/14th and 2/16th battalions as well as the 21st Brigade Headquarters.

With this, all troops at Kagi were issued with extra ammunition and rations; the movement south was helped with the arrival of 50 Papuan carriers who arrived from Myola. These men would be used the next day to help carry ammunition and food back to Efogi from the dump at Kagi.[2]

Meanwhile, continuing north to join the rest of the 2/27th Battalion was White's force who began to arrive in the Efogi area. By the late afternoon, it was increasingly clear that the tactical situation was deteriorating, and with darkness, it looked very desperate indeed. The diarist with the 2/27th Battalion recorded in relation to White and his men: '0630: Supplies dropped on NAURO from air. Camp staff failed to notify OC Party of danger area'.[3] Unfortunately, this resulted in the death of 22-year-old Private Patrick Dynon from Adelaide when he was struck by falling supplies. Four others were injured – two seriously. His two companies reached Menari before 1.30 pm, and from there the men of 'C' Company and elements of 'D' Company pushed on. The war diary records those who reached Efogi included the adjutant of the battalion, Captain Ronald Johnson, a 26-year-old solicitor from Prospect in South Australia, and his men of 'C' Company, along with some men of 'D' Company. The remainder stayed overnight at Menari. A hot meal was served. It also reports that the regimental, 'Padre and RAP Sgt remained at NAURO with injured to bury Pte Dynon. 1330: QM [quarter master] remained at NAURO to check unit stores'.[4]

On hearing that the first two companies of the 2/27th Battalion had reached Kagi, Potts thought he might yet be able to hold Myola, with the remaining two battalions quickly closing in. Reality, however, soon set in – he did not have enough men to hold both tracks and the rugged terrain in between, let alone his flanks beyond. Confirming his decision to evacuate Myola was the knowledge that the men of the 2/16th Battalion were now falling

CHAPTER 5

back rapidly, and the rearguard would not be able to hold back the advancing Japanese for long.[5]

Potts informed Major General Arthur Allen of his decision that night: 'Strong enemy attack driven in 2/16 Bn 1½ hrs [from] MYOLA. Am supporting defence with 2/14 Bn but country utterly unsuitable to defend …. Regret necessity [to] abandon MYOLA. Intend withdrawing [to] EFOGI. No reserves or counterattack. Men full of fight but utterly weary. Remaining coys 2/27 Bn too late to assist. Will keep you informed'.[6]

The war diary of the 21st Brigade provides a vivid account of the rapidly deteriorating tactical situation throughout the day:

0730 hrs: 2/16 Bn reported all quiet, but 10 Japs said to have been seen halfway between river Xing and MYOLA. Bde Comd ordered a patrol of 39 Bn to KAGI – TEMPLETON'S – MYOLA rds junction and a standing patrol from HQ Coy 2/16 Bn to rd junction half an hour North of MYOLA

0900 hrs: Message from 2/27 Bn reported 1 killed and 2 seriously injured at NAURO when planes dropped sups at 0600 hrs. No warning was given. Bde Comd, Bn. Lt-Col CAMERON, 53 Bn, and Lieut. [sic] KIENZLE went back to recce a position for firm base in the vicinity of EFOGI.

0930 hrs: Lt Col CARO, 2/16 Bn, began recce of DRY LAKE and LAKE MYOLA area. An attempt to communicate ground to air by 101 set was unsuccessful.

1129 hrs: BM and Lieut. KIENZLE returned. The plan was changed as there were not enough men to hold the high ground fwd of Efogi.

1130 hrs: Cpl [Ted] NORRIS and 15 ORs 2/14 bn, six of whom were wounded, came in round the right flank of 2/16 Bn. They reported that no enemy and no tracks were seen. 2/16 Bn sent back 1 officer and 32 ORs with dysentery.

SAVING PORT MORESBY

1200 hrs: An order was despatched to KAGI to send back 9 Pl HQ Guard Bn to guard FSD [forward supply depot] at EFOGI.

1300 hrs: Lieut. CAIRNS, LO. was sent to contact 2/27 Bn at KAGI.

1515 hrs: 2/16 Bn reported approx. 100 Japs attacking on left flank. A runner was despatched to inform Bde Comd and BM who were on recce of DRY LAKE area.

1525 hrs: 2/16 Bn reported 200 Japs on left flank. The Japs came across river NW of 2/16 Bn position and laid a telephone line.

1555 hrs: 2/16 Bn estimated 500 Japs were now attacking.

1600 hrs: A message was sent to 2/27 and 39 Bns to stand by.

1615 hrs: Bde Comd ordered one Coy 2/27 Bn to proceed immediately to MYOLA. Possibly would ask 2/27 Bn to put in an attack on TEMPLETON'S tomorrow with one coy.

1550 hrs: Line to 2/16 Bn was out.

1650 hrs: 2/16 Bn reported moving back to high ground with D Coy fighting the rearguard.

1905 hrs: Bde Comd instructed 2/27 Bn to send a coy to raid TEMPLETONS to take the strain off 2/16 Bn. If there were no contact at TEMPLETONS the coy was to go along the river towards 2/16 Bn position and take the Japs in the rear.

1725 hrs: 2/16 Bn reported only just getting 2 coys on high ground before the Japs attacked. Capt. SUBLET, Adm Comd, did not think he could hold.

1800 hrs: Bde Comd sent a message to Gen ALLEN, GOC 7 Division, to say he would have to withdraw to EFOGI. 2/16 Bn reported having fallen back to next ridge with all coys in position.

CHAPTER 5

1825 hrs: 2/27 and 39 Bns were ordered back to EFOGI and the raid on TEMPLETONS cancelled.

1830 hrs: 2/16 Bn reported moving back to a position for the night. Capt. [John] O'NEILL, Adjt 2/16 Bn, arrived Bde HQ to make a report on the situation. During the night all supplies at MYOLA were destroyed and ammunition and sig equipment buried. A hot meal and new clothing was taken to the trail junction to give to 2/16 Bn.[7]

* * *

Still in the Kagi area was Lieutenant Colonel Cameron, Captain Joseph Gilmore, Lieutenant Alfred Parsons and his intelligence officer, 34-year-old Captain Bruce MacCormick from Uralla in New South Wales, and the most fit 40 or so men of the 53rd Battalion. The war diary of the battalion records Gilmore and MacCormick were ordered to move to Nauro, while Parsons, with 23 men, was to search for signal cable that had been dropped by aircraft; before leaving, they were to collect some Thomson submachine guns at a carrier food dump just south of Kagi. Meanwhile, the composite company under Captain James Bryce was to carry supplies from Myola to Kagi. Soon, new orders arrived from Cameron that the adjutant, 34-year-old Captain William Reid from Macksville in New South Wales, along with Lieutenant Parsons and all available men were to report to Bryce at the Efogi–Myola Track junction adjacent to the food dump by dawn the next morning: 'All personnel to be fully armed and grenades issued'.[8]

* * *

Having left Myola the day before, 40-year-old Major Albert Moore and 36-year-old Corporal Jock Inglis of the Salvation Army had woken up early at Menari; it had been a marathon effort by the two men who were in a hurry to establish their Red Shield Post further south to assist the Australian sick and wounded as they fell back towards Port

SAVING PORT MORESBY

Moresby. They were especially anxious about their mates of the 2/14th Battalion; they had a long affiliation with these men, having served with the battalion during the Middle East Campaign and following them to New Guinea. Moore recalled years later:

> During breakfast we heard the drone of aircraft engines and raced outside the hut to hide. We had only just left when a case of bully beef crashed through the roof of the hut where we had been sitting. From tree to tree we ran trying to escape the falling cases of supplies! We returned to the hut to finish breakfast, and then made another hasty exit to escape the flying cases that poured from the air. We thought we should get out of that hot spot, so took our departure from Menari. We reached Nauro for lunch and sent an urgent signal to the Red Shield base in Moresby asking that supplies and comforts be made available for us when we reached the head of the road. I repeated the request wherever I found a signal station along the Trail.
>
> After lunch we set off for Ioribaiwa, where we arrived at 3.30 pm to find the village crammed with troops, many going forward and lots of sick and wounded on their way back. There was no hope of getting under cover, so we managed to secure a groundsheet and a blanket and sought one of the few spots available, even in the open, to rest.
>
> I had just settled down when a chap asked if he and his mate could share my spot. I was alongside the dump where all the waste of the area was tossed out. I agreed and the men introduced themselves as Chester Wilmot and Osmar White. They were war correspondents … and were hurrying back to base to get their story … through to Australia. The night was long, yet when one thought back to those soldiers in the frontline, one felt fortunate to be sleeping, even on a rubbish heap.[9]

CHAPTER 5

* * *

Heading in the opposite direction, but also on a mission of mercy, was Captain 'Doc' Geoffrey Vernon who was bringing with him medical supplies for the aid posts he had established along the track. Doc must have come across the two salvationists on the track and likely had a brief chat. He recorded for 4 September in his official report to ANGAU: 'Reached Nauro at midday, inspected hospital and prepared the staff for more strenuous work, as the Japanese were now advancing towards Menari'.[10]

* * *

Meanwhile, north of Myola, Captain Sydney Buckler and his lost men of the 2/14th Battalion remained in Eora Creek Valley east of the Kokoda Track. It had been five days since they were forced into the valley during the battle for the Isurava Rest House position.[11] That day, while reconnoitring an abandoned enemy camp, they encountered nine dead Australians – likely executed by the Japanese. Buckler wrote in his report: 'Had the unpleasant task of removing identity discs from the week-old corpses'.[12] Buckler was hoping to find a suitable place to lay up for a few days to rest and gather any food they could find. Lieutenant Maurice Treacy with Buckler's party later recorded in his report:

> After passing through a bivouac area in which at least 500 enemy troops had camped; the party reached a suitable bivouac area at 1600 hrs. Shelters were built to keep off the torrential rain which had started to fall .... Despite the lack of food and sleep and the rain; they had carried the stretchers through the virgin jungle and along steep, narrow tracks without one word of complaint. The wounded had shown great fortitude. Despite serious wounds and the continued jolting, they remained cheerful and demanded little. Cpl Metson ('B' Coy) (wounded in ankle) had actually crawled the whole distance for five days

and was continually wet through. He was admired by all. Pte Fletcher continued to do good work among the wounded; unfortunately, the last of the medical supplies had been used that day. Capt. Buckler, as OC party, was an outstanding leader and showed great devotion to duty .... Confidence in his ability to lead the party back to the unit was felt by all.[13]

Lieutenant Treacy was now ordered by Buckler to advance ahead of the main party to find the Australian lines. The lieutenant recalled this OC's orders: 'I want you to leave tomorrow with two men; travel to Myola to contact the unit, obtain medical supplies, food and native carriers, and return here with them. If the unit is not at Myola push on till you find it. This party will remain here for four days after your departure and search for native foods. If at the end of that time you do not return; I shall leave two men here for a further two days, and the rest will make for the coast, after moving NORTH to the KOKODA Valley, then EAST through the less mountainous area where it will be easier going, and native gardens more plentiful. I want you to follow the enemy tracks for a time to make observations'.[14]

* * *

Meanwhile, west of the Kokoda Track was Lieutenant Robert McIlroy and his lost patrol of the 2/14th Battalion. They had been forced west of the pre-war track between Templeton's Crossing and Kagi and could not regain the track. They started off early that morning, following the creek they had camped beside the night before, but they were soon forced to forgo this decision: 'During the day it became impossible to travel along the creek, as it had developed into a series of small waterfalls. We therefore cut our way up onto the ridges alongside and followed the general fall of the country'.[15]

6
'I WAS BLOODY LUCKY THAT TIME'

During the early hours of 5 September, Major Koiwai finally got word at 2.40 am that his men were holding the former Australian position and were digging-in because they expected an Australian counterattack at dawn; *No. 8 Company, II/41st Regiment* were now advancing to support them. In the last few days of fighting, the Japanese had suffered 43 killed and 58 wounded.[1] The Japanese official history records: 'The Japanese greeted the morning at the pass over the highest peaks of the Owen Stanley Range. It was at least 2000 metres above sea level, and all the trees were covered in moss. All the officers and men shouted out *Banzai* together'.[2]

Meanwhile, the men of the 2/16th Battalion had resumed their journey to Myola with no intention of conducting a counterattack. The battalion war diary records that at 3.30 am, the battalion moved by moonlight but stopped shortly soon after, as there was a risk of being fired on by a 2/14th Battalion ambush party. By 8 am, these men had passed through the holding position of the 2/14th Battalion and had arrived at Myola, where hot tea, stew and clean clothes were available.

Among these men was Captain Henry 'Blue' Steward, a 30-year-old physician from Melbourne and RMO with the 2/16th Battalion, who recalled: 'At Myola, after our hard, slow grind through darkness and lashing rain we got our reward – a hot meal, the first for ten days. The sight and smell of food simmering away in dixies made us salivate like Pavlov's dogs, and the comforting glow warmed us from the stomach outward'.[3] Even so, within 30 minutes, Steward and the rest of the battalion were heading south to Efogi, 'very tired and foot sore'.[4] They had in front of them another four-hour hard slog, crawling, sliding and edging their way south through the rain towards Efogi.[5]

Earlier that morning, Brigadier Arold Potts sent Brigade Major Hugh Challen on another reconnaissance mission to identify the positions to be occupied on the high ground south of Efogi along Mission Ridge. Captain Sims recalled getting the word to fall back to Efogi from Challen: 'Early morning A Coy moved off towards Myola. One hour distant they contacted Bde HQ withdrawing to Efogi. Brigade Major put the Company in the picture and Coy about turned and moved back to Efogi'.[6] Lieutenant Colonel Geoffrey Cooper was then still at Kagi with two of his companies – Captain Arthur Lee, a 30-year-old company director from Glenelg in South Australia, and his men of 'B' Company, and Captain Johnson, with his men of 'C' Company. Moving up from Menari was Captain Thomas Gill, a 28-year-old chartered accountant from Largs Bay in South Australia, and his men of 'D' Company who had advanced elements of this company to Efogi.[7]

Challen and his party arrived at Efogi at around 9.30 am and contacted Lieutenant Colonel Cooper. Lieutenant Pearce and his small party – which included the wounded Lieutenant Valentine 'Vincent' Gardner, a 22-year-old grocer from Mildura in Victoria, who had just days before arrived at the Australian lines at Sublet's former position near the track junction a few kilometres south of Templeton's Crossing – were also there.

Within days, Damian Parer would use the last of his film to take one of the most defining images of the Kokoda Track, filming the young, wounded Australian officer being cared for by Salvation Army Major Albert Moore. After conferring with Pearce, Challen pushed further south to reconnoitre the defensive positions along Mission Ridge; he remained there to direct the men into their positions as they arrived.[8]

Earlier that morning, at around 5 am, the adjutant of the 53rd Battalion, Captain William Reid, along with Lieutenant Parsons and

CHAPTER 6

27 ORs had, as ordered by Lieutenant Colonel Cameron the previous day, moved forward from Efogi to assist Captain Bryce in his holding action. At 7 am, Bryce, who was just north of Kagi, had received word from Cameron that he and his composite company were to hold their position astride the Kagi–Efogi Track junction until they received further orders, and that he would soon be joined by others of the battalion. Bryce and his men witnessed the withdrew of Potts force throughout the day.[9]

Meanwhile, Lieutenant Applebee and his men of 10 Platoon, 2/27th Battalion passed through Kagi at 7.30 am, along with the Corps Guard Platoon about an hour later. Kagi was cleared by 9.30 am, except for the brigade liaison officer and the 2/27th Battalion intelligence officer who left at around midday. All ammunition had been carried out of the village, and as much food as possible was sent on to Efogi by Papuan carriers. The remaining balance – 20 large tins of egg powder, 20 tins of dehydrated onions, 40 tins of biscuits, 20 bags of rice, 20 bags of dried fruits, 100 tins of milk powder, quantities of sugar, tea, jam, flour and salt – was all destroyed.

Still forward of Efogi, but south of Kagi, was Lieutenant Bryce and his platoon from the 53rd Battalion, who were manning an observation post to warn of any Japanese advance from Kagi.[10]

At around 11 am, Lieutenant Colonel Honner reported to Brigade Headquarters that he and his men of the exhausted veterans of the 39th Battalion were covering the track just south of Efogi along the main track on the ridgeline. Honner was instructed to fall back when the 2/27th Battalion was in position. By 11.30 am, the brigade war diary records that a reconnaissance party had proceeded to Mission Ridge to reconnoitre the positions to be occupied by the men of the 2/14th and 2/16th battalions, while the 2/27th Battalion took over from 39th Battalion, which would soon be heading south for Port Moresby.[11]

In all, Cooper's battalion number 28 officers and 560 other ranks. They were to take up defensive positions on the middle slopes of Mission Ridge by 2 pm, the rest of Maroubra Force was to fall back on his position. Cooper was to also take over the automatic weapons and other equipment of the 39th Battalion, as Honner and his men would not need them because they were now ordered back to Port Moresby for some well-earned rest and reorganisation.[12]

By 2 pm, the 2/27th Battalion Headquarters' staff, along with 'A' and 'B' companies had rejoined 'C', 'D', and the Headquarters companies located on high ground south of Efogi. Rain began to fall from 3 pm, with the Brigade Headquarters' personnel passing through to their new position. The 2/27th Battalion war diary concludes that the battalion took up a defensive position astride the track, while the forward elements of the 39th Battalion, who had been in position on the track junction around Kagi and Efogi, now passed through and joined the balance of the 39th Battalion, who had been holding the position recently occupied by the 2/27th Battalion. These militiamen now handed their automatic weapons and ammunition over to the men of the 2/27th Battalion and began their long trek back to Port Moresby. Each company of the 2/27th Battalion patrolled its own front throughout the night to prevent infiltration by the Japanese. No contact was made with the enemy, and as usual, it rained throughout.[13]

At around noon, leading elements of the 2/16th Battalion began to arrive exhausted at Efogi, and permission was granted by brigade to stay in the village for the night instead of moving back to Mission Ridge.[14] A few hours behind them were the men of the 2/14th Battalion, having vacated their forward-holding position. By 1 pm, these men had started marching south through heavy rain, which lasted until they reached Efogi at 4.30 pm. Here they bivouacked in the village with the men of the 2/16th Battalion, with patrols sent forward to guard against any Japanese advance.[15] Even so, when the men were on the point of collapse, Captain Clifford Thompson recalled how on countless times the sight of their brigadier inspired confidence and renewed vigour: 'The call, "Pottsy ahead!" made men straighten up'.[16] Undoubtedly, this occurred when the brigadier and his staff pushed through Efogi that afternoon.

CHAPTER 6

With the men of the 2/16th Battalion was Lance Corporal William Cousens and Private William 'Nugget' Goode, a 34-year-old miner from Boulder in Western Australia, who had with a few others been lost in the jungle for almost a week and had stumbled into Myola the previous day. Cousens had spent the night guarding a small creek crossing near Myola, while Goode had fallen asleep. The young NCO later wrote: 'When daylight broke, we were up and away to Efogi on a forced march to beat the Japs there, and to meet up with a detachment of our troops who had approached Efogi by way of Kagi, where they had kept the Japs at bay through that terrain to deny them a free run to Moresby. We were given some respite here for a rest and a good meal which was the first real break we'd had since meeting the enemy many days ago and at last we were to be reinforced by our own 2/27th Battalion who had been kept back from this campaign until now, pending the outcome of the Milne Bay conflict'.[17]

By 5.30 pm, Potts and his senior staff officers had established their headquarters on a small hillock just to the rear of the summit of Mission Ridge – this position would become known as 'Brigade Hill'. The brigade diarist recorded: 'The condition of the feet of the 2/14 and 2/16 Bn personnel was very bad and the men physically exhausted'.[18] By 6 pm, Reid and his forward patrol reported back to Cameron. Two hours later, a standing patrol of the 53rd Battalion reported seeing lights in the direction of Kagi.[19] Indeed, forward of Bryce's men, this patrol reported enemy troops about 300 metres forward of their position, but the Japanese were shortly after observed to withdraw. These were likely the Japanese of *No. 4 Company, II/144th Regiment*. Just after 8 pm, messages arrived from Captain Bryce, saying that he could see enemy movement near Kagi; lanterns were being used. Cameron issued orders for Bryce and his men to fallback south of Efogi at 8 am the next morning.[20]

* * *

No doubt, Honner and his men were keen to get back to Port Moresby, as for the first month of the Japanese invasion, they – along with their mates of the 1st Papuan Infantry Battalion (1st PIB) and the Papuan Royal Constabulary (RPC) – had borne the brunt of the fighting and remained in the fight even when reinforcements from the 2/14th and 2/16th battalions finally arrived in late August.[21]

Sergeant Major Cops with the 39th Battalion had returned to Efogi from his standing patrol the previous night and recorded in his diary: 'Today was the last day of action for the 39th after six long weeks of delaying the enemy. We were then relieved by the 2/27th Bn, the remaining unit of the gallant 21st Brigade to whom we must give thanks for our survival; they will never be forgotten by the 39th Bn. After handing over our automatic weapons, grenades, blankets and medical supplies we set off for Menari Village'.[22] Private Howson, still suffering from a piercing ringing in his ears caused by a near-death experience from a bomb dropped by Allied airmen at Kagi, recalled the march south to Menari: 'At Mission Ridge we were relieved and handed over our automatic weapons and other stores to the 2/27th, but I kept four grenades. I was at the RAP when the order came to move out. Walking along the track I was half-crazy with pain – no rifle or automatic weapon, just four grenades, two in pouches and two hanging from my webbing ready to rip off'.[23]

Among the last to come into Menari that night with the battalion was Sergeant William Guest, a 21-year-old clerk from East Brunswick in Victoria, with 'A' Company. During the second Battle for Kokoda Plateau, Guest had got his men off the plateau as it was being stormed by the Japanese – weeks later, he was again fighting with his men at Isurava.[24] He recalled coming across an individual who was almost certainly Damian Parer:

> As we filed out, we had to hand our weapons and blankets over to the 2/27th and then we went back to Port Moresby. The next village, which was called Menari, I was one of the last, and we

CHAPTER 6

came up out of the river, wet, miserable and hungry and it was getting dark. All I had on was what I started off with which was a ragged pair of shorts, and old shirt and my equipment and hat. All of a sudden, I was conscious and suddenly someone had appeared walking beside me on the track. I looked across the track and there was this figure dressed in old khaki full shirt, and a full khaki trouser, wet and dirty and muddy and he yelled out to me and his words I've never forgotten, 'I was bloody lucky that time' and I thought, 'Who the hell is that?' He wasn't a soldier. He said, 'I was filming, and the Japs came into the village back there' …. I still didn't know what he was talking about and who he was … he disappeared in the mist, and I struggled up. That night we all stayed huddled underneath huts waiting for dawn to come.[25]

Now also at Menari was Captain 'Doc' Vernon who was finally told that Myola had been evacuated and that the Australians were planning to make a stand just south of Efogi: 'Went forward to Menari meeting the 39th Battalion on its way to the base for a rest. I arrived at the village about midday and was immediately involved in heavy hospital duties. Numbers of sick and injured carriers from our frontline on the next ridge were received and we evacuated the worst cases, together with such medical supplies as we could spare, without delay to Nauro'.[26]

* * *

That day, the leading elements of the *144th Regiment* arrived at Myola; they all searched the area keen to retrieve any supplies that they could find. Lieutenant Sakamoto Atsushi, commanding *No. 2 Machinegun Company, II/144th Regiment*, and his men were among these Japanese who searched the former Australian supply dump for food. The rain was penetrating the punctured tins of bully beef, and much of the food was spoiled from the day's heat. Even so, Sakamoto was pleased to discover some unpunctured cans of bully beef, jam and milk;

however, there was nowhere near enough for the troops now scouring the immediate area. Within days, the dry lakes would be searched thoroughly for all available supplies that had been dropped by the biscuit bombers, just beyond the main area of the drop zones and missed by the Australians.[27]

Also arriving at Myola that day was Private Teruoka Akira with the *Tanaka Unit, 55th Division Medical Unit* who had come down with a fever. He recalled receiving two letters from home informing him of the preparations for the harvest and stock breeding. He remarked in his diary that he wanted to return to 'normal physical condition and be strong and healthy'.[28]

With hours, Kusunose's men of the *144th Regiment* had passed through the men of the *41st Regiment* to continue the advance. Indeed, Kusunose's men wasted no time as 1st Lieutenant Horibe, *No. 6 Company, II/144th Regiment*, recorded being just four kilometres north of Kagi by day's end. Taking the lead were their comrades of *No. 4 Company*. Horibe recorded that they encountered no enemy troops that day. These men were using the old pre-war track. He recorded in his diary: 'Eora – prepared for battle – over taking YAZAWA BUTAI – advanced and occupied a high Hill … 4 km north of KAGI'.[29]

Also heading south towards Kagi was 2nd Lieutenant Noda Hidetaka with his men of *No. 8 Company, III/144th Regiment*. The young officer was clearly in a reflective mood, recording in his diary: 'I think it's about two years since I left school. Everybody very pleased to find a river flowing at a point which we thought to be the highest in STANLEY. Long rest at a point of KAGI'.[30] He was only days away from meeting his fate at Brigade Hill.

7

'... THE JAP ALWAYS GOES FOR THE HIGH GROUND'

Having established their medical aid post at Nauro, the medical personnel – now commanded by Major Oldham – were still having troubles in getting sufficient supplies. More wounded men were coming down the line, with most also suffering from dysentery. Supplies of sulphaguanidine were required to treat those suffering from this debilitating disease, but no supplies were available. It was now that Captain McLaren, who had been treating the sick and wounded for weeks, came down with a bad case of malaria, which required his evacuation to Port Moresby.

Meanwhile, Major Magarey was again making his way north to discuss the situation with Brigadier Arnold Potts.[1] The senior medical officer had already decided that no surgical units should be within two days journey from the frontline. He recognised that this decision meant assured death to those suffering a stomach wound, but it was a price that had to be paid, as he wrote in his report: 'This means the loss of all abdominal wounds, but it seems inevitable that this should occur in this type of warfare. If these wounds are to be saved, a surgeon would usually need to be placed with, or in front of, the RAP'.[2] Indeed, 45 years later, Magarey lamented: 'If you got an abdominal wound on the Kokoda Trail, you might as well have given up. You never told the troops that, but you knew bloody well that that was what would happen … so you gave them a shot of morphine'.[3]

SAVING PORT MORESBY

* * *

Just north of Port Moresby, the Australian militiamen of the 3rd Militia Battalion, 14th Brigade, commanded by Great War veteran Lieutenant Colonel Albert Paul, a 53-year-old dental mechanic from Bankstown in New South Wales, received orders in the early hours of 5 September that they were to advance into the Owen Stanleys and take up a defensive position along Ioribaiwa Ridge, just north of Port Moresby. These militiamen, like those of the 39th Battalion, would clearly demonstrate that well led Australia militiamen were just as good man-for-man as any in the 2nd AIF. The Australian 7th Division issued orders to Paul that he was to hold Ioribaiwa and secure lines of communication for Potts. Paul would only have hours to initiate his orders and get his officers and men ready for the move; however, he had been warned on 2 September to be ready to move out at a moment's notice.

The battalion war diary records that the movement order was received and the Battalion Headquarters and three rifle companies, along with a portion of Headquarters Company moved out at 9 am. The Battalion Headquarters and 'A' Company would occupy Ioribaiwa village; 'C' Company would be positioned at NAURO; while 'B' Company would take up a position at Edwalla, north-west of Ioribaiwa. The task of 'C' Company was to search for tracks and protect Nauro from the northern approach, while 'B' Company would patrol tracks coming into Edwalla from the northern and eastern approach.[4] The orders also stated: 'If 21 Bde is overwhelmed – and this is not thought probable – 3 Bn is to be prepared to hold both positions, NAURO and IORIBAIWA' (the main KOKODA track).[5]

By 9.30 am, the men of the battalion, less the men of 'D' Company, the machinegun unit of 'E' Company, and the Battalion Headquarters Company, were transported by trucks to Ilolo, arriving their two hours later. From there, they began their trek towards Uberi. The leading elements reached the bivouac area at around 4.30 pm, with most of

CHAPTER 7

the remainder arriving within the next hour. It was then that a tropical downpour broke out, not ceasing until around midnight.[6]

Sergeant Colin Kennedy, a 21-year-old public servant from Canberra, with 'C' Company recalled many years later: 'The vehicles which took us towards the Owen Stanley ranges were American "Blitz buggies" which to negotiate a hairpin bend up the steep Hombron Bluff had to back-and-fill on one occasion. Up on top more rapid progress was made through the cool-looking rubber plantations of the Sogeri Tableland. Soon the troops had to start walking on a very muddy road, where only bipeds, quadrupeds and the sturdy little four-wheel drive "Jeep" could advance further. Less about 200 personnel left behind, the battalion started a trek which for about 25 per cent of them was to continue for more than three months.[7]

* * *

That morning, salvationists Major Albert Moore and Corporal William 'Jock' Inglis were up early and preparing to move south to collect the required supplies for their aid station. 'Up at 5.30 am we managed, even under those conditions, to get a bite to eat, and set off. We left Chester and Osmar, but their pace was too fast for Jock and me. There was a long, downward grade from Ioribaiwa with thousands of steps. Each step was about a foot deep and almost wrecked one's leg muscles. On arrival at the foot of the grade we found one of the war correspondents stretched out on the broad of his back. Jock and I stepped past and went on our merry way, feeling that the tortoise had triumphed over the hares!'[8] Indeed, Osmar White recalled their hasty trip back to Port Moresby: 'Wilmot and I went back to Moresby in the second week of September. Parer stayed with the troops to film the retreat. It was a hard, fast trip – 60 hours from Myola to the roadhead with the first "eye-witnesses". At the end of it, I wasn't much use to myself or anyone else. From a professional point of view, I had timed my run poorly. The climax of the campaign was still to come. The battles that would be remembered were still to be fought'.[9] Major Albert Moore continued his narrative:

SAVING PORT MORESBY

We were at Uberi at 10 am, thrilled with our progress, and managed to get something to eat before pushing on to the head of the road. There we ran into some of the worst conditions we had encountered. Mules getting into Uberi with supplies had turned the tracks into deep mud. On arrival at the roadhead, transport was not available, so we plodded through mud over our boot-tops and sometimes nearly to our knees.

On arrival at 'Ned' Kelly's,[viii] after getting a lift over the last couple of miles, I rang Major Groves, who had an education centre near the Laloki River. Red Shield officers from Moresby were there and I was able to speak to them. They came out, picked us up and we journeyed to Major Groves. Jock and I waited there whilst [George] Woodland and [Charles] Geddes went into Moresby to secure the equipment and supplies I needed back on the trail. When Woodland and Geddes met us on the roadside, they asked me if I had had any mail. I replied that I had not. They then extended their hands, offered congratulations and told me I had been appointed deputy commissioner of Red Shield services. That news shocked me, as it would mean leaving the 21st Brigade and, incidentally, the men of the 2/14th Battalion. I later sent a request to Salvation Army headquarters asking that I be left with my present appointment, but this was refused. I had to set up my headquarters at New Guinea force headquarters .... I also sent a signal to the rear echelon of the 2/14th requesting that a tent fly be sent out for use when we set up a Red Shield post in the jungle.

Sugar was in short supply, so in desperation I went to a ration depot and sought help. The officer-in-charge assured me that he was sympathetic but unable to make any sugar available, I continued talking and pleading. He scratched his head, called

---

viii Forty-six-year-old Captain Eric Kelly from Caulfield North in Victoria was a former officer with the 39th Battalion.

CHAPTER 7

his sergeant, said that he had to go out on a job and asked the sergeant to do what he could for me. The sergeant was a good type and soon I left with ample supplies of the commodity I was seeking.

We then set off for Captain Kelly's, where we got transport as far as trucks could go, then by jeep a few miles further. There we loaded our goods on three mules which took the load to Uberi. That was Saturday, September 5, 1942.

We needed native carriers from Uberi to the place at the foot of [Ioribaiwa Ridge] where I planned to set up a coffee post, so we decided to stay at Uberi overnight and set off on the final stage on the Sunday morning. But there was a major problem. Informed that the carriers would not carry on Sundays, I was not happy with the news and retired disappointed.[10]

* * *

Meanwhile, still lost further north, Lieutenant Maurice Treacy had left Captain Sydney 'Ben' Buckler's party that morning to reach Myola and find help. With him was with Private Frederick Rockliffe, a 28-year-old farmhand from Sassafras, and Private William Avent, a 28-year-old railway employee from Ararat in Victoria. Treacy wrote in his report: 'The party carried some [canned] salmon and rice as rations. After setting off, this party travelled in a SOUTH to S.W. direction; the intention being to cut the track which runs from MYOLA to TEMPLETON'S CROSSING then move along this track to MYOLA. Unfortunately, maps of the area were not available and accurate planning of the route could not be done. On the afternoon of this day, heavy rain began to fall, and a fog blanketed the mountains. This rain and fog continued for two days and nights'.[11]

Lieutenant Robert McIlroy and his patrol of the 2/14th Battalion were still lost in the mountainous area west of the main track; he wrote: 'Late in the afternoon we came out of the heavy growth onto a spur covered with Kunai grass. Across the front of the spur the mist had just fallen in a large re-entrant'.[12] Unknown to these men, Potts had already evacuated Myola and even Efogi further south.

* * *

By now, Major General Arthur Allen, who was based in Port Moresby, had been informed of the tactical situation along the track. The following orders were issued: 'GOC stressed that the Jap always goes for the high ground; the man who holds the high ground will always win. We do not know what a standing patrol is, as far as the fwd area is concerned – they are all fighting patrols. When a man goes out, he is to look for the enemy and fight him. Any rallying point is to be in the form of an ambush and the Jap is to be lured into it'.[13]

* * *

Meanwhile, on the other side of the Owen Stanley Range at the Japanese beachhead at Sanananda was 57-year-old Yorkshireman Father James Benson, who had been a prisoner of the Japanese for almost a month. Unknown to him, he was the sole survivor of the Gona and Sangara mission parties who had tried to flee from the Japanese since the invasion on 21 July 1942. Within weeks, all – including women, children and wounded Australian troops and American airmen – had either been hunted down and speared and clubbed to death by Papuan collaborators or had been captured and handed over to the Japanese marines of *No. 5 Special Naval Landing Party*, based at Buna, who quickly beheaded any prisoners who fell into their hands.[14]

Benson was still a prisoner of the Japanese Military Police at Sanananda, which was commanded by Captain Nakaiama, who had apparently taken a liking to the English priest. The Japanese captain had already spoken on Benson's behalf, almost certainly saving him from a swift beheading as a suspected spy.[15] Benson, after the war, described the Japanese captain who held sway daily over whether he lived or died:

CHAPTER 7

Nakaiama was a great home lover. He would often speak to me with deep feelings of his family and their home in an outer suburb of Tokyo. There was a son, I gathered, almost ready for the university and devoted to teaching, as indeed was Nakaiama himself. I gathered that Nakaiama specialised in mathematics, and during his early military service he was with the engineers. Later he had lectured at a big military academy, and on his retirement, he became headmaster of a secondary school. The war in the Pacific brought him back into the Army; and because of his age, fifty-seven, he was given a more or less sedentary post with the Military Police – the 'Kempetai'. He was also a poet and was very interested in some of our hymns which I recited to him from memory, for I had no hymn book …. I spent several uneventful months at Sanananda. I was, I suppose, a prisoner, though I could move about quite freely, and little attempt was made to limit the extent of my wanderings along the beach strip and into the jungle.[16]

For now, it may have seemed an uneventful place on internment, but within months, Sanananda would become a ferocious and bloody battlefield – a fight to the death between Japanese and Australian and American soldiers. Indeed, the battles for the Japanese beachheads at Gona, Sanananda and Buna, in late 1942 and early 1943, would become the largest battles fought by Australian troops in the Pacific, in terms of troops committed and casualties sustained.[17]

SAVING PORT MORESBY

PART 2
THE BATTLE FOR MISSION RIDGE

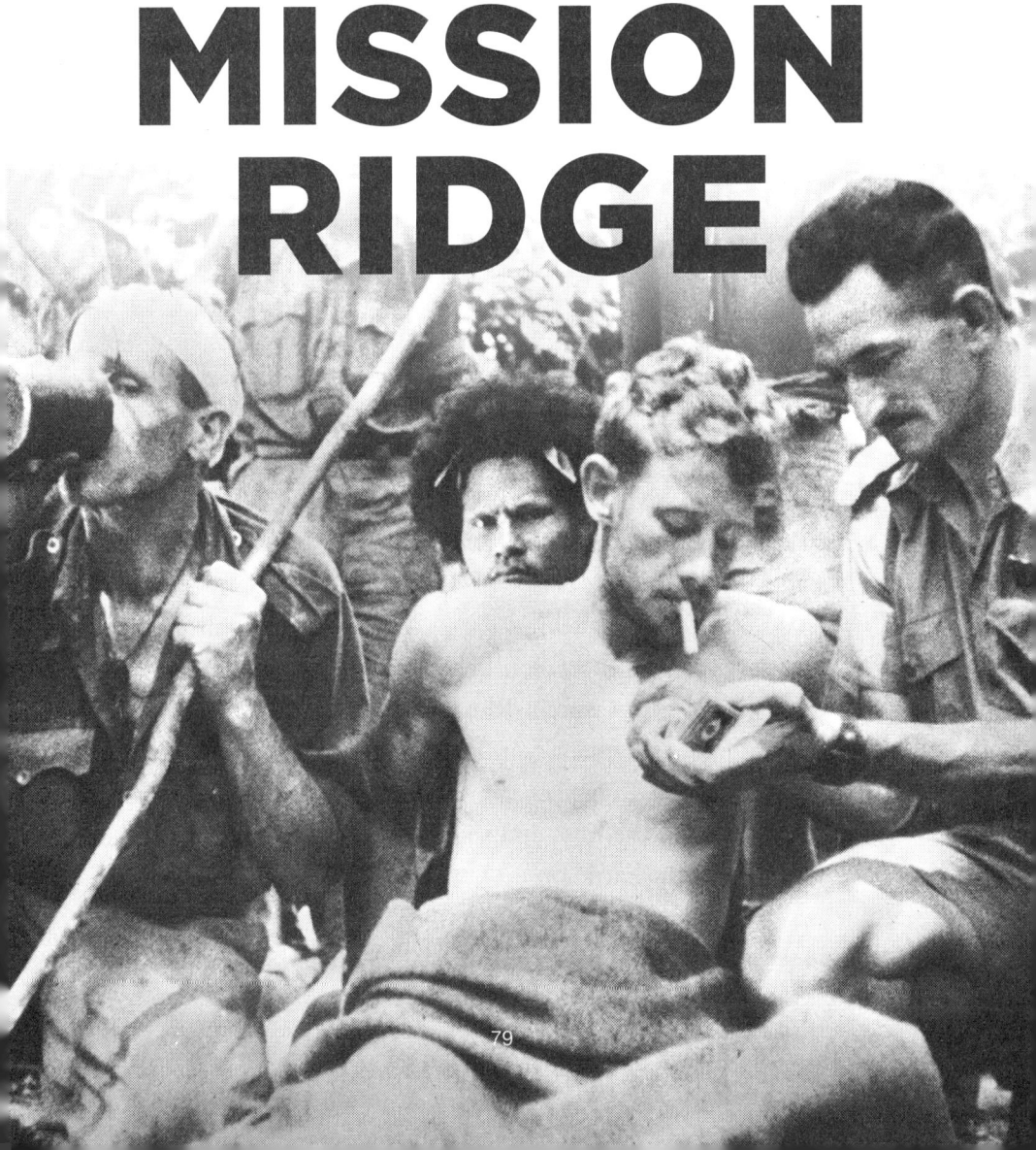

8
'... CONTINUOUS STREAM OF LIGHTS'

On the morning of 6 September, Lieutenant Colonel Caro, commanding the 2/16th Battalion, along with the temporary commander of the 2/14th Battalion, Captain Phillip Rhoden, led their weary men out of Efogi village and through the fresh 2/27th Battalion, who were now on the mid-slopes of Mission Ridge. As they passed through, their cobbers of the 2/27th Battalion shared their remaining cigarettes and tobacco with the tired veterans. For many, it was the first smoke in over a week.[1]

The men of the 2/14th and 2/16th battalions good naturedly taunted their newly arrived cobbers: '"Better late than never" or "You'd be late for yer [sic] own bloody funerals" ... there were weary grins as they threw the banter at the Croweaters [South Australians]. It was good to have them up at last; it was good to know someone else would take the initial shock when the Nips showed up again'.[2]

These exhausted men now took up positions to the rear of the newly arrived men of the 2/27th Battalion. The 150 or so men of the 2/14th Battalion were just behind the 2/27th Battalion with the 250 or so men of the 2/16th Battalion behind them, on the highest part of the slope to the rear. Captain Rhoden, with the 2/14th Battalion intelligence officer, Lieutenant Stan Bisset, were both standing on the mid-slope of Mission Ridge, near a thick tree line. Rhoden turned to Bisset, pointing to the trees and thick vegetation either side of the track: 'Potts wants the battalion up in there. Brigade HQ's about half a mile back [800 metres]. I wouldn't mind having your boys forward through, to keep an eye on things up front'.[3] Stan and his intelligence

CHAPTER 8

section would be located forward of the battalion to observe and report on Japanese movements.

Meanwhile, the 2/16th Battalion war diary records that at 5.30 am, 'A' Company had moved south back along Mission Ridge, followed at half-hour intervals by the Battalion Headquarters' staff, the Headquarters Company, then 'B', 'C', and 'D' companies. By 8 am, the battalion had taken up positions behind 2/14th Battalion, with the men of the 2/27th Battalion holding the forward defensive positions. Captain Sublet later described the position: 'The re-entrants east and west of Mission Ridge spur were thickly forested and carried creeks which joined the stream flowing from east to west at the foot of the spur. On the lower slopes of the Ridge grew Kunai grass growing four to five feet high [about two metres], while the remainder, including the flanks, carried forest of varying density ... water would be available only in the streams flowing down the flanking re-entrants, and the position could easily be outflanked'.[4] The battalion war diary records that at 10 am the arrival of the supreme morale booster occurred: 'Mail arrived for Bn – first since left ITIKI.'[5] Men quickly scribbled letters for home, but these letters were not 'enclosed' in envelopes. Which had long since become pulp. The sticky paper from packets of army biscuits was used to 'seal' the letters'.[6]

The medical officer with the battalion, Captain Steward, also recalled arriving on Mission Ridge: 'Here we got the ... great morale-lifter for a soldier – mail. No one had had a letter for five weeks, and the envelops were torn open eagerly and letters read at once .... Almost every man sat down at once to write his reply. Paper and envelopes were scarce, and many used biscuit packets instead. They wrote on the wrapper and used the sticky ends to seal them.[7] Among those to get mail was Captain Rhoden, who received a cable from his girlfriend Patricia in Melbourne, but he was too busy to read it at the time; he recalled years later:

> It wasn't until I got on board the troop ship at Brisbane, on August, on the way to New Guinea, and I wrote a letter to her,

a very short one, sitting out in the hold on frozen mutton just before the ship sailed out of Brisbane, 'Dear Pat will you marry me?' I leant over the side and gave the letter to a wharfie, I said, 'Will you post this for me digger?' He said, 'Okay mate, I will.' Leaping forward, in leaps and bounds, at Efogi on the way, when we were fighting a withdrawal, some mail filtered through and somebody handed me what was, in those days, an urgent telegram. We were engaged with the Japanese in the last struggle at that point, it was after Isurava. I put the thing in my pocket and thought no more about [it]. Wasn't until we got relieved, a fortnight later, that I was going through things and sorting myself out, I saw this urgent telegram. I opened it and it said, 'Yes, love Pat.' So, I was really engaged in the Owen Stanley Mountains.[8]

At 7 am, Lieutenant Colonel Geoffrey Cooper had sent Lieutenant Frederick Bell, a 28-year-old grazer from Denman in New South Wales, leading 8 Platoon, 'A' Company out on a patrol to the junction of the Kagi–Myola Track. He was to replace Captain Bryce and his militiamen of the 53rd Battalion; unknow to him, Bryce had already got orders to fall back before Bell arrived at their position. Bryce and his men were already on their way back to Port Moresby to rejoin the rest of the 53rd Battalion.[9]

At 11 am, Cooper sent out another patrol, this one led by Lieutenant Sidney Hewitt, a 23-year-old shop assistant from Adelaide, with 'D' Company who with his 25 men were to conduct a deep patrol to the east to check for any Japanese movements in the area covering the new track being cut by Captain Burt Kienzle and his men. Hewitt and his patrol were also to take up a position on the high ground covering the new track.[10] Kienzle was now just south of Mission Ridge and later wrote in his report:

On the new track from MYOLA to MENARI WO, [Joseph] Rae overtook me having been sent by Brig Potter. Fargher advising

CHAPTER 8

the evacuation of MYOLA ... and to proceed to MENARI and not to report back. We contacted the main track south of Efogi near Seventh Day Adventist Mission station at about 1300 hrs. and noted the arrival of 2/27th Bn taking up defensive positions to allow the already exhausted Bns to withdraw through. At 1700 hrs., we arrived at MENARI very tired after the long journey and had the responsibility of organising the ordered withdrawal of personnel not ... wanted to maintain supplies to forward troops back to NAURO and IORIBAIWA. News received late at night that the Japs were at EFOGI with lamps. WOs Davies and Preece were doing a grand job in charge of forward supply camp and holding stretcher-bearers ready to evacuate wounded. Both of these men volunteered for the job. WO Preece was placed in the most forward camp with 36 carriers.[11]

* * *

At around 2.30 pm, Japanese mortar and machinegun fire was heard north of Efogi. It was here that Lieutenant Bell's patrol was ambushed. The young officer was among the wounded, as reported in the 2/27th Battalion war diary: '1530 Lieut. Bell's patrol returned. Reported that Japs ambushed them by watching them get into position and then opening up at close range'.[12] The *General Narrative of Events* records the ambush occurred at around 2 pm. Just before, Bell and his men had observed at least 300 enemy troops moving in close formation down the main track about 1500 metres away. Just as Bell was taking up an ambush position, they were fired on by a Japanese advance guard. The enemy fire was at close range against Bell's right flank, which included fire from a Japanese heavy machine gun and at least one mortar. This fire 'was returned with probable casualties to enemy, but the patrol was forced to fall back after suffering casualties – four killed – four (incl Pl Comd) wounded'.[13]

By 5 pm, the Japanese were observed to have occupied Efogi, and Japanese long-range heavy machine guns targeted the Forward

Defence Lines (FDL) of the 2/27th Battalion on Mission Ridge. At the same time, enemy movement to the right of the ridge was observed. Tactically, the position the brigade occupied was described by the diarist of the brigade as being: 'a strong one though the feature required a full Bde strength and could be held for an indefinite period. Water was the main problem, which might have been overcome but here, as elsewhere, it was obvious if the enemy wished he could bypass either to the left or right with ease. Large concentrations of enemy tps were seen in EFOGI village at dusk and an urgent request was sent back for air action on EFOGI and on track back to MYOLA and MYOLA camp'.[14] Brigadier Arnold Potts message to Major General Arthur Allen requested: 'Turn on all you've got from [the] air on Efogi and Myola at earliest tomorrow morning. Include track Efogi–Myola. Our Forward Defence Lines Mission Spur and high ground 200 yards south [of] Efogi'.[15] Lieutenant Colonel Cooper recalled meeting Potts that day and how calm and composed he was, given the enormous responsibility on his shoulders:

> When we met Arnold Potts at Efogi he didn't show the strain or pressure he'd been under. He was quiet and gentlemanly and had a sort of gruff farmer's approach to people. He was quietly ebullient, if that's not a contradiction ... a little bubbling up all the time; never depressed, always pleasant to people. I never heard him ever dress anyone down; I never heard him make any aggressive comments about anyone in his own force. He was not an aggressive man, but he had a great deal of drive. He tended to jolly people along, to help them. He might make a suggestion. 'Well, have you thought of doing it this other way?' .... 'What about this proposition?' .... 'Do you think we could pull it off if we did this?' .... Never overbearing. There was, however, never any question that he was the boss.
>
> He was a strong man but never had the attitude, I'm at the top; I'm leading ... this will be a big deal for me – which a

CHAPTER 8

lot of commanders [had]. He was one of the commanders who was never looking to his own professional future. He was just doing his job with the brigade ... he knew all his company commanders; he got around, he was always there in the thick of it ... a very competent commander and a pleasure to be under. He was inclined to be humorous rather than dull about any difficult situation.

He certainly was a good planner .... He didn't take a risk at Brigade Hill. He had to sit down and shoot it out at every point usable for that purpose.[16]

Adjutant to the 2/27th Battalion, Captain Harry Katekar also recalled Brigadier Potts' decision to make a stand at Brigade Hill: 'If there is any question about stopping to fight at Brigade Hill it should be completely ignored. He was the man on the spot, what else could he do? He had two battalions which had been ... torn to bits really, done a hell of a good job; weary but still willing to do their best for their Brigadier. The 2/16th worshipped him, and it brushed off on the 2/14th and 2/27th'.[17]

An unnamed officer with the brigade recalled Potts was a 'magnificent bloke with a marvellous sense of humour. He was the greatest orator you ever heard, and he'd lash you with noble sentiments and tell you how good you were, and everyone clapped and cheered. He'd promise you the earth, too. A week's leave, extra rations, and we'd cheer him again – and never believe a bloody word of it! We really loved that bloke'.[19]

* * *

Just after 4 pm, Potts ordered Cooper to withdraw his men to a better position with less exposed flanks, south of the Mission Hut area. Potts gave a spirited address to these men telling them that 'every man must hold out as long as ammunition lasted'.[18] Thirty-seven-year-old Private Raymond Baldwin, from Claremont in Western Australian, with the

2/27th Battalion recalled: 'Brigadier Arnold Potts ... suggested that we pull back further up the hill because we are far too exposed and indeed, we came under heavy calibre machinegun fire there and we were vulnerable. We withdrew up the hill a bit further and scraped out a bit of a hole in the ground as best as we could'.[20]

Upfront, the men of 'C' Company, 2/27th Battalion covered the withdrawal as the rest of the battalion moved up the slope, rejoining the rest of the battalion at around 6.30 pm. About an hour before, a strong force of Japanese was seen passing through Efogi village. The battalion war diary records that by 5 pm, large numbers of enemy troops were visible in Efogi village; it was observed that these men appeared to be eating their rations. They were too distant for this battalion to open fire against them. Arrangements were made, however, for air support at first light the next morning.[21] The diarist of the 2/16th Battalion recorded the Japanese were closing in on the ridge, and they were ordered to take up positions astride the track at first light.[22]

It was now that one of the most famous incidents occurred during the whole Kokoda campaign, which became known as the Japanese 'lantern parade'.

While Lieutenant Bryce had observed something similar the night before – this was on a much larger scale. The 2/27th Battalion war diary records a procession of Japanese were observed moving with lights from Myola to Efogi between 9 pm that night until 5 am the next morning.[23] The brigade war diary concludes for the day: 'From dusk till 0200 hrs a continuous stream of lights ... were reported moving into Efogi'.[24] The historian of the 2/14th Battalion records that, 'Lieuts [Alan] McGavin and [Jack] Clements, maintaining an observation post, were greatly amused by the sight of lanterns flying through the air as the Japanese slithered and toppled on the slimy, sloping roots that crossed the "track"'.[25]

Twenty-three-year-old Private Alec Little, from Magill in South Australia, with the 2/27th Battalion recalled: 'That night we dug in at Efogi. I'll never forget that night as long as I live! We dug in, in

CHAPTER 8

a bit of open Kunai grass and we dug a hole with a bayonet and tin hat, and that's where we slept that night. And coming down the other hill towards the river were Japs, hordes of Japs with lanterns, and you could hear them jabbering and carrying on. And no noise! I mean there was us, no lighting a cigarette, no making any noise. We had to sit there watching them come up. It was pretty nerve-racking'.[26]

Lieutenant Grayden, 23-year-old mechanical engineer from Como in Western Australia, later learnt that the Japanese 'lanterns' were burning lengths of rubber-insulated signal wire.[27] The Australians, not having any Vickers machine guns, could do nothing to target the Japanese, as they were beyond range, and they could only watch in frustrated in not being able to retaliate, but also mesmerised by the trail of 'lanterns' slowly snaking its way towards Efogi. Indeed, Sergeant Robert Thompson with the 2/14th Battalion recalled: 'At night we saw them coming around the hills carrying lights and if we'd had even a [Vickers] machine gun, we could have picked them off, but we had nothing at all to hit them with. They were too far away, but anyway we watched them and knew that it was going to happen again'.[28]

* * *

Sometime that morning, Potts took time off to scribble a note to his wife Doreen on some crumpled, tissue-thin paper. He obviously completed this letter before the mail arrived. He told her nothing of the day's events, or what the next day might bring, but commented on his two battalion commanders, who had been in the thick of the fighting over the last two weeks, and the admiration for his men:

Sweetheart,

Have I neglected you so very badly? Sorry Hearts Delight, but it has been a busy and not very successful period. Not that land sales haven't been at a premium figure, but that was not my anticipation. Hell, but I'm tired and sleepy, down to under 12 stone [76 kilograms], as hard as nails and as dirty. Key's crowd (he's missing) and Caro's have done a marvellous job, soldiers to their fingertips .... I'm tremendously proud of the Brigade.

Sorry I can't talk more witchlet, but the main thing is that I love you always. Don't dream much as I don't sleep much, but the old carcase is still well on its feet Belovedest, no tummy of any kind and I can still lump 40 lbs [18 kilograms] over the most impossible tracks ....

No mail since I left Australia, but odd bags are coming up and I'm hoping in a day or two to have a real treat. Did I comment on the possibility of bringing back a grass skirt? It's quite attractive. Judy would look ripping in one, and so would her mother.

Sorry for such uninformative letters my darling, but NG isn't conducive to long letters and the opportunity to write ... rare.

It rains every day and generally catches you in transit with the little yellow devil being a perfect nuisance. So, one gets wet and more or less stays wet. Sleeping wet never did appeal to me and days of cold diet, no tea (no fires) gets monotonous. We do see life, however, and last night was hilarious as it was cold, and we are all sharing one blanket. It took an hour to quieten down the team and get to sleep. Must shave and barge off again.

All my love Hearts Delight, give my love to Muzzy and the Girls, just can't make time to write to them. I love you Witch and do be good to yourself, extra good and my love to Dad and Kiddies.

Your Bill.[29]

9
'THEY SHOULD BE VERY PROUD OF THEIR PERFORMANCE'

Colonel Kusunose Masao, commanding the *144th Regiment*, was pleased to see that the Australians were apparently going to make a stand on the spur to the south of Efogi. Missionary Ridge provided excellent opportunities for his force to conduct flanking manoeuvrers; he was determined to finally surround and decimate the Australian force concentrated in depth along the ridge. The Japanese of the *41st Regiment* would play no role in the fighting here, as they were still 20 kilometres north at Templeton's Crossing, as were the gunners of the *55th Mountain Artillery*. Each battalion, however, would go into battle with its attached mountain gun platoon. The Australians and Japanese who had fought each other at Isurava would again confront each other in a desperate and ferocious action.[1]

The Japanese official history indicates that on reaching Kagi, Kusunose's men were dismayed, as they expected to be able to see the southern coastline of Papua, and maybe even Port Moresby itself – however: 'The pursuit party went down the slope and passed through Kagi on 6 September. Soon after, they stumbled across an Australian camp south of Efogi. The expectations of the Japanese troops that they would soon reach the plains of Port Moresby after crossing the path were soon dashed. Mountain after mountain stretched in front of them as far as the eye could see'.[2]

Among these men was 1st Lieutenant Horibe, *No. 6 Company, II/144th Regiment*, who merely recorded in his diary for the night of 6 September one word: 'Efogi' (AWM 52 8/3/78 39th Battalion Appendix). Coming up behind their comrades of the *II Battalion*, was 2nd Lieutenant Noda Hidetaka with his men of *No. 8 Company, III/144th Regiment*, who recorded his second-last entry in his diary that day: 'Distributing the food supply left by the enemy at KAGI …. Lt TOKUHIRA and Lt FUJITA were wounded'.[3] Not far behind would have been the unknown leader of the signals unit with *No. 2 Infantry Section, No. 1 Company, I/144th Regiment* who also took some time off that day to record in his diary: 'The sec worked on the communication line which connects HQ with the Officer's Patrol'.[4]

Now approaching Kagi was the Japanese war correspondent, Okoda Seizo, who recorded in his unpublished manuscript his experience of pushing deeply into the heart of the Owen Stanley Range: 'The Horii detachment crossed mountain after mountain and was now marching through what seemed to be the heart of the Owen Stanley Range. The jungle became thicker and thicker, and even in mid-day we walked in the half-light of dusk. The humidity was almost unbearable. The surface of the ground was made up of decayed leaves and was covered with thick growths of soft, velvety green moss. We felt as if we were treading on some living animal. We were walled in by enormous mountains. The vapor that rose from the jungle hung heavy over us, rain falling almost all day and all night. The soldiers got wet to the skin through their boots and the undercloth round their bellies, and shivered with cold, especially at night and in the early morning'.[5]

Back at Myola were the men of the *I/144th Regiment* assigned to collecting any supplies left behind by the Australians, as well as helping carry the gravely wounded back to the field hospital on Kokoda Plateau. Among these men was 2nd Lieutenant Onogawa, *No. 2 Platoon, No. 3 Company, I/144th Regiment* who recalled the dry lake area being targeted by Allied airmen that morning; this was becoming a regular occurrence. During this raid, the lieutenant recorded that his battalion commanding officer, Lieutenant Colonel Tsukamoto Hatsuo, had been wounded while several of his men had also become casualties.[6]

CHAPTER 9

*　*　*

Just like the Japanese 'lantern' party, another famous episode occurred that day just a few hours south of Mission Ridge; the parade of the gallant militiamen of the 39th Battalion at Menari. Honner approached his adjutant: 'I want you to put the battalion on parade for me. I want to talk to them. I haven't met a lot of them. I don't know them. I don't know them by name. They should be very proud of their performance, and I'd like them to know that I think they're some of the best soldiers that I've had anything to do with and they've got no fears about being able to hold their heads up'.[7]

The subsequent address by the battalion CO was filmed by Damien Parer and would feature in Parer's Academy Award-winning film *Kokoda Front Line*. The battalion war diary records: 'The Bn was paraded, and the CO delivered a message of appreciation from Brig Potts for the splendid service the Bn had given while under comd 21 Bde. During this parade, a cinematograph film was taken to record the return from action of the first militia unit to meet the enemy'.[8]

Sergeant Major Cops, who was with the battalion, wrote in his diary that day: 'After our arrival, Colonel Honner called a parade of the survivors of the 39th and read a message from Brigadier Arnold Potts comd 21st Brigade thanks and appreciation for the splendid service the 39th Bn had given under his command of the 21st Brigade. We then proceeded back to Koitaki to recuperate and enjoy some comfort for a change'.[9] Honner later wrote of this parade: 'When I glanced along the steady lines of pallid and emaciated men with sunken eyes and shrunken frames that testified to the hardships, they had long endured, I saw no hangdog look – only the proud bearing of tired veterans who had looked death and disaster in the face and had not failed'.[10]

While at Menari, Sergeant Major Josph Dawson, a 21-year-old salesman from Footscray in Victoria, with 'B' Company, 39th Battalion – who had been in the thick of the fighting since the Japanese invasion

– recalled: 'I took advantage of the opportunity to have a bit of a clean-up. Due to being constantly wet and not being able to remove my boots, my feet felt dead, as though they did not belong to me. When I removed my boots, which were falling to pieces, and then my stinking wet socks, all the skin on the soles of my feet came off, stuck to the socks, leaving my feet raw. After a rummage in my haversack, I came up with another pair of socks and a couple of khaki hankies. All I could do was wrap the hanky over each foot then pulled my socks on over the top of the hankies, followed by the old worn boots. And move on'.[11]

Soon after, these men were heading south from Menari, which required negotiating a very steep treacherous ascent and descent to Nauro. Among them was acting platoon militia sergeant with 14 Platoon, 'C' Company, 29-nine-year-old Private Harry 'Bluey' Jardine, from Tanwood in Victoria. Like everyone else in the battalion, he was suffering from exhaustion and sickness. He had also been wounded at Deniki the month before. Jardine recalled: 'I was running a temperature at Menari, my boots were busted and in the heap of boots there the only ones fit to wear were a size 8 and a size 7 (so I wasn't the only one with an odd pair of boots!). Les Burnett of the 2/16th went crawling past (he had been shot in both calves and had walked and crawled all the way from Abuari). When we reached Nauro that evening my head was throbbing and I was sweating freely. Next day, as we reached Ioribaiwa, I flaked out. Keith McKinnis and Jim Cashmore brought a meal to me because I couldn't stand up. But next morning I was on my feet heading for Uberi – the thought of getting back was spurring us all on'.[12]

Staff Sergeant James Cowey, a 52-year-old orchardist from Belgrave in Victoria, was a veteran of Gallipoli and the Western Front. During the night of 10 August, during the fighting for Kokoda Plateau, he volunteered to stay behind to help get the men of 9 Platoon, 'A' Company off the plateau; the rest of 'A' Company had already withdrawn. Not only did he get these men out, which was then being

CHAPTER 9

stormed by the Japanese, but he managed to get them back to Isurava. Now here he was focused on getting one of his sick men back to Port Moresby, Corporal Allan Smith, a 21-year-old clerk from Kilsyth in Victoria. Almost all considered Cowey to be one of the bravest men of the battalion, and the men worshiped him.

Corporal Smith had figured in the fighting for the plateau, risking life and limb to get a hot meal and ammunition to the men. He had also led a small party of lost men from the battle out of the jungle onto the main track, soon coming across his CO Lieutenant Colonel Ralph Honner.[13] Smith recalled stumbling on his way to Uberi and Ilolo: 'After three days at Ioribaiwa, with the encouragement of Jim Cowey (who reminisced about the 1914–18 War and kept me going when I had lost interest in life), I finally made it to Uberi and onto Ilolo. I had scrub typhus (95 per cent fatal then), hookworm, both malarias and weighed five stone [32 kilograms]'.[14]

Most of these men would arrive back in the Port Moresby at the 30th Militia Brigade bivouac area on 9 September, almost two months to the day when 41-year-old Captain Samuel 'Uncle Sam' Templeton led the first men of the battalion, 'B' Company, across the track to reach Kokoda a week before the Japanese invasion.[15]

Corporal Donald Daniels, a 19-year-old interior decorator from Caulfield in Victoria, with 'A' Company, 39th Battalion had fought at Kokoda during the second battle for the plateau. Years later, when he was asked by a school student what it was like when the bullets started flying at Kokoda, he paused for a moment and then answered: 'Well, what I was thinking is, if you look to the north, you see nothing but Japanese. If you look to the south, you see Australia, and the only thing in between the two is the 39th Battalion'.[16]

How true he was.

* * *

Meanwhile, having cleared his aid post of the sick and wounded at Menari back to Nauro the day before, Captain 'Doc' Vernon, who oversaw the health of the Papuan carriers, was also preparing to assist, if required, the Australian Army medical teams. All were aware of the fighting that was about to break out around Efogi, and with it would be the inevitable avalanche of wounded. He recorded that his carriers suffered the first known casualties associated with the fighting along the track: 'On the 6th of the month, as far as I can recall the actual date, two carriers received slight but multiple wounds from bomb fragments dropped by Allied planes, and on the following day another carrier was killed instantly (fractured skull) by a [mortar] base plate falling from one of the dropping planes. For some time, these were the only cases of war injuries that I knew of in our [carrier] force. If only the incidence of sickness had remained at the low level of wounds the carrier force would have come through the campaign with trivial losses'.[17]

* * *

Early that morning, further south at Uberi, salvationists Major Albert Moore and Corporal Jock Inglis awoke to the good news that the carriers had volunteered to work that day: 'I learned that the carriers had volunteered to carry our gear and comforts, so we loaded the supplies and set off, while Jock awaited the arrival of supplies still to come to Uberi. He arrived later and by nightfall we had set up our post, happy that the morrow would see the Red Shield with a coffee post on the Kokoda Trail'.[18]

* * *

Early that morning, the militiamen and officers of the 3rd Militia Battalion continued their march north from Uberi to Ioribaiwa – their ultimate destination was Nauro, one day's march from Ioribaiwa. The battalion diarist recorded the men were on the move at 7.30 am, and by 6 pm, the leading elements arrived at Ioribaiwa during another all-too-familiar downpour. The officers of the battalion, however, soon got word that their CO, Lieutenant Colonel Albert Paul, was having difficulties and was still just two hours out from Uberi and unable to

CHAPTER 9

continue the march; the decorated officer of the Great War was forced to returned to Uberi.[19] Sergeant Kennedy recalled:

> The track was narrow and hemmed in on each side by high trees and a dense mass of undergrowth. Here and there beside the track were very large trees which were supported by a multiple flanging of the lower trunk, spreading over a large area at ground level for support in the soft jungle soil …. Bang the axe against the flanges and they would resonate a little.
>
> Frequently heavy vines as thick as a man's wrist, and thicker, could be seen hanging down like huge python snakes. It was said that they dripped water freely if cut, but experience did not always prove this to be the case. The canopy some 20 metres above permitted a dim light only to enter making the jungle gloomy. It was not impenetrable but tough rambling vines snaking through at all heights delayed movement. There was a new generation of trees thrusting upwards, straight as ramrods. These were cut down with (sharpened) bayonets for walking sticks. Noticeable after a few days was the silence of the jungle. Apart from man's activity, the only sounds came from running water, white cockatoos greeting the new day and saying farewell to the old. Occasionally a large hornbill might be heard above, whirring in flight to or from some tree with edible fruit …
>
> Few fires could be made from the damp rotting vegetable on the ground. Lieut. Alex Palmer, of Mittagong [with 'B' Company] by persistence, however, made one, feeding twigs unto a very small flame slowly improving the fire until he got a jam-tin boiling for a cup of tea (tea was regularly included in the rations). It was a good lesson for the observers. Most troops soon salvaged a tin for tea-making or heating meat. The tin supplemented a heavy metal dixie on issue to each soldier. Night revealed an eerie glow of phosphorescence in the fallen leaves and fallen tree

trunks beside the track. It was not generally known until about 18 months later that the rotting leaves were the home of the scrub typhus tick. The soldier who received a scab about the size of a five-cent piece on his body had reason to fear. It was said that there was no treatment. Some died, others recovered with no aftereffects. The jungle beside the track (necessarily) became used as though it were a latrine (few proper latrines were constructed along the track to Kokoda. The blowfly population 'mushroomed'; no soap was available, personal hygiene was virtually impossible, and very soon diarrhoea and dysentery among troops became a real problem).[20]

* * *

Still trying to make their way to Myola, not knowing the dry lakes were now well behind the Japanese lines, was Lieutenant Maurice Treacy and his two men from Captain Sydney Buckler's party. That morning they had crossed a creek and on its western branch they soon came across a large Japanese bivouac area. Treacy later wrote: 'This place appeared to be unoccupied except for two JAPS – one of these was shot. Later an improved track, which now roughly [ran] NORTH and SOUTH, was encountered. At this stage, the party had no rations and Pte Rockliff was suffering from dysentery. It was apparent that our troops had withdrawn SOUTH along this track; sounds of fighting could be heard [the ambush of Bell's platoon]. The party set off to attempt to reach the battle area and our own lines'.[21]

Meanwhile, Lieutenant Robert McIlroy and his men were on the mountain spur west of the track, having eaten the last of their rations. Later that day, two men who had become separated from Lieutenant Bell's patrol of the 2/27th Battalion came across them. McIlroy wrote in his report:

> Breakfast finished our rations. As soon as the mist rose a small native village could be seen on the far side of the re-entrant.

CHAPTER 9

Therefore, leaving the main body of the patrol to dry out their clothes and blankets which had been wet since the first night out, I proceeded across to the village with two men. We contacted one old native at the village who undertook to guide us to KAGI. Unfortunately, he gave up the idea and left us on the way back to pick up the main body.

In the meantime, the patrol had discovered a track leading in the general direction we required, and we followed that all day arriving at a native garden as the afternoon rain started, took some yams, and moved on to a partly cleared patch. About this time, we heard rifle fire to our left, but was unable to get observation from a nearby ridge. During the evening meal we were joined by two men from 2/27th Battalion who had been cut off from a patrol that day between KAGI and EFOFI. They had some tobacco with them and were very popular as our supplies had run out the previous night.[22]

* * *

Meanwhile, General MacArthur, still comfortably ensconced in Australia, sent a communique to the American Army Chief of Staff, General George Marshall, back in Washington. As succinctly assessed by historian David Horner, MacArthur saw an opportunity to ensure that any potential military defeats or disasters in Papua would fall at the feet of the Australians, despite the key Australian victory at Milne Bay, which he also took credit for. MacArthur cabled Marshall: 'The Australians have proven themselves unable to match the enemy in jungle fighting. Aggressive leadership is lacking. The enemy's defeat at Milne Bay must not be accepted as a measure of relative fighting capacity of the troops involved. The decisive factor was the complete surprise obtained over him by our preliminary concentration of superior forces'.[23]

10

'THE AIRCRAFT THUNDERED UP THE VALLEY'

During the early morning hours of 7 September, the Japanese patrol that Lieutenant Bell and his men of 8 Platoon had encountered the day before had now blown into an all-out Japanese advance with two battalions, two-gun platoons, and supporting elements, including an engineer platoon, closing in on the deplete Australian brigade's position strung out on Mission Ridge. Overall, Colonel Kusunose's force totalled around 1450 men, while Brigadier Potts had a slightly smaller force numbering around 1300 men.[1] Kusunose, however, had a significant advantage in terms of medium machinegun, artillery and mortar support, and he was on the offensive, and as such, he was well placed to flank the strung-out Australian force at any one point of his choosing. He could mass his troops for a knockout blow, and at this point significantly overwhelm Potts' defenders. This was the same tactic used so successfully at Isurava and Eora Creek just weeks before. Potts would have to dance to Kusunose's tune.

Kusunose divided his force into three main elements, an artillery group, machinegun group, and his two infantry battalions. He would pin down the Australian frontline with several feints by the men of the *III/144th Regiment*, while those of the *II/144th Regiment* would represent the main attacking force seeking out a weak point in the Australian left flank. The artillery, totalling two 75 mm guns, two 70 mm guns and two 37 mm guns would support both battalions. These guns were in a bamboo grove on a small hillock 1500 metres

CHAPTER 10

from the Australian frontline. From here, they could also site Potts' headquarters' position on Brigade Hill, around 3000 metres distant. The medium machinegun company of the *III/144th Regiment* was equipped with telescopic sights and was placed forward of the artillery group, about 1000 metres from the Australian frontlines.[2]

Even so, of ongoing concern for Horii was the state of his men of the *144th Regiment* who had being fighting the Australians since the invasion of 21 July, as malaria and dysentery was endemic. He was determined to reach Port Moresby, issuing the following orders to his officers: 'Those who were required to be hospitalised should be sent forward according to the capacity of each company instead of being sent to the rear and should be hospitalised at the field hospital that was scheduled to be built shortly at Port Moresby'.[3]

* * *

Earlier at 5 am, 'D' Company, 2/16th Battalion moved out on the left to a position along the main ridgeline above a significant cliff face. An hour later, 'B' Company moved astride the main track and along the ridge to connect with the right of 'D' Company. At 7 am, 'A' Company moved to a position along the ridge east of the track overlooking the position held by the men of the 2/14th Battalion. Thirty minutes later, the men of 'C' Company moved out to cover the extreme right flank, occupying the ridge astride a subsidiary track leading east from the Kokoda Track – this track would later become significant in the survival of the brigade.

Indeed, Lieutenant Colonel Caro had already judged that any withdraw would be along this narrow eastern subsidy path originating at this spur from the main ridge. By 10 am, Battalion Headquarters had established itself just to the left of the Kokoda Track on a low, thickly timbered transverse spur separating the 2/16th Battalion from Brigade Headquarters. By 11 am, the battalion CO, major, adjutant, and intelligence officer had moved up to reconnoitrer the high ground on their battalion's right flank.[4]

At 8.30 am, 'C' Company, 2/27th Battalion reported a party of what appeared to be Papuan police to the west; this was confirmed by Brigade.[5] This Papuan force was led by Captain Thomas Grahamslaw, a 40-year-old civil servant from Port Moresby. Grahamslaw had been the senior ANGAU officer in the Kokoda before the Japanese invasion. Indeed, he had a narrow escape during the first week of the fighting and was critical in getting two shot-down American airmen back to Port Moresby during this period. He was now in charge of a small Papuan force tasked with seeking out and eliminating Japanese infiltration parties along the track.[6]

The battalion war diary of the 2/27th Battalion reported at this time a party of 12 enemy troops observed in Mission Hut who soon moved out towards the western flank of the battalion. They were targeted by the men of 'B' Company, but it soon became clear that the party consisted of at least 30 men. Behind them, another party of Japanese arrived at the Mission Hut area, attempting to set up a mortar; however, they too were targeted with rifle and machinegun fire forcing them to withdraw. These were patrols from the Japanese *III/144th Regiment* who were trying to identify the Australian positions.

Within minutes, Australians holding the frontline were targeted by a Japanese mountain gun that cut a signal wire near 'C' Company's position, disrupting communication for a time with Brigade Headquarters and wounding in the head 26-year-old Captain Ronald Johnson, from Adelaide, commanding 'C' Company. The track behind the battalion's position was also constantly swept with Japanese machinegun fire. Cooper's men requested mortar fire support at 9.15 am on a concentration of 50 Japanese to their front, but the mortar team ran out of bombs before ranging was completed.[7] Lieutenant Grayden recalled the Japanese mountain guns could be 'aimed with almost pinpoint accuracy. Australian troops moving on the ridge stood out like ducks silhouetted in a shooting gallery and were quickly targeted. Casualties, particularly among the 2/27th Battalion, rapidly mounted'.[8] Corporal John Burns also commented on the effect of the

CHAPTER 10

close-range shelling by the mountain gun: 'The disconcerting and damaging effect of the Japanese quick-firing mountain gun was due to the fact that being fired at such close range the shells had actually struck and exploded before the noise of discharge could be heard'.[9]

At around 10 am, it was reported by 'A' Company, 2/27th Battalion that another Japanese patrol was moving around their right flank. The brigade now ordered 29-year-old Lieutenant Lindsay Schwind, from Wangaratta in Victoria, with the 2/14th Battalion and commanding the brigade mortar battery to 'take action'.[10] Thirty minutes later, the forward observation post of the battalion reported seeing additional Japanese, at company strength, moving out of Efogi towards their position. All companies now sent out patrols, each consisting of ten men, to observe and report Japanese forward movements.[11]

* * *

At 8 am, the Australians witnessed eight B-26 Marauder twin-engine bombers – from the 89th Bomb Squadron, United States Army Air Force 3rd Bombardment Group – escorted by four American Airacobras from the 80th Fighter Squadron strafe and bomb the Japanese positions at Efogi and further north.[12] An unknown soldier who witnessed the attack wrote: 'The aircraft thundered up the valley below them, banked steeply around the northern toe of the feature and dropped their bombs on Efogi. Smoke and flames billowed as huts exploded and pieces of timber and bamboo flew into the air. The aircraft climbed away and swooped back repeatedly, machine-gunning the village and the track to Kagi and Myola'.[13]

Medical officer Captain Steward with the 2/16th Battalion, also recalled the aerial attack: 'On 7 September, the morning sun shone from a sky of cloudless blue – an event in itself rare enough. We heard the drone of aircraft approaching and the usual question sprang to every mouth: "Ours or theirs?" As eight bombers and four fighters zoomed in over Efogi village the roar went up: "They're ours!" "You bloody beauties!" "Give it to the bastards!"'[14] Fires were seen in

Efogi, and several explosions were heard, which were believed to be ammunition dumps that had been hit.

About two hours later, another sortie of Airacobra fighters made an appearance and swept down strafing the track and ridges between the marked Australian frontline positions and Efogi.[15] Corporal Burns also recorded: 'This was a most cheering sight and the planes swept very low past the battalion position in carrying out their mission. It later proved to have been most effective. The battalion used air-recognition strips and smoke grenades fired from a rifle discharger to indicate its FDLs to the aircraft'.[16]

One Japanese report stated these attacks resulted in 'heavy casualties'.[17] That day, Major General Horii reported to *17th Army Headquarters*: 'Enemy planes made three passes this morning, 7 September, from 0630 hrs to 1000 hrs. Some tens of enemy planes inflicted approximately 100 casualties and deaths in a sustained attack. The lack of Japanese fighter activity means that enemy planes have complete control of the air. It is felt that these attacks will certainly continue in this way. I wish for steps to be taken to despatch Japanese fighters'.[18] His report was not received until 10 September. While the Japanese suffered casualties from these sorties, it is doubtful the casualties described by Horii are accurate, and he may have inflated the casualty figures in an attempted to emphasise the need for Japanese air support – if that was his expectation, he would be disappointed because the Japanese strung out along the track would receive little if any direct support from the Imperial Japanese Air Force during the Kokoda Campaign.

* * *

At around 9 am, Brigadier Potts sent a signal to Major General Arthur Allen in Port Moresby, informing him that he intended to make a stand on Mission Ridge, but also appraising him that he was in no condition to undertake offensive operations – indeed, the Japanese could readily flank his current position: 'Will not give ground if you

CHAPTER 10

guarantee my line of communications ... [men] full of fight, but physically below par, mainly feet and dysentery, naturally affects most. Every endeavour being made to contain enemy on line approach but nothing to prevent him bypassing, hence line of communication problem .... Respectfully consider offensive ops require more than one Bde task. Air action this morning most successful and heartening'.[19]

The area held by the 2/27th Battalion was described by Corporal Burns as being covered in high grass just above a small clearing near the mission ruins that were represented by an old wooden building. It was bounded at each side by 'forest-covered gullies which fell steeply away on its flanks. Timber in the actual battalion area was scarce, but owing to the high grass and undergrowth, there were no fields of fire, and visibility was limited to 15 yards or less. The battalion took up a compact diamond-shaped defensive position astride the track; "A" Company was right forward, "B" Company left forward, "C" Company left rear, "D" Company right rear, with Battalion HQ centre rear on the reverse side of a knoll in the centre of the battalion area. The troops hastily constructed weapon pits using tin hats and bayonets and cut fields of fire in the long Kunai grass, using the cut grass and boughs for camouflage'.[20] The battalion intelligence section set up its observation post in 'C' Company's sector, as did the three-inch mortar crew. Intense and accurate Japanese mountain gun fire, however, would soon force these posts to move.[21]

Meanwhile, located on the centre of Mission Ridge, the men of the 2/14th Battalion had an excellent view of the aerial attacks. The men of 'B' Company were positioned upfront covering the ground left of the track, with 'D' Company in support just behind them. The men of 'A' Company were in the centre covering the forward position to the right of the track, while covering the right flank was 'C' Company, with Battalion Headquarters Company close by in support of 'A' and 'C' companies. During the day, it was estimated by the men at the battalion's observation post that around 600 Japanese had entered Efogi village at the foot of the ridge, who were subjected to heavy raids by Allied aircraft throughout the morning. The Japanese responded throughout the day with intermittent mortar and machinegun fire.[22]

* * *

At 11 am, both sides were bombing each other with intermittent mortar fire, resulting in few casualties on either side. It was now that the 2/14th Battalion reported to brigade that a force of around 100 Japanese was carrying heavy loads into Efogi village.[23] The brigade diarist at this time also recorded: 'A runner from 2/27 Bn reported [to] … BHQ and requested water to be sent with the rations. Also asked for more signal cable. Casualties were light at this stage, but the Japs were working up on the right and left. 11.15 hrs 2/27 Bn were ordered to clear flanks with strong patrols'.[24] At this point, another sortie of Allied airmen appeared with nine Airacobras strafing and bombing the Japanese in Efogi village.[25] Lieutenant Schwind, resupplied with mortar-bombs, was ordered to put down fire against Mission Hut and beyond. He commenced firing at 11.30 am, but the first mortar-bomb failed to clear the timber and several men of the 2/27th Battalion were wounded. Soon, the mortar was being targeted by a Japanese field gun, and not long after, Schwind's observation post – which was directing fire by using a telephone cable connected to the mortar team – was also targeted by the same gun. Later that day, this mortar team suffered a direct hit, killing Private John 'Norm' Wilkinson, a 26-year-old labourer from Essendon in Victoria.[26]

By noon, the Japanese of the *III/144th Regiment* were still pushing their patrols south to pinpoint the Australian frontline positions. The 2/14th Battalion reported to brigade at 1.30 pm that around 200 enemy troops were observed moving in close formation along the Myola–Efogi Track. The 2/27th Battalion, holding the northern forward perimeter, sent out several fighting patrols to disrupt the Japanese; its war diary records: 'Three patrols sent out by A Coy – one

CHAPTER 10

patrol reported Japs in Hut in clearing on RIGHT flank. Fighting patrol under comd Sgt [Robert] Johns sent out at 1530 to contact these Japs. Two patrols from B Coy out for 2½ hrs. Nothing to report. Lieut. [Aldebaran] Trenerry and patrol left at 1530 to move in on MISSION HUT from flank. Nothing to report. Failed to reach HUT. C Coy sent out fighting patrol and snipers to contact enemy moving on LEFT Flank. Nothing to report. No enemy located'.[27] At around 3 pm, Captain Grahamslaw and his patrol of Papuan constables reported back to brigade having also not encountered any Japanese.[28]

Meanwhile, the 2/16th Battalion further up the slope had sent 31-year-old Captain George Wright, from Perth, and a platoon from 'B' Company up to the ridge east of the Battalion Headquarters to relieve 'C' Company, which was to move further east to relieve a patrol from the 2/27th Battalion. Tasked with helping to defend the Brigade Headquarters and the brigade supply dump was 'D' Company, 2/16th Battalion, moving south. An hour later, word came into brigade from the 2/14th Battalion that more Japanese troops were observed 'carrying and dumping gear halfway between' Efogi and the Australian forward defence lines.[29]

At 5 pm, the patrol – led by 24-year-old Sergeant Robert Johns from Adelaide with 'A' company, which had moved out to contact the Japanese near the Mission Hut – had returned. As recorded in the battalion war diary: 'Johns reported having killed six Japanese and captured one light machine gun and a grenade launcher without loss'.[30] The 2/27th Battalion historian recorded: 'The patrol was in high spirits and came over the skyline [with] ... Sergeant Johns waving the LMG in the air and Private H. "Snowy" Hill the grenade discharger'.[31] Johns himself later recalled:

> I took out a patrol which caught a group of the enemy by surprise and killed six of them. Returning without casualties of our own and when members of our patrol were safely back in their foxholes, I remembered the enemy weapons left behind

with the dead Japanese soldiers. At that stage of our conflict with the Japanese, we had captured no enemy weapons and the thought occurred to me that it would be a good idea to go back to the location of our small victory and bring back a sample of the enemy's weaponry .... I called for a volunteer to come back with me so we could provide cover for each other, and one man came forward ... 'Snow', which was his nickname. I could not have wished for a better mate for the job in hand. We made our way back to the scene of our conflict. Snow covered me as I carefully moved the last 50 yards or so into the spot where the bodies of the dead Japanese lay undisturbed. Without wasting time in selecting which weapons to take, I picked up a light machine gun and a small mortar and rejoined Snow as rapidly as possible. Wasting no time, we were soon climbing the steep track which led up to the beginning of Efogi Ridge, where the Corporal and his section were watching us from their foxholes. I must admit that Snow and I were very relieved to be on the way back to relative safety.[32]

Thirty minutes later, the patrol – led by Lieutenant Hewitt with 'D' Company, 2/27th Battalion – covering the right flank came in, bringing back with them 11 large biscuit tins of water, which was much appreciated because the water situation was bad. Along with these tins of water, the usual afternoon downpour enabled more water to be collect in groundsheets. It was now that 'A' Company's sector was again targeted by at least one Japanese mountain gun resulting in two killed and five wounded, the shelling lasted for around 30 minutes.[33] The brigade war diary reports the situation as darkness began to fall: '1730 hrs 2/14 Bn reported having tied up with 2/27 Bn patrols. 100 Japs reported going into EFOGI. Water position was now better as two patrols had filled water bottles at the creek unmolested and further it was reported that a water point had been located in the Bn area'.[34]

CHAPTER 10

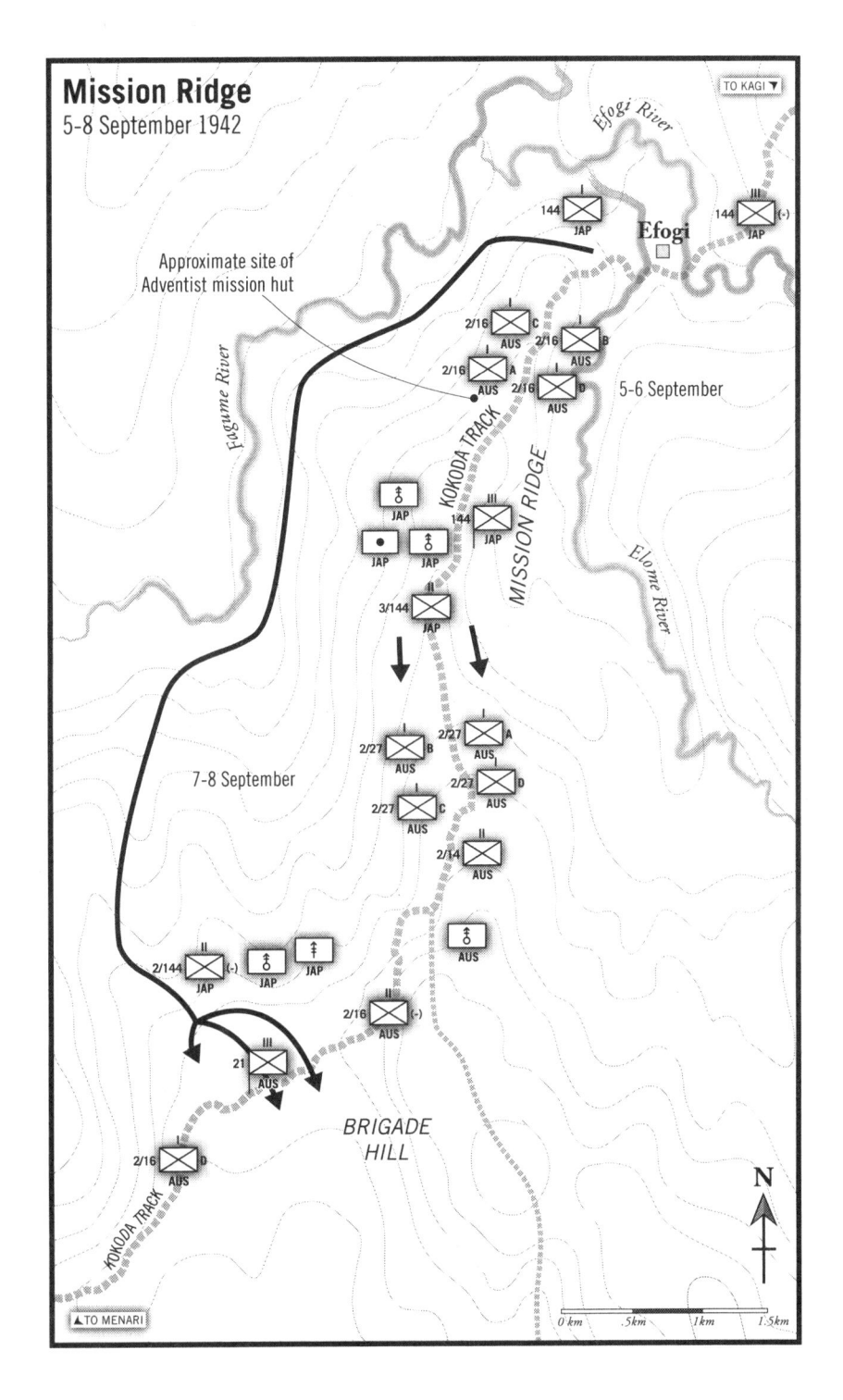

Mission Ridge
5-8 September 1942

By 5.45 pm, a large Japanese force was observed moving around both flanks. Lieutenant Colonel Caro reported that his men of the 2/16th Battalion had observed over 100 enemy troops moving down from Mission Hut on his left flank.[35] Indeed, the men of the 2/14th Battalion also observed more Japanese bringing up supplies closer to the front, reporting another large party of heavily laden Japanese moving down the main track towards the Australian northern perimeter. On hearing this, 'D' Company, 2/27th Battalion sent out a patrol led by 24-year-old Lieutenant William Egerton-Warburton, from Kensington Park in South Australia.[36]

The brigade war diary concludes for the late afternoon of 7 September that one of the Japanese mountain guns began to shell the sector held by the 2/16th Battalion, and the brigade's mortars retaliated against this field piece. An hour later, the Japanese responded with a more intense bombardment from at least one mountain gun and several mortars targeting the positions of the 2/27th Battalion; while their frontline was also swept with machinegun fire.[37] The Brigade's *General Narrative of Events* concludes for 7 September at dusk, the firing on the forward positions by mortar, heavy machine guns and quick firing mountain guns increased in power, and skirmishing continued throughout the night.[38] Corporal John Burns concluded his narrative of the 2/27th Battalion, recording, 'the battalion settled down for whatever might be in store. Burying parties in each company worked hard until 2200 hours, quietly laying to rest those who had fallen during the day'.[39] Most expected the Japanese to launch their attack against Mission Ridge early the next morning.

* * *

Just days before, Sergeant Makings with the 2/6th Field Ambulance had evacuated their aid post at Efogi and was now located south of Mission Ridge, at Menari. He recalled: 'Our planes over for quite a while dropping supplies. Later our bombers and fighters bombed and strafed Efogi, the Japs were in there. A little later three planes suddenly

CHAPTER 10

dived over the hills and dropped a few bombs and machine [gunned] our village, the planes happened to be USA who had mistaken the village, fortunately very little damage. Was very fortunate because there was quite a crowd about and didn't take much notice of the planes. This was the first time I had been in close bombing. In the afternoon we were warned of the seriousness of [the] position and were given emergency posts and were given a rifle and grenades each. I hadn't handled either before but soon got the drift of both'.[40]

11

'AT ABOUT MIDDAY ANOTHER JAP WAS DISCOVERED'

During the late afternoon of 7 September, Japanese Major Horie Tadashi and his men of the *II/144th Regiment* set out on a wide flanking manoeuvre to the west to cut off the Australians, and to attack them from the rear at first light the next morning.[1] With these men would have been Lieutenant Horibe, with *No. 6 Company*, who recorded in his diary: 'EFOGI – 0710, fought against approximately 100 of the enemy. Casualties, five wounded and two killed. The enemy is occupying a position along the crest. Thirteen men were wounded, and one man killed by heavy-mortar fire. From 1700, commenced to advance to attack. Because of uncertainty of enemy situation, dispatched scouts to reconnoitrer before advance. Prepared to attack at dawn'.[2]

Also appearing to have been attached to Kusunose's headquarters during this period was the unknown leader of the signals unit with *No. 1 Company, I/144th Regiment*, who recorded that day: 'Maintained the communication line between Regt HQ and the No. 2 Bn. NISHIMURA and NISHIORI, together with some natives, worked on the communication lines'.[3] Most of his comrades of the *I/44th Regiment* were still scouring the dry lakes around Myola for dropped Allied supplies or were tasked with carrying the seriously wounded back to the field hospital at Kokoda.[4]

That night, Major General Horii, who was located just north of Efogi, issued Nankai Shitai Operational Order 'A' No. 113:

CHAPTER 11

Enemy forces equipped with trench mortars have occupied a succession of positions on the highlands on the western [south] side of OOD-O [Efogi?]. Intense strafing and bombing by enemy planes is expected tomorrow morning between 0600 and 1100 hrs.

The SHITAI will endeavour to defeat this enemy force between tonight and dawn tomorrow, following this up with a pursuit attack.

The pursuit force will capture by means of night attack the enemy position confronting our arty position, while the main strength of our advanced guard, at latest by dawn tomorrow, will defeat the enemy in the vicinity of the straight road, and follow this up by a pursuit attack. In view of strafing and bombing raids by enemy aircraft, every effort must be made to change our action to a pursuit attack by 0530 hrs at latest, and have each TAI open up the vicinity of the mountain arty positions.

The various TAI(s) forming the main body will alter their dispositions in their present area by dawn, in accordance with requirements for safety from intense enemy strafing and bombing, in particular taking up wide deploying formation. Their further movements will be covered by a separate order.

I will remain at my present location until dawn, and then go forward at the head of the main force.

Personnel to receive order will be sent here at 0500 hrs tomorrow, the 8th.

HORII Tomitarô
SHIOTAI Comdr.[5]

Sometime that night, 2nd Lieutenant Noda Hidetaka, with his men of *No. 8 Company, III/144th Regiment,* recorded his last entry in his diary: 'The village above the ravine was raided from the air (twice before 0900 hrs). Two of the company were killed and one wounded. For the first time I experienced the honour of an air raid. "Death is fate". No good being pessimistic. Advance to Port Moresby believing in final victory'.[6] The young Japanese officer was likely one of the ten Japanese killed sometime the next day (another 45 were wounded) while fighting the Australians of the 2/27th Battalion on Missionary Ridge.[7] His diary was recovered a month later by Australian troops as they pushed north in their offensive.

*　*　*

Further south at Menari, Captain 'Doc' Vernon of ANGAU had established his aid post; he recorded in his ANGAU report for that day: 'There was great activity at Menari during this period but no signs of panic or even confusion when it became obvious that a further retreat would be forced on us. Carrier parties were sent to the frontline right up to the day of withdrawal, often returning with stretcher-cases, but though some were later cut off by the Japs our losses were light'.[8] He also recorded in his personal diary[ix]: 'The excitement at Menari gradually increased. I got what sick I could back to Nauro and prepared for instant evacuation of the rest. Graver and graver news come in from the front, we slept with our packs ready to hand, ready to move, and the nights were disturbed by distant firing and officers arriving for meals at all hours, in fact there was a continuous restaurant service run by ANGAU which functioned well right to the end. Nevertheless, there was no confusion or panic. Quantities of tobacco, chocolates, boots and clothing dropped from the air were spread out in the village square for anyone to take'.[9]

[ix] A copy of Doc Vernon's personal 'diary' has been published by the 39th Battalion Association. This 'diary' was clearly written after the events, as Doc refers to places by name before they were then designated as such (e.g., Templeton's Crossing) and subsequent events. It is likely a combination of records of his personal diary and additional information gained likely in early 1943 – it is best considered a narrative of his experiences of the campaign, as opposed to a strict diary.

CHAPTER 11

Earlier, Doc Vernon's good friend Captain Burt Kienzle left Menari during the morning to report to Brigadier Potts at Brigade Hill. Kienzle later wrote: 'I reported to Brig HQ. Brig Potts and advised him about new track discovered from MYOLA to south of EFOGI. Whilst at BHQ, saw the Japs advancing on mission and the action of our 3" mortars bursting on them. These seemed to cause many casualties. I also saw a splendid action by our Air Force in strafing and bombing the Kagi–Efogi Track and depots. After inspecting the native camps and carriers just behind BHQ on MANARI track I returned to MENARI. Capt. Vernon had returned from Port Moresby to take up medical duties in forward area but owing to the danger of being outflanked advised him to return to NAURO'.[10]

* * *

Still at Naruo, medical officers Major Oldham and Captain Wallman were trying to obtain sufficient supplies to treat the sick and wounded. That day it was decided that the medical personnel would need to withdraw to Uberi and from there to Ilolo before finally reaching Port Moresby. However, Captain Wallman remained on the track and continued to perform surgery at Uberi for another ten days before he too was forced to return to Port Moresby with the Japanese advance to Ioribaiwa.[11]

* * *

Further south at Ioribaiwa, the militiamen of the 3rd Militia Battalion were still waiting on their rear elements to reach their position. It was around 8 am that the officers were informed by Major General Arthur Allen that Lieutenant Colonel Albert Paul was to be temporarily replaced by the battalion's 2i/c – Major Robert Grantham, a 43-year-old shire clerk from Moruya in New South Wales. The rear elements of the battalion did not arrive at Ioribaiwa until just after midday. The men of 'C' Company, led by Captain Thomas Boag – a 28-year-old public servant from Canberra, who had been tasked with reaching

Nauro – with two of his officers and 81 other ranks, moved out at around 2 pm; they would not reach their destination until noon the next day. The men of 'B' Company were to move out to Edwalla, north-west of Ioribaiwa at 6 am the next morning to conduct patrols, while those of 'A' Company would remain at Ioribaiwa.[12]

* * *

Just behind these men, on the southern slopes of Ioribaiwa Ridge, Major Albert Moore and Corporal Jock Inglis were up early having established their post near Ua-Ule Creek. Both were keen to get the coffee and tea brewing – troops were already passing through north and south. A few years later, Moore remembered with some pride the Red Shield Banner next to his makeshift shelter of timber and canvas:

> I awoke on the morning of Monday, September 7, full of hope that the day would see us meeting a great need on the Kokoda Trail. We did not have proper facilities for heating water, only kerosene tins and a couple of urns. Wood we cut from the steep mountain along the side of the creek, and from the creek itself came our water supplies.
>
> By 8.30 am coffee was flowing. The steady stream of sick and wounded were goggle-eyed as they beheld the Red Shield sign some hundreds of yards distant along the track and round a bend in the Trail. Most of them thought some chap had hung the sign from a tree for a joke, but as they approached the post, they discovered that it was for real. The men would slump on the ground and, with a cup of coffee and some biscuits, would be refreshed. That, without a doubt, was to be the most compensating of any service I tried to render to the men of the forces. That day we had 39 stretcher-cases as well as many wounded and sick men and supplied a total of 40 gallons [180 litres] of coffee.

CHAPTER 11

There was practically no one on the trail after 4 pm as none would be able to make Uberi by dark if late at this post. During later days, as the invader shortened the line between us, there were latecomers, but they had to camp somewhere along the Trail when dark overtook them. I had been instructed that no one be allowed to stay at our post.

It had been my joy to erect it in many places and under varying circumstances, but never before or since has the erecting of any shield given me a greater thrill than this one. We placed it so that it was visible before the tent could be seen – every man knew what stood behind that emblem.[13]

Among the men returning to Port Moresby that day was Private Jack Howson, with the 39th Battalion, who recalled reaching Ioribaiwa Ridge: 'Somehow, I kept walking. I remember going down … [Ioribaiwa] Ridge with no boots, no shirt, half a pair of trousers, starving, crook (if that's a strong enough word) and those grenades hanging and swinging on my webbing. I was nearly at the bottom of the stairs when I see a RED flag. In a flash I've got a grenade off my webbing, pulled the pin out and was just about to throw it when I realised: That's not a Jap "Rising Sun", it's the "Salvos' Red Shield" flag. And there they were with mugs of hot coffee and packets of "Log Cabin" tobacco. You couldn't believe it, but it was true – the "Salvos" were nearly as close to the "front" as the troops'.[14] Soon, Moore would be famous among those who served on the Kokoda Track, known to many as 'the Simpson of the Owen Stanleys'.[15]

***

Lieutenant Maurice Treacy and his two men from Captain Sydney Buckler's party were still pushing south to reach Myola, not knowing they were already south of the dry lakes:

> During the morning of 7th Sept., the party moved through several bivouac areas which the enemy was in the habit of building

as shelter from the torrential afternoon rains. These bivouac areas consisted of a varying number of shelters constructed of palm branches and will accommodate four to six men. Some had a raised floor. It was subsequently noted that enemy parties moving up to the frontline, always used these places at night and during the day sick JAP troops occupied them. At about midday another JAP was discovered – his uncooked rice provided a meal for the party. Sounds of mortar fire could be faintly heard during the afternoon and the party continued along the track, hoping to gain contact with our own troops in the shortest possible time. Three of the four days given by Capt. Buckler had almost elapsed. Darkness eventually prevented movement, the party halted and consumed some dry rice.[16]

Meanwhile, to the west, Lieutenant Robert McIlroy and the men of the 2/14th Battalion – having that morning taken a well-earned drag from rolled cigarettes supplied by the two lost men of Lieutenant Bell's patrol of the 27th Battalion – now tried to break out from the jungle to the main track somewhere close to Efogi. He late wrote: 'With 2/27th chaps as guides headed for EFOGI but the track we followed along a ridge took us too far to the west'.[17]

* * *

General Douglas MacArthur – who had made ongoing derogatory remarks to the American Army Chief of Staff back in Washington regarding the Australian efforts to halt the Japanese advance – now conveyed similar sentiments directly to Australian Major General George Vasey with a request that he inform Lieutenant General Sydney Rowell of the necessity to 'energise combat action'.[18] Vasey wrote to his good friend Rowell and also referred to a complaint that Rowell had made about the incorrect news of the fighting in New Guinea. Vasey said the misreading of the situation was due to the 'time-lag between the actual events and the issue of communiques', and because

CHAPTER 11

of 'MacArthur's own personal outlook and actions'.[19] Of the latter, Vasey wrote: 'My information . . . comes from Howard our Press Relations Liaison Officer at GHQ. He says that MacArthur will not admit that any serious operations are going on in New Guinea and, as you probably know, all press articles must bear out the tone of the official communique . . .. The reasons for this attitude of MacArthur I do not know – nor does Howard'.[20] One can only read this as an attempt to be diplomatic – Rowell's response went unrecorded.

12

'... HE WAS STRUCK IN THE CENTRE OF HIS THROAT BY A BULLET'

Throughout the early morning hours of 8 September, the men of the brigade were distracted by the Japanese to their front who were keen to make their presence known. All this noise was likely intentional to draw Australian attention away from any concerns about their left flank that was then being skirted by the Japanese *II/144th Regiment*, and/or as part of their panache for psychological warfare. All seemed to hear the digging and incessant chatter by the men of the *III/144th Regiment*, who had obviously crept forward during the cover of darkness.

Sergeant John Scott with 'C' Company, 2/16th Battalion recalled the Japanese activity throughout the night and early morning hours: 'Couldn't see the Japs, but we could hear them. We were given sacks of grenades. At first, we would pull the pin, throw the lever away and roll the grenade over the edge down the slope towards the noise we'd heard. Trouble was they had seven-second fuses and they'd bounce right down to the bottom before they'd explode. That wasn't any good, so we counted to three or four and threw them over. It was like Russian Roulette in the dark, hoping there weren't any four-second fuses among them. It seemed to do the trick, though'.[1]

Sergeant Allen Gittos, a 20-year-old assays assistant from Pinjarra in Western Australia, with 'A' Company, 2/16th Battalion also recalled: 'For the first time in over a week we had a chance to recuperate – and it was wonderful to know, with the 2/27th Battalion below now taking the brunt of the Japanese drive, we had support at last. But all the

CHAPTER 12

time we knew the enemy would be working around us and preparing. The Japs were funny fellows. Through the night we could hear them chopping at the vegetation and moving around yelling to each other as though there was no need to be careful'.[2]

In response to this enemy activity, by 3.30 am, Lieutenant Egerton-Warburton, 'D' Company, 2/27th Battalion led another patrol forward of the Australian lines. By 5 am, reports came into Brigade Headquarters that heavy skirmishing had broken out in the forward sector held by the men of the battalion, with the Japanese focusing on Cooper's right flank. These were the Japanese of *No. 8 Company, III/144th Regiment* that were tasked with keeping the Australians pinned down. To their right were the men of *No. 7 Company, III/144th Regiment* that were tasked with attacking the centre of the Australian frontline. During the dark hours, Major Horie Tadashi and his men of the *II/144th Regiment* had continued their wide flanking manoeuvre, for a time hugging the banks of the Fagume River, before swinging east to attack the Australians at dawn.[3]

Holding the northern frontline perimeter was Captain Sims and his men of 'A' Company right of the track with 'D' Company in support, while Captain Lee and his men of 'B' Company were left of the track with 'C' Company behind in support. Sims' men beat back several Japanese assaults, pouring fire into the advancing Japanese who were using fire and movement, and carefully using the terrain and vegetation to their advantage – the assaulting troops on *No. 7* and *No. 8 companies, III/144th Regiment* were determined to seek out and destroy the Australian positions.[4] One unknown Australian recalled: 'Along they came blaring bugles and making all the weird noises under the sun'.[5] It is likely that Japanese Lieutenant Noda, with *No. 8 Company, III/144th Regiment*, was killed here.

The battalion war diary records at 4.30 am that the enemy attacked strongly along the track targeting 'A' Company's front. These men replied with 'intense small arms fire and liberal use of hand grenades. Japs were stopped but a few were considered to have penetrated our

FDLs between A and B coys and to be hiding in long grass. Japs concentrated on A Coy front and made attack after attack, but small arms fire and grenades effectively stopped any move fwd. Enemy strength now estimated to be at least five coys in addition to patrols of unknown strength'.[6] Corporal Clive Edwards, a 25-year-old salesman from Pinnaroo in South Australia, likely with 'D' Company recalled: 'At about 4 am hell broke loose, and the Japs attacked A Coy's front. We had no attacks on our particular front but there was a terrific amount of muck whizzing over our heads. Well, the racket continued with slight lulls, all the morning'.[7]

The historian of the battalion recorded it was a rifle section and a section commander's fight: 'Each section post had to be a veritable fort in the line of posts forming the battalion perimeter, relying on its own ability to conceal and camouflage its position and to resist enemy attacks and infiltration, without hope of calling for outside fire support; it was also self-contained and self-reliant so far as supplies were concerned, as it was impossible for anyone to move safely to and from these posts in hours of daylight'.[8]

* * *

Before dawn, Major Horie Tadashi and his men of the *II/144th Regiment* were ready to launch their attack against the rear of the Australian position close to Brigadier Potts' Headquarters. They had spent eleven hours trekking along the banks of the Fagume River and then eastwards through the jungle and up the 45-degree western incline of Mission Ridge.[9]

Private Nishimura Kokichi with *No. 2 Platoon, No. 5 Company* recalled the advance many years later: 'The march took a night and a bit. We left before sunset and arrived just as the sun was about to come up. 6th Company led and I was in 5th Company just behind. Leading 6th Company were friendly local natives. It was not difficult going along the track system, but the last bit was difficult. Towards the end we went up a steep track'.[10] Close by was Lieutenant Sakamoto

CHAPTER 12

Atsushi, commanding *No. 2 Machinegun Company, II/144th Regiment*, who also recalled the advance here: 'Started to climb a steep mountain which takes 11 hours. Detoured in order to come out at enemy's rear … slashed through the jungle aided by engineer Tai'.[11] Indeed, Private Albert Ward, a 24-year-old shop assistant from Adelaide in South Australia, with the 2/27th Battalion recalled the difficulty of the Japanese flanking approach: 'Still, when they encircled us at Efogi in an area like that …. Well, you'd have to be a qualified mountain goat to be able to do physically what they did – to be able to get right around the battalion; around Brigade Hill'.[12]

Having reached their objective, Nishimura recalled: 'The Australians were very close to us, perhaps not more than twenty metres away, but we could not see them'.[13] Nishimura and his eight-man squad quietly dug several foxholes in a small area forming a crescent. With him was his officer commanding (OC), Lieutenant Inoue, with his head still wrapped in bandages from a friendly fire wound inflicted weeks before by a Japanese medic during the fighting along Misima Ridge east of Isurava. From their position, they could see the Kokoda Track and a huge tree with a three-metre-thick trunk; they would have been close to the position of the men of the 2/16th Battalion.[14]

* * *

Brigadier Potts was up early having heard the crescendo of fire as Horii redoubled his assault against Cooper's front. He and his staff took mental notes of its onset, received Cooper's report, and went about their duties: 'Potts walked across the clearing soon after dawn to a newly excavated pit [latrine] near the jungle's edge. A sentry – Pte. Gill[x], one of the 'old and Bold' of the Guard Platoon – manned a Bren-gun post on the forward slope of the knoll overlooking the point where the track broke from the jungle saddle. The two men exchange pleasantries of the morning. Returning after a few minutes, Potts had

---

x   The author has unfortunately not been able to identify any details regarding Private Gill.

barely reached the roofless shack when a sudden shot rang out across the clearing. He turned to see the sentry fall ... the dead man lay sprawled over the shallow hole of his post'.[15] Potts had a lucky escape.

Lieutenant Norman Cairns, the brigade liaison officer, now collected Corporal Beveridge[xi], of the Brigade Transport Platoon, and both crossed the clearing, passing Gill's position on the knoll approaching its north-western corner. Here the corporal said quietly to his officer: 'Look, there's a Jap'. Cairns told Beveridge, 'Give him a grenade'. It was now that Cairn saw another Japanese soldier in the shadows of the jungle with a light machine gun aimed at them. He yelled to Beveridge to get down, just as the Japanese opened fire. Beveridge had just tossed his grenade as he heard his officer yell, [and] the corporal crumpled as a hail of bullets tore into his body; moments later his grenade exploded, killing both Japanese. Cairns tried to drag the mortally wounded corporal back but now another Japanese soldier leapt from the scrub and charged at him. Cairns scrambled to his feet and just then a shot rang out, [and] the Japanese soldier fell dead'.[16] Lieutenant Burnham Fraser, a 41-year-old grazier from Caulfield in Victoria, announced to a thankful Cairns: 'Couldn't miss him'.[17]

* * *

The Australians now realised they had been flanked. Lieutenant Stan Bisset and his intelligence section were still located forward of the 2/14th Battalion at an observation post focusing on the Japanese activity to their front along the perimeter held by the men of the 2/27th Battalion when they realised the Japanese were behind them. He and his men could hear from the Brigade Headquarters' area concentrated fire; Bisset recalled one of his men saying: 'Bloody hell, what the heck is *that*?'[18] Years later, Bisset recalled: 'At Brigade Hill in particular, we were just desperate, but we had ... three-inch mortars, and short term we caused a lot of damage there with the three-inch mortars but they

---

xi The author has unfortunately not been able to identify any details regarding Corporal Beveridge.

CHAPTER 12

only had a limited number of rounds, but we would loved [to have had] our Vickers guns there because they were all within range of our Vickers and that would have made an enormous difference'.[19]

The diaries of the 2/16th Battalion also record the unwelcomed and unexpected appearance of a large Japanese force to their rear: '0600 Two shots fired down road towards Bde HQ on investigating it was discovered that japs had got between Bde & Bn'.[20] Medical officer Captain Steward, with the battalion, recalled hearing the first shot of the Japanese to their rear: 'How could one shot reveal so much? Even the RAP men, not accustomed to the daily use of firearms, learned to distinguish the sounds of the different weapons used in close jungle fighting. It only took a little practice to recognise the sharp snap of a rifle, the clatter of the submachine gun, the flatter burst of a light machine gun or the heavy thud of a weapon like the Japanese Juki gun. But most important of all, one can tell the difference between "ours" and "theirs"'.[21] With this, Lieutenant William Grayden, commanding 15 Platoon, 'C' Company, 2/16th Battalion, was sent out along the left flank to conduct a patrol; his task was to identify the strength and location of the Japanese who had flanked their position. The young lieutenant provides a fascinating account of his patrol:

> I was tasked with forming and leading the patrol. Doug Burgess had been wounded at Alola; Warrant Officer Tommy Foy had been appointed to replace him as Platoon Sergeant of 15 Platoon. He was one of the older members of the battalion and had served with the British Army in India before the war. I chose him and four others as members of the reconnoitring party. The patrol moved off down the extremely steep slope … with a sense of anticipation, hoping for a speedy clarification of the threat that the new development posed.
>
> As the officer in charge, I intended to proceed in a wide semi-circle to intercept the saddle behind where the Japanese were likely to be established, at a point 150 to 200 metres along its length.

After descending 100 or 150 metres, we unexpectedly intercepted a newly made single file track in the soft, moist rainforest floor. It was apparent that a large number of troops – a company of approximately 120 or possibly more – had moved along this track during the night. We had found by our own experience that the only practicable way for troops to move in the complete blackness of the rainforest at night was in single file with each man following the man immediately in front of him.

There was no way the significance of this track could be misconstrued. It was clearly only a few hours old, and a large number of troops had passed over it. Our patrol would now follow the track to find where it led. We proceeded with extreme caution, not knowing when or where we might intercept the force that had created it. Apart from the track itself, there was no sound or other indication of the enemy as we followed in their footsteps of the previous night. As we cautiously advanced, the track approached the crest of the saddle. The most likely place to make contact with the enemy would be at or near the crest.

Sharing the risk equally among all ranks, the six members of our patrol formed an extended line – each about two metres apart – and commenced a slow and careful crawl to the top of the saddle. I would then have to decide whether we would move to the left or the right. Press as flat as possible on the sloping rainforest floor we were mostly concealed by 750 mm or so of light undergrowth. To my right, however, there was a small clear patch perhaps two metres wide. Our platoon sergeant Tommy Foy crawled across this clearing. With his head raised he was looking intently at the ridge, scanning it for any sign of the enemy.

I turned to look at Tommy and, as I watched, he was struck in the centre of his throat by a bullet fired from six metres or less in front of us. Being on a steep slope and having been fired from above, the bullet travelled the full length of his body killing him

CHAPTER 12

instantly.[xii] Tommy Foy had joined 15 Platoon only a few days earlier and had been welcomed because of his experience and his rank of WO2. He had willingly taken a step down in rank to be appointed platoon sergeant of 15 Platoon. Because of our regard for Tommy, we were all thankful for the small mercy that at least his death had been instantaneous. Tommy's death shocked our patrol. To lose a person of his status so suddenly and so unexpectedly in such circumstances was a bitter blow, especially as we had approached the lip of the ridge so cautiously.

We had made a completely silent approach and, except for that tiny clear patch, had been totally concealed from the enemy. Prior to that single shot we had no indication of the Japanese presence. The fact that no other member of our patrol came under fire indicated that the enemy could not see … the rest of us in the undergrowth. From the numbers we had estimated from the track it was apparent that the Japanese had sufficient strength to man all points of the saddle. Our brief contact also made clear that they were well dug-in, camouflaged and alert. The Japanese were masters of camouflage and, even though present in numbers, could maintain complete silence and stillness for extended periods.

We had not seen the enemy, but it was obvious that anyone who showed themselves would have suffered instant death. The slightest movement in the low undergrowth could betray our position. Minutes passed. As a reconnoitring patrol we were charged with obtaining information and reporting back to HQ. We now had all the information we required. With the continued lack of activity and it being impossible to recover Tommy's body at that stage, we carefully and noiselessly slipped backwards down the steep slope and out of contact with the

---

[xii] Forty-one-year-old Sergeant Thomas Foy, from Narrogin in Western Australian, is listed as Killed in Action, 8 September 1942.

Japanese. Retracing our steps along the Japanese track of the previous night, we reported the existence, disposition, and our estimate of the approximate strength of the Japanese forces to our Commanding Officer, Lt Col Albert Caro. That information, limited as it was, formed the basis for the subsequent bayonet attack to clear the track.[22]

Lieutenant Grayden and his men had verified that a substantial Japanese force had established itself between Brigade Headquarters located just behind a western spur on top of the Mission Ridge on a small hill – they were now cut off from the three battalions positioned further north.

13

'AN ATTEMPT IS TO BE MADE TO FORCE OUR WAY THROUGH'

Earlier at 6.30 am, Major Horie Tadashi and his men had launched their attack against the 2/16th Battalion Headquarters' position about 400 metres forward of Brigade Headquarters. Corporal John Corbett, a 24-year-old tailor from Subiaco in Western Australia, with the Battalion Headquarters recalled the fighting here:

> We called it 'Butcher's Hill' because so many people were killed there, we had terrific fighting there and I finished up … with about a dozen blokes on a very steep long hill, it was a hill but a valley, the side of a valley and … we saw several boxes of grenades which had been one of our earlier ammunition dumps and as we raced across the top, we grabbed a couple of boxes of grenades each and flung ourselves down and the Japanese were able to fire thousands of rounds at us. You know they had stuff to spare, and we would not have fired a round unless we had a target and all they did was strip all the trees in the background, but as long as we lay flat they couldn't hit us, and we were just, every time a group of Japanese got close to us, whoever they were closest to just dropped a grenade on them and bang! That was the end of them. And then the next group would try and come up here and when they got up there, bang! …
>
> I remember the funniest thing that happened there was the padre came wriggling in alongside of me and [said], 'Give me

one of those, Johnny,' and I was a bit of a churchy bloke and I said, 'Oh Padre, you can't have one of these.' 'Oh, I suppose not old boy. I must think of the cloth and all that,' and he wriggled out and smacked me on the bum as he went down to the next chap. There was humour in different places. That was a terrible place, and we lost a terrible lot of fellas there at Butcher's Hill.[1]

Soon after, the Japanese were attacking the Brigade Headquarters itself. Horie's men quickly secured a foothold astride the main track between Brigade Headquarters and the 2/16th Battalion, requiring a hurried exit by their Captain Henry Steward, who was with the Regimental Aid Post (RAP) that was towards the top of the ridge. The Japanese of *No. 5* and *No. 6 companies, II/144th Regiment* bore the brunt of the fighting. The men of *No. 5 Company* attacked north of the Brigade Headquarters, isolating it from the rest of the brigade, while those of *No. 6 Company*, which included 1st Lieutenant Horibe, attacked the Australian Brigade Headquarters itself. Between both companies on the western crest was the machinegun company, and just behind them was the Battalion Headquarters, with the men of *No. 4 Company* in reserve.[2] The Australian 21st Brigade diarist recoded Japanese machinegun fire targeting the 2/16th Battalion Headquarters' area, while the Brigade Headquarters' staff had to stand-to; it was then that 'the Japs attacked Bde HQ'.[3]

Now the Australians, in the headquarters' area, from brigadier on down, were exchanging steady fire with the Japanese, who were attempting to infiltrate their perimeter; they had occupied the saddle, cutting the main track, and with it the telephone cable linking them to the Battalion Headquarters further north. Meanwhile, Lieutenant Norman Cairns had acquired a two-inch mortar and 'proceeded to bombard the jungle, sometimes with more enthusiasm than judgement'.[4] This attack by the Japanese of *No. 6 Company*, however, was checked by the Battalion Headquarters Company, supported by Lieutenant Bert Madigan leading 12 Platoon, 'B' Company and two

CHAPTER 13

platoons from 'C' Company north of their position, forcing these Japanese to fall back about 100 metres.[5]

In the fighting here was Lamce Corporal William Cousens, who recalled: 'It was decided that we would attack the heavily entrench Japs dug in on either side of the track and try to clear the way through to evacuate the wounded. Captain Bert Madigan and an OR led this attack and made it through before the main body entered this cauldron of hell and were cut to pieces. It was just plain suicide, akin to the charge of the Light Brigade. I was very fortunate in this attack, as my role was to fight the rear guard with my Tommy gun'.[6]

Brigade Headquarters and the survivors of 'D' Company, 2/16th Battalion were now isolated from the rest of the brigade. Soon after, Cairns – who was leading the defence of the Brigade Headquarters – discarded the mortar for a Thompson submachine gun to extracted revenge upon the Japanese for killing Private Gill, Corporal Beveridge and others of the brigade team by knocking out a Japanese light machinegun position and pouring a hail of bullets into the gun's crew.[7]

Also involved in the defence here was Brigade Intelligence Officer, 28-year-old Captain Ronald Cowan, from Adelaide, as recalled by Corporal John Burns: 'Captain Cowan … played an effective part in this action and must have been glad of his experience as a section commander in "A" Company of the 2/27th Battalion'.[8] Of major concern was that the brigade supplies were now out of reach from the bulk of the brigade, and the men of the 2/27th Battalion were especially feeling the strain because they were constantly under attack. Burns recorded: 'There was … no chance of receiving the much-needed supplies of ammunition and food. The ammunition position was serious in the battalion and food was short. Only a ration of water had been available for two days'.[9]

The diarist of the 2/14th Battalion also recorded that their Battalion Headquarters was forced to move its location due to the Japanese attack to their rear. Even so, the observation post manned by men of 'C' Company maintained their position to report on enemy movement through Efogi village. It was estimated that during the previous 48 hours, around 1500 enemy troops had entered the village, which was again bombed and strafed during the morning by Allied airmen. The enemy, however, kept up continuous pressure on the front perimeter of the 2/27 Battalion, and mortar fire was falling on their forward area, which resulted in several casualties.[10]

While the Japanese of *No. 8 Company, III/144th Regiment* had been charging 'A' Company's position with great determination, they had also been employing their all-too-familiar tactic of trying to flank enemy positions. The 2/27th Battalion war dairy records that Japanese machinegun fire was sweeping their whole battalion area, making it difficult for the men to move, even though ammunition was brought forward to 'A' Company.

At 7.15 am, Airacobras were overhead; however, hopes of the promised 'close air support' were in vain, as they targeted the rear of the Japanese position. It was now the Japanese of *No. 7 Company, III/144th Regiment* put in a relatively light attack on 'B' Company's position, but this assault was readily repulsed, and most considered this attack to be a feint.[11]

It was clear to all that a Japanese force to the west was attacking Potts' Brigade Headquarters' position to the rear just below the reverse slope of Mission Ridge on a small hill. While this became known as Brigade Hill, most who fought there called it 'Butcher's Corner'. The 2/14th Battalion war diary records the Japanese were applying pressure against the frontline positions, coupled with flanking movements, which became increasingly evident, as to the rear, Australians could be heard from higher positions directing fire against enemy infiltration parties now between the 2/16th Battalion and the Brigade Headquarters' positions. A 108-wireless set with the 2/14th Battalion was soon put into use, and intermittent communications was re-established with brigade. For some unknown reason, brigade would only be contactable using this radio from 8.30 to 10.30 am and 1.45 to 4 pm. Apart from this, no communication was possible.[12]

CHAPTER 13

Messages received from Brigade Headquarters could be re-laid from the 2/14th Battalion to the 2/16th and 2/27th battalions by telephone cables.[13] The 21st Brigade signals later reported that the performance of the 108 MK 2 Wireless Set was by no means satisfactory. They would sometimes work well up to nine kilometres, but at other times failed to work over much shorter distances.[14]

The diarist of the 2/27th Battalion also recorded the attack to the rear, writing it was found that 'Japs had attacked 2/16 Bn and succeeded in breaking through and taking up a position astride track and ridges thus effectively cutting all communications between fwd tps and 21 Bde. 0745 A Coy again called for more grenades after repulsing another Jap fwd move. 0750 A Coy effectively stopped Japs by liberal use of Bren, Tommy, and Rifle fire, and Grenades. A Coy used grenades whenever Japs attempted to move fwd'.[15]

At around 9 am, another sortie of Airacobras strafed the positions held by the *III/144th Regiment* forward of the 2/27th Battalion; however, according to the battalion war diary, they again targeted the track between Efogi to Myola. The Australians fired smoke grenades to attract attention and indicate to the airmen the position of the Australian FDLs, but they either failed to see or just ignored them.[16] At 9.20 am, the aircraft finished strafing and returned to base. It was then that the 'Japs seized opportunity to surge fwd but after fifteen minutes of fruitless endeavour attack repulsed'.[17]

By now, however, six of Sim's Bren gun positions had been knocked out during the early fighting. And while the company had gone into action with 1200 grenades, with each man carrying 100 rounds, all of this was soon used up, along with the whole battalion reserve.[18] The battalion war diary records that food, water and ammunition for 'A' Company became acute at this stage and the company sergeant major was kept busy collecting ammunition from personnel not actively engaged in the fighting. A small reserve of hand grenades was located on the high ground on the track near 'C' Company, but it was effectively covered by 'Jap snipers using LMGs'.[19]

At 9.45 am, word reached the brigade from the 2/16th Battalion that the Japanese were being heavily reinforced – likely those of *No. 4 Company, II/144th Regiment*. The brigade ordered the 2/14th Battalion to help the 2/16th Battalion, while the 2/27th Battalion was to drop back one of their companies to protect the left flank of the 2/14th Battalion.[20] By 10.15 am, word was received from the 2/27th Battalion that the Japanese were cutting a track to the west from the main track. Potts sent orders using the radio for the 2/27th Battalion to 'probe the left flank with a strong patrol'.[21]

At around 10.30 am, just before Potts' radio failed, he instructed Lieutenant Colonel Caro to take command of the brigade if headquarters was wiped out – which then seemed likely. Caro was to withdraw Maroubra Force to Menari. The diarist of the 2/16th Battalion recorded: 'Advice received from Bde that Lt Col Caro would be in charge of ops in the event of Bde HQ being wiped out. Position critical, shortage of food and water and track covered by MGs from rear'.[22] Caro, if possible, was to try and break through to Brigade Hill with his remaining three companies, and if necessary, to commit the 2/14th Battalion to support the attack, while the 2/27th Battalion held the northern perimeter.[23]

Caro and his men of the 2/16th Battalion had been involved in a desperate fight against flanking elements of Horii's force; the fighting here had broken out at 6.30 am, and his own headquarters was in danger of being overrun by the Japanese of *No. 5 Company, II/144th Regiment*. Even so, Caro at this point tried to relieve Potts by pushing Sergeant George Morris down the track. As Morris and his men pushed south, they were forced to take up a defensive position, as they could not break through to the HQ's and 'D' Company's position on Brigade Hill because the whole area was swept with enfilade. The situation was deteriorating with the passing of every minute. If Morris and his men know they could not break through – it was doubtful anyone could.[24]

CHAPTER 13

* * *

At this point, Brigade Major Hugh Challen – to the south at Menari – heard that Potts and his headquarters had been cut off from the rest of the brigade.[25] Challen began to collect as many men as he could to form a company-size relief force. It was now that Captain Geoffrey Lyon, a 28-year-old accountant from Brighton in Victoria, arrived at Menari. Lyon had days before been sent forward by Major General Allen to report on the situation. The young captain conferred with the Brigade Major Challen, who informed him that Potts' headquarters was apparently cut off, and he was about to lead a composite company to Brigade Hill.[26]

* * *

Back along the northern perimeter of Mission Ridge, the Japanese of *No. 8 Company, III/144th Regiment* had launched another attack against 'A' Company's position, with Cooper's men again pouring small arms fire into the advancing Japanese, throwing the last of their grenades into the face of the enemy, supported with fire from the three-inch mortar team. The Japanese were again forced to fall back. With this, the Japanese called a halt to their attacks against Sim's company, as recorded in the 2/27th Battalion war diary: 'Situation quietened down considerably and until midday Japs fired automatics and MORTARS spasmodically with A Coy using small arms fire or grenades on any enemy target that offered'.[27]

At noon, the early morning patrol led by Lieutenant Egerton-Warburton, 'D' Company 2/27th Battalion, was able to make their way back to the Australian perimeter, likely coming in from the east.[28]

It was also now at Brigade Hill that Captain Bret Langridge and his men of 'D' Company, 2/16th Battalion were tasked with guarding the brigade's rear and the brigade supplies. It was here that they found themselves with the brigade personnel heavily engaged; in some places they resulted to hand-to-hand combat with the Japanese of *No. 6*

Company, II/144th Regiment. Meanwhile, the 2/14th Battalion was ordered to withdraw to the 2/16th Battalion's position – to support the charge down the track to break through to Potts' position, while the 2/27th Battalion was to hold the Japanese to the north in check.[29] The battalion war diary for the 2/16th Battalion records: '1145 2/14 Bn ordered to withdraw to our area. An attempt is to be made to force our way through'.[30]

Indeed, the men of the 2/27th Battalion were still holding the northern perimeter expecting at any time a renewal of the Japanese attacks against their thin perimeter. The historian of the 2/27th Battalion recalled: 'No orders from brigade could be received and Cooper therefore consulted the CO of the 2/14th Battalion and agreed to withdraw the battalion up the hill to the 2/14th Battalion position and then, in conjunction with that unit, to force a passage through the Japanese lines to the position at the top of the hill …. The battalion would then be in a position to receive supplies, evacuate the wounded and continue the rearguard action'.[31] The battalion war diary also records that from 1 pm: 'Due to Japs occupying track and ridge between fwd tps and Bde there was no chance of receiving much needed supplies; the food problem was fast becoming serious; the water supply was limited; and ammunition dangerously short. Col Cooper took full advantage of lull in fighting and decided to withdraw Bn to 2/14 Bn position and then in conjunction with that Unit to try and force a passage through the Jap lines to the position occupied by 21 Bde and so be able to receive supplies and continue rearguard action and allow the wounded to be evacuated beyond the battle area'.[32]

The 2/27th Battalion was able to successfully break contact with the Japanese to advance up the slope and occupy the position previously held by the men of the 2/14th Battalion. The route taken was completely covered in thick grass, and the Japanese had no suspicion of their withdrawal, given that the fighting here had subdued. The men moved up a newly cleared track to a position near the top of the hill and crossed a crest of the ridge and moved roughly parallel with

CHAPTER 13

the main track. Stretchers for carrying the wounded were available in 2/14th Battalion area, but the Papuan carriers had disappeared, and Battalion Headquarters' personnel had the task of carrying all the stretcher-cases. At this point, all stretcher-cases and walking wounded were collected in the sector now held by the 2/27th Battalion.[33]

Corporal Clive Edwards with the 2/27th Battalion recalled the retirement further up Mission Ridge: 'After a warning order, we began to withdraw in earnest just after midday and it [was] a nightmare journey. I was ordered to take a platoon up a rough path through the jungle to the left of the main path and that blinking "tree-cutter" M.G. of the Japs chased us every minute of the way but miraculously we got up unscathed. Helping with a stretcher delayed us'.[34] Corporal John Burns also recalled the withdrawal:

> It was most fortunate for the battalion that behind its position there was a certain amount of cover in the trees and scrub, and it was able to withdraw. 'C' Company first withdrew to higher ground then Headquarters Company with the wounded and 'A' and 'B' companies moved back. 'Don' Company had moved forward slightly in rear of 'A' Company's position and in conjunction with 'B' Company covered this withdrawal. The entire battalion was withdrawn through 'C' Company to the 2/14th Battalion area with few casualties. The route used was covered and the enemy evidently had no suspicion of the move, or else had relaxed his vigil and was resting after this morning's activities. Members of the battalion constructed home-made stretchers, comprising two long saplings between which a blanket was fastened by means of jungle vine. Originally four men were detailed to carry each stretcher, but on later days, as their condition weakened and going became more severe, as many as ten men were required to handle one stretcher-case. Headquarters Company was allotted the unenviable task of carrying the stretchers. The battalion deployed near the top

of the hill, while the CO and Lieu-Colonel Caro, CO of the 2/16th Battalion, conferred. It was decided that Lieu-Colonel Caro would attack astride the track towards the hut at the top of the hill where Brigade Headquarters had been.[35]

Most of the 2/14th Battalion had by now withdrawn further up the ridgeline to the area held by the 2/16th Battalion in preparation for their joint attack to break through to the brigade.[36] At 1.45 pm, radio communications with Potts were re-established, with 2/14th Battalion reporting: 'Moving portion of Bn to 2/16 Bn. Required instructions for 2/27 Bn'. The brigade war diary records that 15 minutes later: '2/14 Bn reported having made arrangements with 2/27 Bn and was moving'.[37] As recorded in the brigade's *General Narrative of Events*, Potts informed Caro that if the attack was successful, he would consolidate the ground gained by the men of the 2/16th Battalion; 2/14th Battalion would then take over the position previously occupied by the 2/16th Battalion, with the men of the 2/27th Battalion advancing back to the foot of the main feature facing the Japanese of the *III/144th Regiment*. Since the ground on both sides of the ridge in the rear of the Brigade Headquarters' area fell away steeply, this would require the enemy to bypass the whole position to reach Menari, leaving an intact enemy force to their rear.[38] The fighting at Brigade Hill by the headquarters' personnel was later recorded by the historian of the 2/14th Battalion, who was a veteran of the fighting at and around Mission Ridge, Captain William Russell, a 31-year-old schoolteacher from Highett in Victoria Russell. He commanded the Battalion Headquarters Company and wrote in 1948:

> The fight of Brigade Headquarters 'from commander to cooks' against great superiority of numbers and weapons has been referred to by many writers on New Guinea operations. A platoon of the Corps Guard Battalion was allotted to the defence of Brigade Headquarters. These men, since they were mostly aged

CHAPTER 13

from thirty-five to forty-five years, were affectionately termed the 'old and bold', the 'ruthless and toothless, and the 'rugged and jugged'. They formed the framework of the defence. The signallers, clerks, batmen, cooks, staff officers and others filled the gaps. The enemy commenced his attacks shortly after the first light and contributed at intervals all day. He was within 15 yards of Brigade Headquarters on a number of occasions, only to be forced back. To a certain extent the distraction of close combat was a tonic to a natural fighter like Brigadier Potts; but even when out of contact with his battalions, he still had the dual responsibility of his task and the lives of his Brigade. The man who really enjoyed himself was Lieut. Bernie Fraser, formerly of the 2/2 Pioneer Battalion, and at that time Liaison Officer at Brigade Headquarters. He used two weapons alternatively, throwing an occasional grenade for good measure when the target warranted it. The highest tribute must be paid to every man on the headquarters.[39]

* * *

First Lieutenant Horibe, *No. 6 Company, II/144th Regiment* briefly recorded in his diary the fighting around the Brigade Headquarters' position: 'EFOGI: At 0520, charged the enemy line on the hill and secured a corner. Our casualties in this battle were two killed and eight wounded. By 1030 completely occupied the hill. Collected dead and wounded. Four killed and 17 wounded …. Strength Copy Comd and 91 men'.[40] In all, 43 Japanese would be killed in the fighting around 'Butcher's Corner' and about 40 wounded.[41]

Likely still close to the position occupied by the headquarters of the *II/144th Regiment* near 'Butcher's Corner' was the unknown leader of the signals unit with *No. 2 Infantry Section, No. 1 Company, I/144th Regiment*, who was busy trying to keep Kusunose and Horie's headquarters connected. He recorded in his diary sometime that day that his NCO and two signalmen were killed as they tried to repair the telephone line: 'The road was cut by the bn at a point 50 metres behind a hill. Line maintenance personnel – Cpl OKANOUE and Leading Pvts SHITAMOTO and SOGABI. At 1540, the sec. leader, Cpl OKANOUE, and Leading Pvts KAMOTO and SHITAMOTO were killed'.[42]

14

'EARNED ... THE BLOODY NAME OF BUTCHER'S HILL'

Caro would now attempt to conduct a three-company drive south to break through to Brigade Hill with another company behind in support. Attacking on the right (west) of the Australian flank would be Captain Claude Nye, a 26-year-old railway employee from Ormond in Victoria, and his men of 'B' Company, 2/14th Battalion. Attacking in the centre would be the men of Captain Frank Sublet's 'B' Company, 2/16th Battalion; however, Sublet had earlier been evacuated with a seriously infected hand, and Captain George Wright would lead the men of 'B' Company into the attack. Supporting the centre behind Wright would be Lieutenant Grear McGee, a 29-year-old professional soldier from Launceston in Tasmania, and his men of 'A' Company, 2/16th Battalion. Assaulting the left (east) of the Australian flank would be Captain Douglas Goldsmith, a 24-year-old bank clerk from Northam in Western Australia, and his men of 'C' Company, 2/16th Battalion.[1]

The war diary of the 2/14th Battalion records the situation: 'Two plans offered themselves to 2/16 and 2/14 bns. One was to create a diversion in the vicinity of the 2/16 Bn while the balance of the forces proceeded to MENARI by the detour route to the east of the main track; or two, to fight the way through to Bde thereby clearing the main track. It was decided to adopt the second plan with B Coy 2/14 Bn right [west] of main track, B Coy

2/16 Bn on the main track, and C Coy 2/16 Bn left [east] of main track. Bde HQ were kept informed of developments until wireless transmitter ceased'.[2]

By 2 pm, the plans for the attack were finalised; the attack would be launched before 3 pm.[3] Lieutenant Grayden, who had earlier conducted the patrol identifying the Japanese here, knew the odds were stacked against them. He recalled: 'All of the troops that took part in the bayonet attack on the Brigade Hill saddle (later known as "Butcher's Corner") were aware of the almost suicidal nature of such an attack and of the improbability of a successful outcome'.[4]

He also described the tense situation as the men organised themselves for the attack.

> Attacking roughly southwards, 'C' Company would advance along the ridge and over steeply sloping ground to the left (east) of the track. 'B' Company, supported by 'A' Company, was to attack directly down the track itself. 'A' Company from the 2/14 Battalion was to attack on the ridge and the slope to the right [west] of the track. At this stage, the troops were dispersed on a cleared knoll facing a dense wall of rainforest on the ridge and slopes of the saddle, but there was no means of, or attempt at, concealment. There was no concern at the possibility of the enemy attacking. Indeed, every Australian soldier would have welcomed an attack because they would then be facing an exposed enemy rather than an entrenched one.
>
> Steady rain had begun falling and, to avoid the wet ground, the troops stood around, sheltered to some extent by groundsheets about their shoulders. There was little noise or conversation, and orders and information were passed from one to another in subdued tones. All were aware of the implications of the type of attack that they were about to

CHAPTER 14

undertake and were filled with a deep sense of foreboding. From somewhere in the rear of the companies, desultory [Australian] three-inch mortar fire was being directed to where the Japanese positions were estimated to be. With dull thuds the bombs exploded at intervals in the rainforest about two hundred metres further along the ridge. One bomb dropped short and exploded a short distance in front of our assembled troops. The resulting explosion graphically demonstrated the effect that we hoped the mortars would have on the enemy's positions.

When zero hour approached, the troops were moved into their final formation for the assault. It was all very deliberate and unhurried. There was no hesitancy or reluctance by any individual to do what was expected of him. The purpose of the attack was to enable the wounded to be evacuated. The absence of any other apparent means to achieve this justified an attack of this nature. If ever an example is required of the ultimate in discipline, morale and of Australian soldier's implicit faith in their unit and their comrades, it can be found in each of the 2/16 and 2/14 battalion troops on Brigade Hill as that zero hour approached.

Everyone involved knew that many of them would be walking to their death, yet, as was witnessed time and time again throughout the Kokoda campaign, no one hesitated to accept what was clearly more of a probability than a possibility. Because of the terrain, the type of assault, the necessity to reduce confusion to a minimum and the reduced numbers remaining in the platoon there was no merit in 15 Platoon adopting a formation other than extended line. All members of the platoon formed part of the line. As zero hour approached, the order was given 'fix bayonets.' At the appointed time – taking our line from the adjutant company – the advance commenced.[5]

SAVING PORT MORESBY

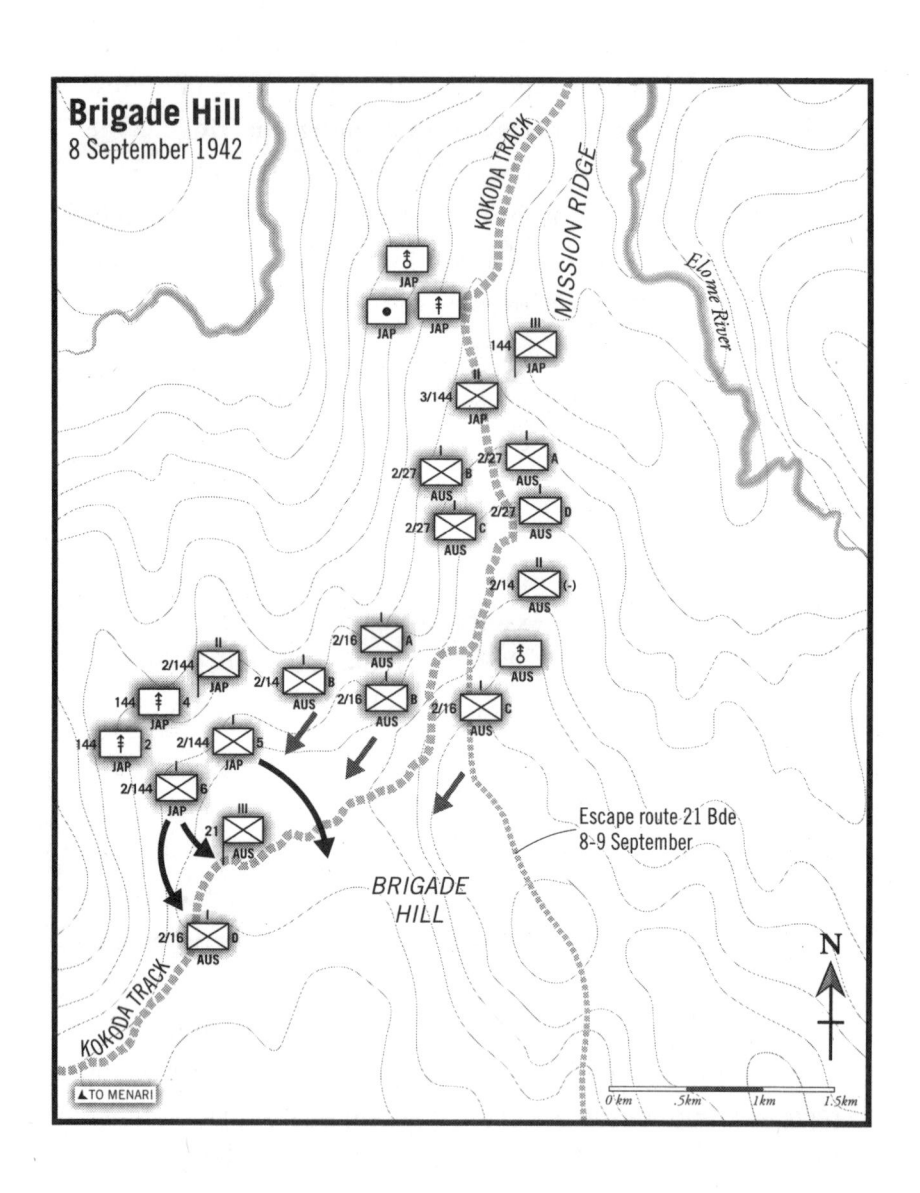

CHAPTER 14

* * *

At 2.45 pm, the main attack south to break through to the Brigade Headquarters went in. Just before the men charged, Lieutenant Stan Bisset recalled his good friend Captain Claude Nye ordering in a booming voice: 'Company, fix bayonets'.[6] The war diary of the 2/16th Battalion records: 'The attack began with a terrific volume of fire'.[7]

On the right, Captain Nye and his men of 'B' Company, 2/14th Battalion struck the heaviest opposition – the Japanese had dug-in to his front with concentrated enfilade to his right, which included machinegun fire, targeting his men as they charged south towards Brigade Hill. Leading the Australian attack were the men of 11 and 12 platoons with 10 Platoon in support.[8] Warrant Officer Wilford 'Wofty' Noble, a 25-year-old salesman from Brighton in South Australia, with his section, some of them wounded, managed to push through the Japanese and reach the brigade area. Seeing the breakthrough, Nye led some of his men through the gab, but he was killed in doing so, along with several of his men. The battalion war diary records: 'Only 8 of B Coy 2/14 Bn, under WOII NOBLE got through to Bde HQ. Our casualties, including Capt. C.C.P. NYE OC B Coy, were heavy because this Coy had the task of attacking the main strength of the enemy in this area. An attempt was made to reinforce the left coys but without success'.[9]

Also killed here was acting Corporal McCallum, a 35-year-old farmer from Tarra Valley in Victoria, who had been in the centre of the fighting at Isurava. With his Bren gun, he held back a determined Japanese assault against his company's position, enabling his men to escape. He just managed to escape himself, with the Japanese literally on top of him, wrenching off his equipment in their hands as he poured fire into them. His men said that he killed 40 Japanese at Isurava and saved a third of the platoon before he himself was able to fall back. He was recommended for the VC but was awarded the DCM for his actions during 29 August. Now he lay dead with several of his men.

Nearby, Sergeant John Matthews, a 24-year-old slipper maker from Fitzroy in Victoria, who now commanded 'B' Company, realised there was no way they could push through the entrenched Japanese. Out of ammunition and with several dead and wounded, he withdrew as best he could to a position near the track so that stretchers could be made, and the wounded evacuated as quickly as possible.[10]

Among the Japanese here was Private Nishimura Kokichi, *No. 5 Company, II/144th Regiment*. Within arm's reach was Lance Corporal Harukazu Oka, who among the firing was head to yell: 'They keep appearing one after another, like *Daruma Otshi* [a popular Japanese children's game]. However, many times we shoot and, however, many we kill. There's just no end to them'.[11] Soon after, Nishimura was horrified to see Harukazu body lying near him with his head blown off from an Australian grenade blast. Several of the men around him were then being picked off by machinegun fire or grenade blasts. He wrote in his diary: 'All remaining soldiers accepted their fate and we tried to defend the current position to the best of our ability'.[12]

Once Lieutenant Inoue exhausted the magazine of his rifle, he asked Nishimura to hand over another from one of the dead or wounded; it was then that his platoon leader was riddled with a burst of submachine gun fire. It was not long after that Nishimura felt bullets piercing his shoulder and he was then fighting hand-to-hand with a young Australian who he described as being 'incredibly young'. The Japanese private managed to grab Lieutenant Inoue's sword and ran it through the young Australian, who sank to the ground in agony – it took him hours to die. Now, wounded, and feeling weak and dizzy, Nishimura collapsed unconscious. He would wake up the next day as the sole survivor of his section.[13]

In the centre, charging down the Kokoda Track, was Captain George Wright and his men of 'B' Company, 2/16th Battalion. They also suffered from intense enfilade, which poured into their ranks as they pushed south. He recalled to Lieutenant Grayden years later that as he 'reached the heavily timbered section of the saddle, he came

CHAPTER 14

upon his batman wrestling with a Japanese soldier. Using his revolver to shoot the Japanese soldier, he looked up to find another Japanese had stepped from behind a sheltering tree just three or four metres ahead and had a rifle aimed at Captain Wright's head. The Japanese fired and although the bullet penetrated the captain's steel helmet, it was deflected and skidded around the inside of the helmet. Dazed and temporarily blinded by a copious flow of blood from a scalp wound but otherwise unhurt, Captain Wright was able to stagger from the fray'.[14]

In the fighting here was Corporal Keith Norrish, a 22-year-old farmer from Cranbrook in Western Australia, with 10 Platoon, 'B' Company, 2/16th Battalion who had a truly lucky escape from almost certain death. Just hours before, he had put in his pocket a stack of mail about 25 millimetres thick, as well as a small mirror. During the bayonet charge, he was targeted by a Japanese machine gun as he tried to retrieve ammunition from the pouch of a dead mate next to a large tree. It was then that a burst from an automatic weapon fired by a Japanese soldier from two metres away who, unknown to Keith, was taking refuge behind the large above-ground roots of the same tree. In an incredible stroke of luck, all five bullets from the Japanese weapon hit the steel mirror and letters in Keith's pocket. Instead of passing through his body and killing him outright, all five bullets were deflected downwards into Keith's chest and abdomen, puncturing a lung and piercing the pericardium surrounding his heart. The battalion medical officer dusted a quantity of sulphonamide powder into the wound, bandaged him, telling him: 'Port Moresby's seven days walk up that trail. When you fall over, someone will pick you up. Now move'.[15] He would have been told to take the eastern track leading back to Menari, which bypassed Brigade Hill.

Close by was Private Jim Moir, a 23-year-old clerk from Bunbury in Western Australia, also with 'B' Company. Jim was a strapping lad who weight around 90 kilograms; he was not overweight, but tall and muscular. He was wounded in the bayonet attack, and when an un-named mate couldn't lift him out on his back, he got next to Jim and rolled him on top of him and then 'crawled with Jim on his back through the mayhem and to the rear'. It would take Jim 31 days to reach Port Moresby as a stretcher-case because he, like many others, would soon become lost in the jungle. Even so, he would eventually make a full recovery.[16]

Wright, having recovered from his near-death experience of his helmet being penetrated by a Japanese bullet, was informed that one, perhaps two, of his platoons had broken through. However, this was a false report, as only a few from the 2/14th Battalion and 'C' Company, 2/16th Battalion got through. When Brigadier Potts informed Lieutenant Colonel Caro of this at 3.15 pm (the radio was working intermittently), Caro replied: 'Enemy being heavily reinforced. Can you assist battalions to get forward?'[17] Potts replied that it was crucial that Caro's men break through to his position and that he would support his attack from their position.[18] The battalion historian recalled the fighting here: 'Using rifle grenades, Tommy guns and captured Japanese light machine guns, a harassing fire was maintained to such good effect that the enemy was prevented from counterattacking. This gave time to gather up the wounded and to take stock of the situation'.[19]

On the left, Captain Douglas Goldsmith's men of 'C' Company, 2/16th Battalion were trying to break through east of the track, while Captain Langridge and his 50 or so men of 'D' Company, 2/16th Battalion, who were helping to defend the Brigade Headquarters, tried to support Goldsmith's attack. Lieutenant Grayden leading 15 Platoon, 'C' Company recalled how he and his men were lucky compared to most others because they avoided much of the enfilade from the right as they advanced on the left to relieve the brigade and their mates of 'D' Company, 2/16th Battalion:

> During the approach to the enemy perimeter, the advance was orderly and unhurried and at normal walking pace. The

CHAPTER 14

extended line formation was maintained without difficulty. We had not progressed far towards the wall of dense rainforest where we expected the Japanese to be entrenched when the first limited exchange of shots erupted into an intense fusillade from both the Japanese and the attacking troops. Even before they could see our troops through the thick undergrowth, the Japanese poured fire in our direction. Initially interspersed with shouting, our troops fired in the direction of the enemy and into anything likely to conceal a Japanese soldier such as rotting logs or suspicious clumps of foliage in the trees.

At such close quarters, the difference between life and death was to see the enemy before he saw you and that was a matter of chance. The alternative was a bayonet thrust or a burst from a submachinegun. The Japanese were concealed in camouflage foxholes and slit trenches, behind trees and in the foliage above. With us advancing through the undergrowth they would always see us before we could see them. Our attacking troops were exposed and disadvantaged at a range of just a few metres.

The barrage of fire from both sides and the need to press ahead with maximum speed meant no thought was given by those in the advancing line to anything but the enemy immediately ahead. I saw no one to my left or right hit or drop back. We did not learn of [the] heavy casualties our troops had suffered until Brigade HQ informed us later in the afternoon. We were still advancing at normal walking pace and the sound of the intense fire from along the line indicated that the extended line formation was being roughly maintained.[20]

The historian of the 2/16th Battalion also recounted the attack by the men of 'C' Company: 'Rain was falling as the attacking force charged down the hill against a withering fire from the dug-in enemy.

SAVING PORT MORESBY

The jungle gloom seemed to vibrate hotly from the terrific hail of machinegun fire. The first fierce momentum of the charge drove the Japanese from their holes, and for a moment it seemed as if the track was clear. When the Japanese retreated, however, they rallied around a bugler and then reformed across the track nearer to Brigade'.[21] Some men of 'C' Company led by Captain Theodore Hodge, a 36-year-old farmer from Warooma in Western Australia, attacked 40 metres east of the track, and about 12 of his men were able to break through to the brigade. Lieutenant Vivian Williams, a 32-year-old accountant from Victoria Park in Western Australia, leading 13 Platoon along with his sergeant, Harry Wilson, a 23-year-old clerk from Geraldton in Western Australia, were among those killed in the fighting here.[22] Among Captain Hodge's men to break through was Sergeant John Scott, who recalled years later:

> The Japs had set up machinegun nests between HQ and the rest of the brigade. The order came through for us to break through if we could. Hec Hodge, our ... commander, was as game as Ned Kelly .... Charlie Harris and I were on the bottom end of the line of C Company on the left, a fair way down the slope from the pathway on top of the saddle. The fellow above me on my right was a married man with children. As soon as we stood up to go, he copped a couple in the chest. As he fell, he said, 'I don't want to die'.
>
> We could see the others on the higher ground above us getting belted. We got though eventually, climbing up to the top through the undergrowth via a re-entrant or sort of gully. As we hauled ourselves over the edge, Brigadier Potts looked over straight at us. I can still see those blue eyes flashing to this day. 'How did you fellas get here?' he asked. We explained. 'Could you make it back.' Jack and I looked at each other, none too keen. 'If it's necessary,' I mumbled.

CHAPTER 14

'No you don't,' Ken Murdoch interjected, 'we need you here!' We breathed a sigh of relief.[23]

Meanwhile, Lieutenant William Grayden and his men of 15 Platoon, 'C' Company managed to also fight their way through the Japanese lines, and on reaching brigade they contacted Lieutenant Henry Lambert, who was leading 18 Platoon, 'D' Company, 2/16th Battalion.[24] Grayden continues his narrative:

> We were more fortunate than those attacking along the actual track itself. The fire from our troops was so intense that many of the Japanese defenders in our path were either dead or had abandoned their positions by the time we reached them. Others concealed in the trees were withholding fire as our troops passed below them to avoid disclosing their positions. The firing finally abated except for an occasional shot or burst of fire from the rear and inexplicably five or six of us on the left [east] of the track had passed unscathed through the maelstrom of the occupied area.
>
> Quite abruptly, we emerged from the rainforest into a partially cleared area on the southern side of the Japanese positions and within a few metres of the track where it led back towards Brigade HQ. A further five or six members of 2/14 Battalion [Noble's men] similarly emerged on the right side of the track. While the area had possibly been a native garden in the past, it was now lightly covered with low regrowth. Although troops were exposed in the area when standing, it also offered cover if required. It seemed unlikely the Japanese would leave their carefully prepared positions to mount a counterattack. As a result, there was no attempt by those of us who had survived the attack to conceal ourselves. The occasional large tree and clumps of waist-high shrubs provided all the cover required. While naturally vigilant, we stood just off the track awaiting developments.

No further Australian troops emerged from the rainforest. Soon, to our astonishment, Brigadier Potts came walking unaccompanied down the track from Brigade HQ, which was over a slight rise to our rear. He had been in radio contact with 2/16 Battalion HQ on the other side of the Japanese incursion and told us that the main force of the attacking troops had been held up. Brigadier Potts assembled the ten or eleven personnel from C Company and from A Company of the 2/14 Battalion who had survived the attack and, as I was the only officer to get through, he placed them under my control.[25]

Meanwhile, just north along the extreme left flank (east) of the attacking force, a small force from 'C' Company, 2/16th Battalion was sent to clear a spur heavily defended by the Japanese; likely the same Japanese that had targeted Captain Hodges' men. With this party was Private Alan Kilner, a 29-year-old labourer from Mont Magnet in Western Australia, who edged his way forward slightly when two of his mates at his elbows were each killed instantly by a burst of machinegun fire. Kilner retired 'inch by inch and, knowing where the enemy was located, he led a small party around the left flank and continued to probe the position. He accounted for several of the enemy and had succeeded in getting astride their line of withdrawal when the platoon was ordered to retire because of enemy pressure right along the line'.[26] Also among these Australians was 26-year-old Private Tom Beard, from Boyup Brook in Western Australia. He and his mates were quickly pinned down in the dense scrub by two dug-in enemy machine guns in the thick jungle undergrowth. A few men moved around the left flank of the enemy position but were immediately targeted with a hail of bullets and grenades thrown from the ridge. Even though those around him had been hit, Beard, 'continued to edge forward in the undergrowth. He seemed disappointed when the order came to retire because the position was too dangerous'.[27]

CHAPTER 14

* * *

At 3.45 pm, Caro radioed Potts again requesting information about those who had reached the brigade position and asked for support in his ongoing attempts to break through to Brigade Hill. Brigade replied five minutes later stating that less than a dozen men from 'C' Company had reached his position. At 4 pm, the brigade informed Caro that Captain Langridge with 17 and 18 platoons would attempt to clear the track from the Brigade Headquarters' sector; they would be assisted by Lieutenant Grayden and his small composite section.[28] Lieutenant Grayden, who had led a charmed life so far, lived to recall the fighting here many years later:

> Brigadier Potts now arranged for Captain Langridge and his group, and I with mine, to attack the enemy from the rear. Apart from diverting the attention of at least some of the Japanese force and relieving the pressure on the main attacking force, there was the possibility that the initial Australian attack had depleted the Japanese position and rendered it vulnerable to an attack from the rear. Now heading in the opposite direction, Captain Bret Langridge, and Lieutenant 'Bluey' Lambert with their two platoons were to engage the enemy on the eastern or right side of the track while facing the enemy. My group would do so on the western or left side of the track through which the 2/14th had previously advanced.
>
> Our two groups moved off quietly and cautiously keeping abreast of each other. Both groups were lightly dispersed in a semi-probing formation; the groups separated only by the area of the track itself. After about 50 metres, but long before we had even reached the edge of the dense rainforest area we came under small arms fire. The enemy fire was controlled and selective. Apart from bursts of automatic fire from ground level, single shots from small arms were being directed at

our troops from the trees above and forward of the enemy positions. When I and the others had passed through the dense rainforest on the saddle earlier, the Japanese positions were to the centre and northern end of the saddle. It appeared now that some had regrouped and were aligned along the southern end of the rainforest.

Exposed in relatively open terrain before reaching the dense rainforest, both of our groups immediately began suffering casualties and someone called out to say that Captain Langridge had been killed. I was about 25 metres from him on the other side of the track. Both groups' advance had been halted and the forward personnel were pinned down in the low undergrowth. Those in the middle and rear of the formation on the other side of the track were also taking cover behind trees which were more numerous on that side of the track. Any movement attracted fire. Casualties continued to occur and soon someone called out that Lieutenant Lambert had also been killed.

I left four or five of my group to assist holding the position and covered by the low undergrowth, took the remainder in a 100-metre semicircle to the left in the hope that the ridge at that point would be undefended and we could get behind the Japanese. Protected by a small though sharp lip on the hill, we crawled to within a few metres of the lip. Our first indication of the close proximity of the enemy was a burst of submachine gun fire cutting a line of holes diagonally downwards through the trunk of a 150-mm thick tree about 30 cm to the left of where I had momentarily halted. The bullets splintered the bark of the tree as they emerged. The 45-degree angle of the bullet holes through the trunk indicated they had been fired by a Japanese soldier concealed in a tree a few metres above and ahead of our position.

CHAPTER 14

Without warning a 2/14 Bn soldier on my left moved over to me and, before I could stop him, slithered over a low rotting log that together with the lip of the hill had been concealing us. He immediately received a burst of machinegun fire from a metre or so beyond the log and was killed instantly. Bursts of fire erupted from other machine guns and automatic weapons also positioned along the ridge. To us it seemed that these were as close as one metre apart. Another of our group was shot from the trees above the crest of the ridge. He died without uttering a word.

We slithered a few metres back down the slope, still concealed from the Japanese positions by the lip of the hill. We were then about ten metres from their positions. The Japanese then began rolling hand grenades down the steep slope. Pressed into the ground, we could see yellow flashes on either side of the rims of our steel helmets as the grenades exploded on the slope a few metres above us. We experienced no casualties from the grenades despite their proximity, although a single Australian grenade would have immobilised our entire group.

We slithered further down the hill and another exchange of fire took place. The Japanese fire was being expended in the trees and undergrowth above our heads and our only targets were their approximate positions on the lip of the hill. The exchange of fire ceased and, the remainder of the Brigade Hill area having already been silent for a prolonged period, we extricated ourselves from the position and made our way back to the track. There we rejoined a handful of Captain Langridge's group that still remained.[29]

Killed alongside Captain Bret Langridge and Lieutenant Henry Lambert, a 29-year-old farmer from Burracoppin in Western Australia, were around 20 of their men, while Lieutenant Grayden suffered at least two men killed from his small section of less than a dozen men. The battalion historian also wrote of the fighting here:

> Meanwhile 17 and 18 platoons, pushing down from Brigade Headquarters, were halted by strong opposition. The enemy were across their path. Captain Bret Langridge jumped to the head of his men and led them in a fierce charge. Hit in the chest he sank slowly to the ground, but as he did so he yelled encouragement to his men. His personal gallantry earned him a recommendation for a Military Cross. This could not be awarded posthumously but Captain Langridge was mentioned in dispatches. About the same time Lieutenant 'Bluey' Lambert and Private Viv Franklin[xiii] were killed. Many of the charging men were wounded, some of them suffering severely. By this action Brigade Headquarters was still intact and able to function. The fierce fighting and the heavy casualties on both sides earned this section of the Kokoda trail the bloody name of Butcher's Hill.[30]

The battalion war diary for the 2/16th Battalion records that the attack failed at this point, although an element of 'C' Company got through to brigade with few casualties, considering the opposition. The men of 'A' and 'B' companies, 2/16th Battalion, and 'B' Company, 2/14th Battalion were on the right flank (west); however, they had a more difficult task in thick country, as they ran into the main Japanese positions held by the Japanese of *No. 5* and *No. 6 companies*, and their supporting machinegun company, *II/144th Regiment*. Due to this, the men of 'B' Company, 2/14th Battalion were forced to withdraw.[31]

Meanwhile, the men of 'B' and 'A' companies, 2/16th Battalion were held up, all companies suffering heavy casualties.[32] As recorded in the *General Narrative of Events*, the attacks to break through to Brigade Hill were unsuccessful, with Captain Bret Langridge, Lieutenant

---

[xiii] Twenty-seven-year-old Private Victor Franklin from Manjimup in Western Australia is listed as Killed in Action, 8 September 1942.

CHAPTER 14

Henry Lambert and '20 other ranks being killed after making approx. 200 yards of ground before being pinned down'.[33] This left the Brigade Headquarters with just five men from 'C' Company, and a weak platoon from 'D' Company (both from the 2/16th Battalion), along with Graden's section and the Brigade Headquarters' personnel to hold their position.[34]

* * *

Still defending the northern perimeter was Lieutenant Colonel Geoffrey Cooper and his men of the 2/27th Battalion. The diarist of the battalion recorded that while one party from the 2/16th Battalion had managed to break through on the eastern flank, the assault had failed. The agreed plan of evacuating Mission Ridge if the attack to break through to the Brigade Headquarters failed was now to be put into effect. They would attempt to break contact with the enemy and withdraw. Leading the way would be the men of the 2/16th Battalion, followed by the men of the 2/14th Battalion carrying the bulk of their wounded using the eastern route. Finally, the 2/27th Battalion would act as the rearguard and would follow as soon as the rest of the brigade had made good their escape.[35]

The men of 'B' and 'D' companies, 2/27th Battalion, now moved back downhill towards the Japanese of the *III/144th Regiment*. These Japanese troops were unaware that the Australians had earlier retired from their forward perimeter – enabling these men to reoccupy the former forward positions. It was now that the Japanese of *No. 8 Company* attacked; however, the men of 'B' Company were able to hold them back while the bulk of the column was formed up and moved along the eastern route south to Menari.[36]

Twenty-two-year-old Private Eric Sambell, from Wayville in South Australia, recalled the fighting here years later: 'I just got me head up and had a look and there's sort of a mob of them a way down and I give a couple of bursts and then I got me head down again. I wasn't going to get hit. But unfortunately, two of our section had been shot dead, one through the head and one in the chest. The poor bloke in the chest, he was alive, and anyway, I won't name him, because it mightn't be wise, and we got the order … to withdraw, and we had to leave him there and he's still conscious you know, it's a terrible thing. But anyway, later on they found his bones, I suppose the Japanese would just, would have just killed him'.[37]

* * *

Meanwhile, Brigade Major Hugh Challen and his composite 'company' finally reached Brigade Hill from the south just after 5 pm. Rather than 100 men, however, his company was at platoon strength with just 38 men. With darkness, Potts, Challen, and their men were able to disengage and fall back to Menari. Potts had been informed earlier that the men of the battalion had been ordered to make their way to Menari by a circuitous route if their attacks failed; as such, he was confident the rest of the brigade would soon make their way to Menari.[38]

Meanwhile, Lieutenant Grayden and his men had returned to the headquarters' area on Brigade Hill, after having helped pushed back the Japanese. All knew a Japanese counterattack would soon follow. On reaching the headquarters' position, however, they found it deserted. Grayden recalled:

> As Brigade had withdrawn, the few of us that still remained followed suit. Without taking cover and unmolested, we made our way leisurely down the very steep and long track to the village of Menari. Although it did not occur to me at the time, in retrospect I would have to regard myself as particularly fortunate in surviving the events of the day. Due to the various roles, I had been required to undertake as a Platoon Commander, I would be the only person in the 21st Brigade to have been engaged in close contact with the Japanese on the four sides of the conflict and the additional aspect when our small group attempted to

CHAPTER 14

get behind the Japanese at the corner of Brigade HQs end of the saddle. Despite being in close contact with the enemy in five different areas of the battlefield and having witnessed the encounter from those five aspects, I emerged unscathed. Such is fate and the precarious nature of the fortunes of war.[39]

* * *

Back at Menari, not having heard from Challen or Potts, Captain Lyon sent an urgent message back to Major General Allen: 'Jap in strength between ABOG [Brigade Headquarters] two platoons 2/16 and remainder. At worst can hold feature as fortress two days, destroy Menari and withdraw Line of Communication troops. At best can hold as fortress four days as NO water on position, forward supplies short. Alternative, withdraw and establish eight days fortress vicinity Nauro. We have NOT sufficient force to prevent outflanking, following troops from Moresby would have to fight into us. Please instruct urgently'.[40]

This was totally inappropriate, as the captain should have not sent this message without discussion with Challen or Potts. He was not merely stating the current situation but outlining three alternatives. Allen was alarmed on receiving this message and replied: 'Ref HQ 21 Aust Inf Bde 0 29 0f 8 Sep from LYON. Has the message been sent with your approval? Is situation stated correct? Advise immediately'.[41] It would be hours before Potts would finally make his way back to Menari and be able to respond.

15

'... GIVE THE BASTARDS A FULL MAG'

Before dusk, Caro, Cooper and Rhoden began their withdraw using the track earlier scouted by 23-year-old Lance Corporal Robert Watson, from North Geelong in Victoria, who forever after was given the title 'The Boy Wonder'.[1] The Japanese were in strength on and around Brigade Hill, and it was decided that the 2/16th and 2/14th battalions would move out in that order, around the eastern flank to reach Menari. Although still under fire, Caro managed to break contact with the Japanese around the Brigade Hill area and by 5.30 pm his men were pushing east along the narrow track. Progress was slow, as the terrain was unforgiving. Coming up behind them were the men of the 2/14th Battalion. Still manning the northern perimeter, the men of the 2/27th Battalion were keeping the Japanese pinned down as the rest of the brigade started their journey along the track.[2]

As the head of the column moved out, the men had to cut and slash at the scrub with their bayonets. Their hearts 'seemed to be bursting as they struggled to help their wounded, hoisting, lowering, pulling and pushing the clumsy stretchers on which the worst cases had to be carried'.[3] Leading the way were the headquarters' staff of the 2/16th Battalion, followed by 'A' and the Headquarters Company with their 1 Platoon helping to carry the stretcher-cases with the battalion medical personnel. These stretcher-cases were mostly casualties of the 2/27th Battalion, who had been wounded during the morning. With his native bearers, Warrant Officer Ronald Preece had carried these men to the 2/16th Battalion Headquarters' position, but the Papuan carriers had 'been badly frightened and had gone bush' when

CHAPTER 15

the Australian counterattack was launched.[4] A hail of bullets flew over the heads of the men of the 2/16th Battalion as they moved out. They were followed 30 minutes later by the men of the 2/14th Battalion, also with several stretcher-cases. It would be some time before the leading elements of the 2/27th Battalion followed.[5]

The historian of the 2/16th Battalion recounted their journey in the darkness: 'This track was one of the worst experiences of the whole campaign. For the first 400 yards the track was narrow but led up a gentle slope. Then the track became steeper and steeper. In the front went the stretcher-bearers, men light on physique but strong in spirit and purpose. Frequently they had to halt while a pathway wide enough to take the stretchers was hacked ahead of them. As night fell the track led down a steep spur and during the next few hours the men literally felt their way ahead, hands outstretched groping ahead, like men without sight. The jungle closed in on them. Even the sky above was only a little less dark. The drizzling rain never stopped'.[6] The medical officer with the battalion, Captain Steward, also recalled the journey: 'The Japanese [now] controlled the main track to the next village, Menari. We withdrew, therefore, along a barely known alternative route to the east, and then southwest towards Menari. We stumbled on through pitch blackness and drenching rain, groping forward with outstretched hands like sightless men. The night of 8 September will not easily be forgotten.[7]

The battalion war diarist also recorded: 'Lieut. Hewitt who had previously been patrolling this area was acting as guide to 2/16 Bn at head of column. Movement of column along track was particularly slow, as much of the track had to be freshly cut and it was found impossible to make any speed with stretcher-cases. As night was overcast and black and drizzling with rain at 2100, it was decided to bivouac until dawn. 30% personnel on guard during night. No alarms'.[8]

Among these men was Corporal William Cousens, 'B' Company, 2/16th Battalion who recalled the assistance provided by an un-named officer of the 1st PIB, likely referring to WO Preece: 'Our position looked really hopeless at that stage until the appearance of a Papuan Army Brigade officer who had approached us from Menari along a side track almost parallel to the Kokoda Trail, it was his purpose to guide us back along this track hopefully before the Japs got to Menari, who were being stoutly resisted by everyone in Brigade HQ as they fell back on the main trail. A lot of our wounded had been sent out earlier in the day, including many stretcher cases, to try to make their escape in a roundabout way in this impossible mountain country. That episode was an epic on its own if the story is ever properly told by those who were actually there .... Darkness was rapidly settling in, which was to stifle all movement for the night, so we all just sat down and lay down where we were, soothing the wounded we had with us throughout the night'.[9]

Now coming up behind the 2/16th Battalion were the men of the 2/14th Battalion, as recorded in the battalion war diary: 'The movement was completed under fire, and contact [with the Japanese] was eventually broken. The unit was collected at the rendezvous some ½ hour away from the main track, and circumstances were explained to Coy Comds. The unit proceeded along its way until nightfall when it was decided that the advance would be continued in the morning at first light'.[10]

*　*　*

Meanwhile, Captain Arthur Lee and his men of 'B' Company, 2/27th Battalion were still holding the northern perimeter to the right of the track and fended off several assaults as the rest of the battalion moved out. At around 5.30 pm, with the Japanese less than 25 metres away, Lee was heard to yell to the men of his small company headquarters' personnel: 'They're getting all the young chaps. We old buggers have something to account for. Come on boys, give the bastards a full mag'.[11] Lee and his men charged the enemy, and before the Japanese had a chance to recover, he led his company off Mission Ridge and onto the

CHAPTER 15

track into the jungles, gullies, and ridges to the east. Also supporting this attack was Captain Justin Skipper, a 28-year-old solicitor from Adelaide. For is actions that day, Captain Arthur Lee was awarded a bar to his MC that he was awarded in Syria; the citation reads: 'Capt. Lee was in command of his Company when it was ordered to cover the withdrawal of the column from Efogi. He displayed exemplary courage in leading his Company under intense fire from the enemy at ranges down to 20 yards. When the Company began to withdraw, Capt. Lee kept behind with two comrades, who emptied their magazines into advancing enemy, causing him to discontinue his attack'.[12]

Captain Justin Skipper was mentioned in dispatches, as was 24-year-old Lance Corporal Albert 'Pompey' Turner from Adelaide, who was Lee's batman and runner, Turner was killed during the counterattack. The battalion history records how Captain Lee 'lost a gallant brother'.[13] Also killed in the attack were the company's two signallers, Private Horace 'Viv' Knott, a 23-year-old salesman from Payneham in South Australia, and 38-year-old Private Edward 'Ted' Churcher from Adelaide; both bodies were recovered and buried by the brigade a month later.[14]

Captain Harry Katekar, a 28-year-old solicitor from Mile End in South Australian and adjutant to the 2/27th Battalion, later recalled the significance of Lee's attack: 'That was a tremendous operation a wonderful action by B Company. They had to buy time some way or other and the way they did it, they counterattacked down towards the Japs. The Japs were so shocked that they broke contact. We didn't see them again that day. They had the impetus, and they were hot on our heels. We were withdrawing with our wounded; they would have known that because they were in physical contact with us, sniping at us and so on. And then B Company was given this job to stop them. Instead of just standing there and firing at them they counterattacked and that must have shocked them considerably'.[15]

The battalion war diary also records that at 5.30 pm, 'B' Company was acting as the rearguard to the battalion. Now these men made a short and sharp counterattack, which caused the Japanese to fall back in momentary confusion. Only then did 'B' Company withdraw to follow the rest of battalion, having successfully broken contact with Japanese. These men now withdrew down the eastern track, acting as the rearguard of the column.[16] In two days of fighting, the 2/27th Battalion suffered 39 men killed, three officers and 43 other ranks wounded and two men missing – close to 15 per cent casualties.[17]

Lieutenant Colonel Cooper and the bulk of his men were now moving out well behind the men of the 2/16th and 2/14th battalions. They found going along the steep and narrow track extremely difficult. Their wounded also slowed them down; 12 men had to be carried on stretchers, mostly by men of the 2/14th and 2/16th battalions who volunteered to stay behind to help with the recently wounded of the 2/27th Battalion. This enabled the bulk of the men of the 2/27th Battalion to remain in the firing line. It was a dark night made worse by the constant rain, adding to their difficulties. At some point, Cooper was forced to bivouac on the track for the night.[18] Corporal Clive Edwards recalled the withdrawal: 'Confusion was the keynote, and no one knew exactly what was happening but when the sounds of battle came from in front we were told that the others were trying to fight their way through. We stood by to help but they failed, and a general withdrawal was ordered …. It was pitiful – the rain was coming down, and there was a long string of dog-tired men straining the last nerve to get the wounded men out and yet save their own lives too. Bewilderment at the turn of events showed on every face, and as the long line faltered and halted, those at the back became affected and sent messages along the line to "keep moving, the Jap is on us"'.[19]

Private Baldwin with the battalion also recalled the situation as they moved out along the eastern track with the wounded: 'We had one boy from the 2/14th Battalion, two from the 2/16th Battalion and the rest were our boys that were stretcher-cases …. Then because of this break in the attack by the Japanese, we were able to pull out from these positions the stretcher-cases, and it was [a] hard slog, believe me. The

CHAPTER 15

stretcher had to be passed hand over hand over this column of men on both sides and pass them up onto the hill and then drop them over the edge and get up to the top again and take over the stretchers again and this went on for quite some time. We knew that some of those boys would not make it, they were so terrible sick and hurt. Come nightfall on the first night on the side of the hill almost at a 45-degree angle, we had to put our feet against the trees to stop from sliding down the hill, and you had to try and brace the stretchers in whatever way you could, so they wouldn't slide down too'.[20]

* * *

Earlier, Brigadier Arnold Potts and his survivors from Butcher's Corner had endured their long and exhausting march back to Menari, which involved a steep descent as recalled by 22-year-old Corporal Alain Thomson, from Adelaide, now a signaller with the 2/27th Battalion attached to the Brigade Headquarters: 'We came back along the track in the pitch-black darkness. And the only way you could go back was by holding the sig, cable in your hand. We had one torch with us and must have gone for about five hours I reckon. We had no idea where we were going, but just kept going down following the cable …. That was a terrible night, I still don't know how we did it'.[21]

Before arriving at Menari, Potts ordered a small composite company to be formed among the survivors from Brigade Hill and those who had arrived at Brigade with Major Hugh Challen. This party numbered around 60 men under the command of Captain Theodore Hodge and were to take up a blocking position forward of Menari. It would appear at some point that Hodge had made his way through to the brigade during the attack of 'C' Company, 2/16th Battalion. These men were to deny the track to the Japanese until first light; they were then to retire to Menari. Potts also ordered Captain William Russell – commanding the 2/14th Battalion Headquarters Company, who appears to have either been with Brigade Hill or Menari at the time of the attack – to lead a composite platoon tasked with occupying the most forward position about halfway between Mission Ridge and Menari. His small force was to act as a warning of any Japanese advance, and they were to be in position before midnight.[22]

The brigade war diary records that by 6.45 pm, the Brigade Headquarters had made its way to Menari. Brigadier Potts now responded to Major General Allen's message; and to Allen's shock the reply was: 'Message Lyon confirmed'.[23] Major General Allen reported to Lieutenant General Sydney Rowell to discuss the situation. Rowell's 21-year-old ADC Captain Gordon Darling, from Melbourne, summed up the situation: 'The brigade had been surrounded and is now cut off from its lines of communication, leaving a few troops this side who were to establish a firm base for the back. It is now race against time. Can 25th Brigade be brought up to restore the situation in time? They are due to arrive tomorrow'.[24]

* * *

Tanaka Kengoro, a veteran of the campaign in his history of Japanese operations in Papua, provides only the barest of details of the fighting to take Missionary Ridge and the condition of the men at this point:

> The 144th Infantry Regiment passed Kagi on September 6, and shortly after confronted the enemy's position to the south of Efogi. The Regiment deployed its main forces, and with repeated attacks through day and night from the evening of September 7, they broke through the enemy's position by the evening of September 8. Strafing and bombing by enemy aircraft became fierce around this time and resulted in heavy casualties for the Japanese. In addition, malaria and diarrhoea plagued us, and many men fell ill. However, it was instructed that those who were required to be hospitalised should be sent forward according to the capacity of each company instead of being sent to the rear and should be hospitalised at the field hospital that was scheduled to be built shortly in Port Moresby.

CHAPTER 15

On the other hand, the provisions supplied to the South Seas Detachment were extremely poor, and the physical strength of the soldiers was exhausted day by day due to the food saving requirements. Nevertheless, their morale was still excellent, and they were anxious to capture Port Moresby as soon as possible in order to receive dinner at the enemy's expense. Then, immediately after capturing Efogi [Mission Ridge], they began to pursue the enemy and confronted the enemy's position in Ioribaiwa on 12 September.[25]

* * *

At Menari, Burt Kienzle was organising supplies to be sent forward, having no idea that the Australian position along Mission Ridge had collapsed. Later that day, he recorded in his diary: 'The planes were dropping supplies by this time, ordinance, chocolates and cigarettes. On the same night Brig Potts arrived at MANARI and all the carriers had been evacuated from the forward camps excepting WO Preece and his carriers, who were cut off when the Japs encircled BHQ and separated the three bns, 2/14, 2/16, and 2/27. We had an uneasy night with no force of strength astride [the] track north of MANARI. There were many scares and I had two trusted police boys standing guard all night to keep the carriers calm'.[26]

Also at Menari was Kienzle's good friend Captain 'Doc' Vernon. Throughout the day, he had been withdrawing his supplies and carriers back to Nauro; however, by nightfall, the Australian Light Horse veteran was still at Menari. He later wrote: 'That night a quantity of army stores was burnt but not before the whole carrier force had received generous supplies of food, tobacco and clothing. The more liberal feeding during the retreat, plus the shortened carrying distance and the concentration of carriers, did much to invigorate and satisfy the carrier force. It might be said that the rapid Japanese advance reacted most beneficially on our men [carriers]'.[27] Vernon also wrote in his diary a more vivid account of that night, including that his assistant Warrant Officer Jack Wilkinson had been based at Brigade Headquarters on Mission Ridge. Wilkinson arrived back at Menari later that night:

> This station [Menari] was now the advance base and was full of activity, yet a few native women still lingered in the villages. It was only two or three hours walk to our frontline on the top of Brigade Hill on the Efogi Ridge, and casualties were constantly arriving on stretchers shouldered by our carriers. Sgt Wilkinson was again at the front, and we had a medical orderly stationed at Brigade Headquarters whom I was anxious to visit but never found time to as our hospital work was heavy. As usual the rough shelters were full of broken-down carriers with fever, pneumonia, and dysentery, but I do not recollect admitting any wounded [carriers] …. The last night was one of some anxiety. The forward medical boy had come in and all the bad cases evacuated. I placed the remaining patients and orderlies in total darkness in the hospital with orders to be ready to move instantly if called. Lieut. [*sic* Captain] Kienzle kept in touch with all parties to ensure the safety of the carriers.[28]

* * *

Further south moving up in support of the 21st Brigade was Captain Boag and his 83 militiamen of 'C' Company, 3rd Battalion. They arrived at Nauro by midday, having left Ioribaiwa the previous afternoon. The men took a well-earned rest, and it was just before 2 pm that Boag got word that his company was now under the command of Potts and the 21st Brigade. He was also instructed that one three-inch mortar and 224 bombs, along with a detachment of mortarmen would arrive at Nauro the following day and these men and their mortar were to dig-in at Nauro.[29] Sergeant Colin Kennedy with Boag's company recalled: 'Our battalion had been part of 14 Militia Brigade, but we were not to know that we would never return to it'.[30]

CHAPTER 15

* * *

Just behind the southern slopes of the Ioribaiwa Ridge, Major Albert Moore and Corporal Jock Inglis had set up their Red Shield post – this was their second day providing refreshments to the men moving up and down the line. Moore recalled that he and Inglis where up at the crack of dawn: 'We had the fires blazing and coffee ready for the early arrivals. I recorded in my diary that "Colonel Norris called in on his way forward and was loud in his praise of our efforts". We also had a call that day from Damien Parer, who filmed the post as we supplied the wounded with coffee and biscuits. Another entry in my diary reads: "This day we had 13 stretcher-cases among the many we catered for ... served 50 gallons [230 litres] of coffee today and supplies are running out ... have sent urgent messages to our officers in Moresby for extra supplies to meet our need"'.[31]

Having left Menari after filming the parade of the 39th Battalion, Damien Parer, who was sick with fever, arrived at Moore's Red Shield Station. It was now that he would use the last of his film to take another image that would help define the image of Australians fighting on the Kokoda Track. This was of Major Albert Moore lighting a cigarette for the wounded Lieutenant Valentine Gardner, with 'D' Company, 2/14th Battalion. In the middle background stands Corporal Inglis looking into the camera with an inquisitive look. Soon after, the sick and weak cameraman returned to Sydney with his precious film, and memories and admiration for those men fighting and dying on the Kokoda Track.[32]

* * *

Now struggling along a mountain track well north of the Australian frontlines, Lieutenant Robert McIlroy and his lost men of the 2/14th Battalion were determined to strike south-east, hoping to reach Menari before the Japanese.[33] He later recorded: 'When the mist rose we got a view of a native village (HAILO [south of Efogi]) on the other side of a deep gully and practically due south of us. At the same time some of our transport planes were seen dropping supplies beyond on a bearing 170°. I therefore decided that as the planes were probably dropping at MENARI we would head for the village [Hailo] and then to MENARI. The day was spent trying to get across to the village'.[34]

Early that morning to the east of McIlroy, Lieutenant Maurice Treacy and his two men who had been sent forward from Captain Sydney Buckler's party of the 2/14th Battalion continued their journey south towards the sound of the fighting, hoping to somehow reach the Australian lines. He later wrote: 'At 0500 hrs the party set off along the track. During the day little distance was covered, it being often necessary to encircle through the jungle enemy bivouac areas to avoid parties occupying them. At night, the party moved for an hour or so using a torch, eventually it was decided that this might attract too much attention. Movement at night without a light was unsuccessful, it being necessary to crawl on hands and knees in some places to find the track'.[35]

* * *

Back in Port Moresby, and having been informed of Potts' situation, Lieutenant General Rowell cabled Major General George Vasey (DCGS) – Potts was to be replaced by Brigadier Selwyn Porter, a 37-year-old banker from Wangaratta in Victoria: 'Potts is at this hour in serious trouble between Efogi and Menari, as a result of enemy penetration. Porter is being sent forward ... to relieve Potts and take charge of all elements in the forward area. I trust you appreciate [the] gravity of the immediate situation and will produce additional troops asked for particularly infantry with minimum delay'.[36] On first hearing that Brigadier Arnold Potts was to be removed from command, Major General Arthur Allen was troubled by the decision and recorded so at the time:

> Commander New Guinea Force G1 [Spry] and self to a conference ... and on arrival he [Rowell] stated that he had

CHAPTER 15

decided to relieve Brig Potts of his Command and that Brig Porter would be sent to relieve him. I was not consulted prior to this decision. I raised no objection because judging from signals from Potts I myself felt he, Potts, was either tired or was losing a grip of the situation. I had intended to send the G1 forward or myself to check up on him. Either the G1 or myself would have gone forward before if I had not the added responsibility of the defence of Moresby area from a sea attack .... When given the dual roles, I expressed doubts to Commander New Guinea Force but at that time there appeared no alternatives. I feel the decision to replace Potts by Porter is a sound one, but in fairness to Potts, I at this stage, at any rate, have an open mind on the matter of whether Potts has shown lack of judgment or has been out-fought. Although it may appear that this is the case, I prefer to hear Potts before condemning him. He took Command at a difficult stage in the fight, and up to Myola, [he] apparently was forced to putty up gaps in the defences. At Myola and Efogi, he seemed to have an opportunity to force aggressive patrolling and adopt the offensive, but there is little evidence of this being done, and without hearing [from] Potts, it is difficult to assess. In any case he is apparently tired, and his relief is in the interest of the service. I, however, refrain at present to express an opinion as to his future employment. His past services have been excellent.[37]

* * *

Australian casualties during the two-day battle for Mission Ridge resulted in 164 casualties: 87 killed, 77 wounded; the Japanese suffered 225 casualties: 60 killed and 165 wounded.

16

'... HOUR AND A HALF LATER THE JAPS WERE SENDING MORTAR-BOMBS AMONG THE HOUSES'

During the morning of 9 September, General Horii pushed his men forward; however, given the casualties suffered by the men of the *II/144th Regiment* during their fight for Brigade Hill (43 killed and around 40 wounded), he pushed the men of the *III/144th Regiment* ahead to lead the advance with their supporting two 37-mm guns.[1]

Among these wounded was Private Nishimura Kokichi, with *No. 3 Platoon, No. 5 Company, II/144th Regiment,* who was the sole survivor of his section that fought at Butcher's Corner. Nearby was the now dead Australian he had run through with a Samurai Sword. That morning, he was laying in pain when he heard a familiar voice: 'Hey, Nishimura, are you all right?'[2] It was Captain Naoma Fujisaki, a machinegun officer from another company who he barely knew; he wondered how an officer from another company knew a lowly private? The officer soon had stretchers brought up to collect the wounded strewn around the battlefield, and Nishimura was carried to an aid post. Here, a medic checked his shoulder and removed a bullet. Nishimura recalled the increasing pain in his shoulder over the next few days and asked if the medic could check his wound again. When the medic did, he saw another bullet coming out of the wound, which he removed. Forty years later, while getting an X-ray, it was observed that another bullet was buried deep in his shoulder. Rather than making his way to the rear, like most wounded, Nishimura advanced, as all expected a field

CHAPTER 16

hospital to be set up for them at Port Moresby.³ Close by would have been Lieutenant Sakamoto Atsushi, commanding *No. 2 Machinegun Company, II/144th Regiment*, who recalled: 'Corpses were piled high … it was a tragic sight'.⁴

As Horii renewed his advance, the *II/144th Regiment* could muster around 525 men, while the *III/144th Regiment* had a similar number at around 540 men. In addition, he had a full engineer company present, which added another 150 men to his force. He also now had nine artillery pieces, five 75-mm mountain guns, two 70-mm battalion guns and two 37-mm light guns.⁵ Moving forward from Templeton's Crossing were the two battalions of the *41st Regiment*; while the men of the *I/144th Regiment*, who had been at Myola, were also now ordered to advance. Finally, that morning, Horii called forward his sole reserve, the *III/41st Regiment* – these men were then at Kokoda and would arrive in the frontline position on 12 September. With their arrival, he would have two regiments, with their supporting elements at his disposal.⁶

First Lieutenant Horibe, *No. 6 Company, II/144th Regiment*, who had been involved in the fighting at Butcher's Corner, recorded in his diary that night that they stopped their advance just '300 metres north of MENARI'.⁷ Meanwhile, earlier that day, Japanese war correspondent Okada Seizo had passed through Brigade Hill and later described the aftermath of the fighting as he witnessed it:

> It was around here in the ravine of Efogi that of all the battles fought in the Owen Stanleys the bitterest hand-to-hand fight took place between the vanguard company and the Australians. It began at daybreak and continued for many hours. Early next morning I hurried through the ravine with staff-Officer Tanaka [Toyonari]. The vanguard company commander told him that there were about 200 bodies, Japanese and Australian, scattered in the ravine. Here and there, on both sides of the path of blood-

red clay which ran through the cypress forest, I saw a great many Japanese and Australian soldiers lying dead. One of them had a twisted neck and broken legs, with his face smeared all over with mud and blood. Another was in a crouching posture, his face resting at the foot of a tree. A third was lying on his back like a fallen tree, and a fourth was shot through his forehead in a prone shooting position, his gun left on the ground in front of him. Another again was hanging on a tree over the edge of a cliff; another lying with his upper body in a ditch; and another leaning against a tree with his body bent halfway forward – men in all postures and conditions showing how desperately they fought and fell.[8]

* * *

Still at Menari, Potts remained unaware that he had been replaced by Brigadier Porter. He was waiting anxiously having heard nothing from his three battalions for many hours. At 8 am, Brigade Major Hugh Challen handed over duties as brigade major to Captain Geoffrey, a 28-year-old accountant from Brighton in Victoria. Challen moved south leading a reconnaissance party to select the next fallback position on the high ground south of Menari. It soon became clear that the ground there was impossible to defend, as it was, 'a razor-back with no depth and easily bypassed but it was the only feature between MENARI and NAURO'.[9]

Having already arrived at Menari was Lieutenant Grayden and his men of 15 Platoon, who – along with a couple of men from the 2/14th Battalion – had earlier broken through to the brigade area. Taking a well-earned rest and advantage of stores carried to Menari, he and his men watched the Japanese further north rummage through the largely destroyed supply dump just south of Brigade Hill, while other Japanese appeared to be preparing positions for their mountain guns. Grayden and his survivors would soon be heading south when the rest of his battalion arrived. He recalled: 'We reached the new position

CHAPTER 16

near Nauro [and] it was evident that once again it was not the sort of defensive site we had expected. Being late in the afternoon, it was decided to remain there for the night and move on in the morning'.[10]

At 8.15 am, the composite company – acting as the rearguard for Brigade Headquarters and led by Captain Hodge – arrived at Menari. Hodge reported all had been quiet after dark and they had gained no word from any of the brigade's three battalions. At 9.30 am, Hodge and his men of 'C' and 'D' companies of the 2/16th Battalion with Brigade Headquarters' personnel were positioned on the ridgeline south of Menari village, while the composite platoon, under Captain William Russell, was now positioned to cover the northern approach to the village and the nearby Emuni River crossing of the track.[11]

By 10.30 am, Warrant Officer Ronald Preece and Staff Captain Peter Smith, a 30-year-old manager from Peppermint Grove in Western Australia, who had been assisting some of the forward stretcher-cases with the 2/16th Battalion, also reached Menari. On arriving in the village, they informed Potts that the lead elements of the 2/16th Battalion were about 30 minutes behind him, and the bulk of the 2/14th Battalion not far behind. Still with these men was Lamce Corporal William Cousens, 'B' Company, 2/16th Battalion, who later wrote:

> Off again at first light on our way to Menari, with our main task being to ease the agony and pain of the wounded as much as possible and to fight off the Japs if they happened to show up along this track. I was mainly concerned with supporting Clarrie Maskiell on this journey. Clarrie had had his arm almost severed, so that it was dangling uselessly by his side, held together by a few sinews ands meat and causing him a lot of pain. "Yet he just grimly struggled on over rough mountain country that would tax the strength of the most able-bodied people, and that was being repeated a hundred-fold by all the wounded who made that journey that day. I met Clarrie Maskiell a few years after the

war and was amazed to find that he had retained his arm, even though he was obliged to keep his hand in his pocket to give the arm some support. The good Lord must have looked after us that day, because we eventually made Menari barely ahead of the japs, with the wounded now assured of some chance of survival with the medical care they could now receive, however inadequate.[12]

Potts was informed that more stretcher-cases were struggling to the rear of the 2/14th Battalion. He soon after ordered Captain Smith back along the eastern track with a stretcher party to help bring in these wounded. Indeed, on reaching the rear of the 2/14th column earlier, and while his men were struggling with their stretcher-cases, Lieutenant Colonel Geoffrey Cooper had ordered them along with the fittest of the walking wounded to push ahead. Cooper and his men would proceed more slowly with the remainder of the stretcher-cases.[13]

Indeed, Cooper and his men were now lumbered with 15 stretcher-cases in all; one would die later that morning. These men were soon cutting their way through the vegetation, pulling themselves painfully up the steep slopes by low-hanging vines and grabbing tree branches, and all too often crawling on hands and knees, dragging the fragile and clumsy stretchers with them.[14] The battalion war diary records: 'Column moved at first light B Coy acting as rearguard. Owing to slowness of 2/27 Bn caused by carrying stretchers 2/14 Bn and 2/16 Bn personnel and walking wounded pushed on ahead and reached MENARI. No further contact was made with them by 2/27 Bn. All through this day HQ Coy did excellent work under most difficult conditions carrying stretchers. The gradients were of steepness necessitating hands and knees climbing and grasping any available bushes, creepers, and trees for support. On reaching a river at about 1100 hrs B Coy watched approaches while column crossed and ascended 1/1 gradient slope on SOUTHERN side. Took over one hour to cross with column and stretchers. At 1330 reached crest of

CHAPTER 16

ridge and struck old track leading down into MENARI. Heavy rain made track a quagmire'.[15]

* * *

Earlier that day, Captain Burt Kienzle, with ANGAU, had evacuated most of his men from Menari and cleared the village of wounded. He ordered the remainder to carry with them south to Nauro as much of the supplies as possible while he remained with a large party of carriers to help bring in the wounded that were thought to be just hours away moving down the eastern track from Mission Ridge. He recorded in his report: 'Before daybreak I despatched all carriers and personnel except 100 natives, who were to act as stretcher-bearers if the bns cut-off should arrive. We cleared the camp of all wounded and took as many supplies and ammunition as could be carried to a depot near NAURO but later transferred to NAURO proper that afternoon. I remained behind with Brig Potts and Maj. Watson, 100 carriers and four Police and troops of BHQ. At 1100 hrs WO Preece reported in with Capt. Smith, S/Capt. 21 Bde that the three bns were on their way in by bush track and should report within the next few hours carrying wounded. I handed over to WO Preece 96 carriers and two Police to help bring in the wounded as quickly as possible before the Japs advanced on MANARI along the main track. We prepared tea and awaited the arrival of the men'.[16]

Captain 'Doc' Vernon was also ordered by Major William Watson, a 55-year-old gold miner based in Papua, who commanded the 1st Papuan Infantry Battalion (PIB), to evacuate Menari and retire to Nauro in the morning. Watson and his Australian officers and Australian and Papuan NCOs commanded a small force of Papuan riflemen that had been the first to confront the Japanese invasion. They had spent the first three weeks fighting with the Australian militiamen of the 39th Battalion in trying to deny the Japanese the airstrip at Kokoda Plateau. Watson and his men of the PIB remain to this day largely unknown heroes of the Kokoda Campaign.[17]

Vernon records that Watson passed on to him orders from the Deputy Assistant Director of Army Medical Services: 'To get out to Nauro at once. This was about 0900 when the retreat was at its height. All staff, patients, and stores were removed in safety. A temporary aid post was formed on the riverbank 2¹/2 miles from Nauro [likely Naoro River] but evacuated later in the day. At Nauro immediate steps were taken to evacuate the sick to Ioribaiwa and two N.M.Ords. [Native Medical Orderlies] went with the bulk of the stores and stretcher-cases. All the forward staff with a few sick stragglers joined us before nightfall'.[18] Again, the old Great War veteran provides a more vivid account in his personal diary:

> About 0930 Major Watson advised me to go, and an hour and a half later the Japs were sending mortar-bombs among the houses. We had saved all the medical stores and practically all the carriers were in or covered by reliable ANGAU officers. As a matter of fact, however, some of the tail end of the retreat from Menari got into difficulties; later I met several carriers who had fallen into the hands of the Japs and had even been impressed into carrying for them. None that I spoke to showed any enthusiasm for their new masters nor for their scale of rations which was nil …. On the way to Nauro the Brigade Major [Challen] asked me to establish my most forward aid post in the nearest part of the Nauro valley. I left a detachment there with WO Davies to put rough shelters and went on myself to see the preparations of the Nauro hospital, but later on these men were ordered back so I did not return to the rendezvous.[19]

17
'THERE'S YOUR BREN BACK, SLIM'

The situation at Menari by noon was recorded in the Australian 21st Brigade's *General Narrative of Events*: 'The Bde Comd had to decide at what time to evacuate MENARI in view of the possibility of 2/27 Bn coming in. The village itself was extremely vulnerable and could not be held. The time fixed for the evacuation was 1430 hrs after which A Coy 2/16 Bn moved out and rejoined its Bn (strength approx. 200) which was to hold main ridge [to the rear] until 0900 hrs the next day'.[1] Indeed, Captain Kienzle recalled:

> At 11.45 hrs. the Japs in possession of an ideal position directly overlooking the camp at MANARI opened up with heavy mortar fire and L.M.G. We flattened out and took cover but were unable to retaliate. With the increasing and more accurate machinegun fire we were forced to evacuate MANARI .... I led a party out over a small bush track to the main track and we assembled there to retire to Nauro. Before retreating I made a reconnaissance along the ridge south of MANARI but could not find a spot overlooking the village to observe enemy movements. A few [men from] 2/14 and 2/16 Bns arrived in very weary after weeks on continuous fighting and took up defensive positions along the track to NAURO.[2]

By 12.30 pm, most of the men of the 2/16th Battalion had arrived at the village, which was then already under a barrage from at least one Japanese mountain gun and several mortars from the high ground to the north; the area was also swept with machinegun fire. After being fed, the men of 'A' Company were placed in a position covering the village, and when the men of the 2/14th Battalion arrived, 'A' Company would be pushed south to Nauro. The diarist of the 2/16th Battalion recorded that at 12.30 pm, advance elements of the 2/14th Battalion and the bulk of the men from the 2/16th Battalion had reached Menari while under Japanese mortar fire from Brigade Hill. The battalion diary of the 2/14th Battalion records that Captain Phillip Rhoden and Lieutenant Stan Bisset checked the men as they passed through Menari, although portions of the Headquarters Company and 'C' Company had lost contact some distance to the rear. The Battalion Headquarters Company was still well behind carrying stretcher-cases with the men of the 2/27th Battalion.[3]

Medical officer Captain Steward with the 2/16th Battalion recalled arriving at Menari at this point: 'We all got an immense issue of ration chocolate, wolfed down at once by every man. Our systems were crying out for the sugar we had almost totally lacked over the last two weeks. Cigarettes, too, were handed out in the unimaginable largess of whole packets. Some strange smokes had been seen (and smelt!) in the last fortnight, tea leaves being a common substitute'.[4] Menari was to be held until '1430 hrs in the hope that stretchers and 2/27 Bn would arrive before then'.[5]

Arriving with the 2/16th Battalion had been two badly wounded privates, Clarrie Maskiell, a 25-year-old mine worker from Perth, and Leslie Dingle, a 30-year-old collector from Boulder in Western Australia. Maskiell had been assisted on the track by his good mate Lance Corporal William Cousens. Dingle was given a shot of whisky and told to start walking south. Maskiell, however, had decided to stay put, even though he was given a liberal doze of alcohol. Brigadier Potts, Lieutenant Colonel Caro and Captain Lyon soon came across the semiconscious man; Potts was determined to get the wounded man moving so goaded him, accusing him of holding up the line. The

CHAPTER 17

original historian of the Kokoda campaign, Raymond Paull, recorded in 1958 the situation:

> Deliberately, because he was determined to save Maskiell, Potts levied a cruel accusation calculated to provoke positive results. 'This ought to shift him,' he remarked. Bending low over the wounded man, he said, 'If you weren't so yellow, you'd do something to help yourself. You're yellow, Maskiell, and you're holding up your mates. Do you want them to be knocked while you lie here?' Maskiell stirred and sat up. He clenched his fist and swung a weak blow at Potts. Caro and Lyon, catching his arm, lifted Maskiell to his feet and started him up the hill. He would not let Potts touch him ... he insisted that he would 'show the grey-haired old bastard' that he was not yellow. Maskiell staggered on to Nauro, walking while consciousness remained, and being carried when it lapsed. He became No. 40 on the list of wounded who received treatment and shelter at Nauro that night.[6]

Potts now sent Captain Grahamslaw and his small force of men from the RPC and a small party of men from the 2/14th Battalion to move out along a south-east track leading to Nauro. They were tasked with reporting back to the brigade any information regarding Japanese movements in that area and to try and contact Cooper and his men of the 2/27th Battalion. Grahamslaw recorded years later:

> The last task assigned to me by Brigadier Potts when I reported to him at Menari, was to investigate a track which it was thought the enemy might be able to use to by-pass our troops. As the task was regarded as hazardous, the brigadier insisted that I take a bodyguard of Australian soldiers (one sergeant and 11 privates) in addition to my police Squad.
>
> It was a useful experience for the troops, particularly as I had to place the party on half rations after the first day. After ranging

far and wide for six days, during which we examined numerous native tracks, none of which would have been of benefit to the enemy, I decided there would be no point in continuing, and in any case, we were short of food. We could not return to Menari as it was now in enemy hands.

By this time, I'd just about had it myself. Lack of a balance diet, constant patrolling in difficult country where it rained almost every day, and mostly without adequate shelter at night, had taken toll. In those days we wore shorts, and shirts with short sleeves. The numerous scratches on my arms and knees had festered, and by the time I reached Iawarere Plantation I was suffering from dysentery, and finally ended in a casualty clearing station.[7]

* * *

The unknown leader of the Japanese signals unit with *No. 2 Infantry Section, No. 1 Company, I/144th Regiment*, who was assigned to the Regimental Headquarters, had just organised the burial of his three dead comrades killed during the fighting for Mission Ridge. Soon after, he was again on the advance with the rest of the regiment. The officer recorded sometime that day in his diary: 'Reported casualties to the coy comdr before evacuation. Asked the WO how to dispose of the bodies. Evacuation started at 0800 and was completed by 1300. The sec was guarded by one section of No. 9 Coy during the evacuation'.[8]

Finally, the orders had arrived for the officers and men of the *I/144th Regiment* to leave Myola to rejoin the regiment's advance further south. Advancing from Myola was 2nd Lieutenant Onogawa and his men of *No. 2 Platoon, No. 3 Company* who had been ordered to pass through the area being held by the *41st Regiment*, as the *144th Regiment* was leading the advance.[9] Close by would have been 1st Lieutenant Hirano, leading *No. 3 Platoon, No. 1 Company, I/144th Regiment* who recorded in his diary that day that he had a severe headache and an upset stomach.[10]

CHAPTER 17

Meanwhile, further south, Lieutenant Sakamoto Atsushi, commanding *No. 2 Machinegun Company, II/144th Regiment* – who had provided supporting machinegun fire for the battalions' attack against Butcher's Corner – recalled that he and his men received rice that day for the first time in three days.[11]

* * *

Still heading for Menari along the eastern track were the stretcher-cases of the 2/14th Battalion who had fallen behind the rest of the column; they were soon joined by the forward elements of the 2/27th Battalion. This stretcher party was led by Lieutenant Alan McGavin, a 22-year-old university student from Black Rock in Victoria, along with the 2i/c of 'C' Company, one of his officers, Lieutenant Jack Clements, a 22-year-old clerk from Melbourne, and his men of 13 Platoon. Also assisting was Lieutenant John Greenwood, a 25-year-old British regular Soldier from Sunderland, who had earlier tried to carry Private Ambrose Mayne, a 37-year-old storeman and packer from Corindap in Victoria, on his back, while both were lost in Eora Creek Valley after the fighting at the Isurava Rest House. With Greenwood were his men of the Pioneer Platoon, 2/14th Battalion.[12]

* * *

Meanwhile, Captain Russell and his composite platoon was still located just north of Menari to warn of any Japanese approach, while McGee and his men of 'A' Company, 2/16th Battalion were holding the village. Lieutenant Colonel Albert Caro placed the rest of his men of the 2/16th Battalion south of Menari, while Captain Phillip Rhoden and the remainder of his men of the 2/14th Battalion moved south to Nauro over a track described as being knee-deep in mud. Potts was trying to hold the village as long as possible to enable the stretcher party of the 2/14th Battalion and the men of the 2/27th Battalion to reach and pass through Menari.[13]

At 2.30 pm, while the Japanese commanded the high ground just north of the village, Potts was forced to withdraw most of his force to Nauro. The men of 'A' Company, 2/16th Battalion now took up a delaying position on some high ground close to Menari, while still in front of these men was Captain Russell and his platoon, consisting mostly of survivors of the 2/14th Battalion, who remained upfront in a holding position.[14] Lieutenant Stan Bisset recalled years later: 'So then we only got the 14th and the 16th into Menari and then the Japs [shelled us] …. Bill Russell our 2iC, who wrote the history … brought a party of our people plus others up and they formed a defence position just forward of Menari, between the Mission Ridge and Menari, and so they held it, delayed the Japs there before they could get it'.[15]

At the time, Potts had ordered Major Magarey that all sick and wounded at Nauro had to be evacuated before dawn. In Nauro, nearly all the stretcher-cases were laying in a hut that served as the medical aid post. The walking wounded were resting nearby. Going to the aid post, the major asked for silence and announced: 'I want every walking case to be away from here by half past five, and every stretcher-case by 6 o'clock. At a quarter to six, I must have a report that the last cases are ready to move'.[16] Magarey recalled watching in admiration as every man capable of walking or hobbling assisted to 'fetch and carry, to feed and prepare the stretcher patients for the day's journey. Before 6 o'clock, all had been fed and dispatched up the track to Ioribaiwa'.[17] It was now that the exhausted and sick Major James Magery was replaced by Major Ronald Humphery, a 20-year-old physician from Fairfield in New South Wales, as the senior medical officer in the forward area.[18]

Later that day, Captain Kienzle and his men of ANGAU had arrived at Nauro. He recorded in his official report: 'We arrived at NAURO about 1600 hrs. and were forced to evacuate all the wounded from NAURO that evening. The First units of the 3 Bn took up defensive posns near NAURO [*sic* Naoro] river to delay the enemy. After sending away stores and withdrew to the ridge north of IORIBAIWA.[19]

CHAPTER 17

* * *

The survivors of the 21st Brigade further north were now forced to evacuate Menari. The diarist with the 2/16th Battalion recorded the men of 'A' Company were to hold the village until 2.45 pm. At 1 pm, the Japanese attack against Menari barred the progress of the stretchers with the 2/27th Battalion who were making it to the village. Just at the bulk of the brigade was forced to withdraw from Menari, Staff Captain Peter Smith and his stretcher party from ANGAU reached Cooper and his men still struggling along the eastern track, who were trying to reach Menari. Smith had brought along 70 Papuan bearers, which greatly eased the situation, although the soldiers themselves still had to carry five of the stretchers. Shortly before the arrival of Smith, Cooper had ordered Captain Sims and his men of 'A' Company and Lieutenant Colin Sandison, a 28-year-old sales superintendent from Glenelg in South Australia, with two platoons from 'C' Company to push forward to Menari and reconnoitre the position; they were to hold the position until the rest of the battalion arrived.[20]

* * *

By 5 pm, the men of the 2/16th Battalion had taken up positions on top of a ridge to the south of Menari with 'C' and 'D' companies on the right of the track with 'A' and 'B' companies on the left with an ambush party from 'D' Company placed just forward of the ridgeline. The Battalion Headquarters was located 50 metres behind the main defensive line.[21]

Among the last to leave Menari had been Lance Corporal William Cousens with 'B' Company, he recalled many years later: 'We moved hastily onto Nauro from Menari, my only pause being my return to Menari when I discovered that I had missed out on a cache of chocolate and tobacco that was earlier deposited there for the benefit of the troops falling back to new positions. I was not going to miss out on those luxuries, Japs or no Japs, so back I went to discover chaff bags full of these wonderful commodities, to load up with all the goodies possible before beating a hasty retreat to our own lines .... it galled me to imagine the picture they would have when they discovered all that lovely booty'.[22]

Still holding the northern perimeter as a rearguard for the brigade was Captain William Russell. He and his 20 or so men now got the welcomed news that they were to fall back in short bounds and pass through Menari. Soon after, they were ambushed by a Japanese infiltration party and forced off the track, losing two men killed. While some elements arrived back at the various units, some were forced to go bush, including Russell and Lieutenant David Rainey, a 22-year-old farmer from Pyalong in Victoria.[23] Russell provided an account of his time forward of Menari. It is likely the log bridge referred to spanned Enili Creek about one kilometre north of Menari:

> At 10.30 am the enemy sent three men dressed as Australians to the Efogi end of the log bridge in an attempt to draw fire. There wish would have been gratified but for the possibility that they were some of our own troops cut off in earlier fighting. The enemy then commenced infiltration by crossing upstream from the bridge. The first patrol at 11 am was sighted by Pte Joe Attard[xiv], who killed two and dispersed the patrol. This remarkable soldier was a native of Malta, who though unable to read or write English and barely able to speak it, had been a successful racehorse owner. He had served in the First World War and understated his age to serve in the Second. Only his gameness, and a determination that he would kill at least one Jap had carried him over the trail. He was himself killed in action later in the day.
>
> At the same time as the enemy made his first attempt to cross the creek, he brought heavy fire from woodpeckers and mountain

---

xiv Private Joseph Attard, a 37-year-old wharf labourer from Melbourne, is listed as Killed in Action, 9 September 1942.

CHAPTER 17

guns to bear on Menari village. It was obvious to Brigadier Potts that Menari could not be held since it was overlooked from the high Efogi track. He therefore ordered it to be evacuated for new positions on the low angle between Menari and Nauro. Lieut. Lex Tillyman of Brigade Headquarter, and others, made repeated trips into the bullet-swept village to secure tins of salmon, bully beef and cheese for the 2/14th and 2/16th as they came into their covered rendezvous near the village. After eating, the two battalions made a detour round the village to the new positions – tactically better than Menari, though far from satisfactory:

Meanwhile the enemy, having been held at the creek crossing had moved up the creek and come on the track being used by the main body. They established positions here at about 3 pm forcing the stretcher parties of the 2/14th, 2/16th and 2/27th to detour to the east …. At 4.30 the patrol in accordance with earlier instructions from Brigadier Potts commenced to move back to the village, moving by short bounds. About 150 yards from the crossing the track was swept by crossfire from across the creek and from ambush positions on the left of the track, so that the patrol was forced to take cover on the opposite side of the track.

Ptes Joe Attard and Eric Vial[xv] had been killed and Pte Pierce Allen wounded. Bullets had struck Pte Basil Quarrel's equipment but had been deflected the right way. Sgt Thorne and three men were instructed to move off immediately to the new Battalion area through the jungle, as the Japanese were now in great numbers and preparing to camp for the night. Sgt Thorne's group, suffering from exposure and exhaustion, came out at Uberi nine days later. The remainder of the standing patrol

---

[xv] Private Eric Vial, a 29-year-old baker from Abbotsford in Victoria is listed as Killed in Action, 9 September 1942.

left at dusk. After moving all night, they came up to the track between Menari and Nauro about the middle of the morning to find the Japanese in possession. The same thing happened two days later at the Nauro [sic] River crossing. The patrol reached our own lines at Ioribaiwa after seven days' climbing through the jungle.[24]

* * *

By now, Brigadier Arnold Potts was moving south towards Nauro when he came across a couple of men from the 2/14th Battalion taking a rest beside the track. Twent-six-year-old Corporal Leonard Parker, from Geelong in Victoria recalled: 'We had just passed through Menari .... We – that is my section which included Joe Royals and Sandy McAlister – must have decided to have a bit of a rest and a bit to eat halfway up a long hill when along came Brigadier Potts and his single escort, both carrying a fair bit of gear. We didn't recognise him until he was close. He said, "Come on you men, keep moving, the Japs aren't far behind." We started to move, a bit unwillingly. We were tired and hungry, so he said, "Give me your Bren, I'll carry it up the hill for you." Which he did, then gave a wave when he reached the top before making off. We greatly appreciated this at the time, but thinking it over later, with the problems he had on his mind – he probably knew the 2/27th had been cut off, how low in numbers were the 2/14th and 2/16th and how close the Japs were – this was the act of a great man who we all felt privileged to know'.[25]

* * *

Captain Sims and his men of 'A' Company, 2/27th Battalion were now leading the way forward towards Menari, unaware it was already in Japanese hands. Not far behind them was Lieutenant Sanderson and his men of 'C' Company. As the men of 'A' Company approached the village, 24-year-old Lance Corporal James 'Jim' Mannion, from Broken Hill in New South Wales, patrolled to the edge of the Menari

CHAPTER 17

and observed from behind a patch of cane grass the Japanese and 'gained a most useful and accurate estimate of their numbers'.[26] Also, with Sims's company was 24-year-old Platoon Sergeant Edgar, from Orroroo in South Australia. He recalled their plight at this point: 'In my platoon, 9 Platoon, after the seven or eight attacks in the couple of days before we came out of Efogi and went up the hill, I went around and said: "Well how much ammunition have we got?" We had very little, very little …. I know I struggled out from Efogi with Slim's gun, and I got up the top and said, "There's your Bren back, Slim," And he said, "What's the use of that bloody thing! I've got no ammunition for it anyway!" So, the stretchers weren't the only problem Cooper had. It was lack of food, and A, and possibly B Company, were practically out of ammunition.[27]

After conferring with Sims, Lieutenant Sandison with 'C' Company detached Lieutenant Ernest Caddy, a 30-year-old civil servant from Adelaide, along with a section of 14 Platoon as a rearguard. He was also to inform Cooper of developments up ahead and conduct patrols between Cooper's split force. Soon after, 'A' Company and the two platoons of 'C' Company were attacked by the Japanese who forced them to withdraw away from the Menari area. Caddy and his section soon again joined Cooper and the bulk of the 2/27th Battalion.[28]

Captain Sims recalled moving back to Brown River to try and locate Lieutenant Colonel Cooper: 'This icy cold water …. There was no bridge or anything to get across on and you had to wade up to your chest and hold whatever weapons you had over your head. We were stuck down in this great canyon! The main track was well away to the right. I told them to look for any sign of a track coming down, but mainly leading up on our right-hand side. And I said: "Well if you can see that, that's presumably where the rest of the battalion's gone. That's their advance party, going to establish a firm base somewhere on the left flank of Menari"'.[29] While the track was used by troops, they were not used by Cooper and the main party; indeed, they were still further north. Sims continues: 'And at that stage, that day or the next day, we had stragglers coming in from the 2/14th from all over the place. Poor beggars! One bloke started off with half his behind shot away. He'd had a Jap grenade tossed at him. He had all the guts in the world, struggling along with two blokes helping him'.[30]

Meanwhile, Cooper and the remainder of his men were awaiting word from Sims, but hearing nothing from him and seeing from their position that the Japanese were already at Menari in force, these men set out to skirt around the village. Cooper left Caddy and his men to instruct Sims and his party which way they had gone. It was likely soon after that the fighting broke out between the Japanese and Sim's men just east of the village. With this, Cooper sent forward Sergeant Paul Robertson, a 30-year-old clerk from Hawthorne in South Australia, and a small party towards Menari to try and contact Sims, but Robertson was forced to return finding no sign of 'A' or 'C' companies.[31] The battalion war diary records the situation of the 2/27th Battalion from 3 pm onwards: 'Jap patrols were active, and the enemy were felling trees and making a lot of noise generally .... ANGAU representative [Captain Smith] commenced cutting [a] track, roughly parallel to main track, but at 1830 [hours] due to darkness and proximity of Jap patrol, Bn was forced to halt and bivouac for the night'.[32]

The adjutant of the 2/27th Battalion, Captain Harry Katekar, recalled the decision by Cooper not to attack Menari: 'There was only one option for Cooper and the rest of us, and that was to get our wounded out, and not create more. We thought of that; we thought of attacking Menari, that was deliberated on not only by Cooper but by all the senior officers that were there including myself. And the decision was clearly made; we would have been utterly foolish [to attack]'.[33]

* * *

The *General Narrative of Events* concludes for 9 September that the remnants of the 2/14th Battalion (just 120 strong) were to take up a position the next morning just forward of Nauro Ridge as a rearguard

CHAPTER 17

until the 2/16th Battalion passed through while companies of 3rd Militia Battalion would establish themselves on top of the ridge. That night, while on Nauro Ridge, Potts was in telephone contact with Brigadier Porter, who had originally commanded the militia 30th Brigade and Maroubra Force, before Potts had replaced him. Porter was now moving forward and was then located south of Ioribaiwa village. Potts was told that Porter had important information to pass on to him. Unknown to Potts, Porter had been sent by Major General Arthur Allen, commander of the Australian 7th Infantry Division, to take over command of Maroubra Force. Porter was instructed that after he took command, Potts was to make his way back to Port Moresby to report to Allen. It was arranged that next day, 10 September, Porter would meet Potts just south of Nauro Ridge at the Royal Australian Engineers camp. Events, however, would overtake them, and the two brigadiers would meet the next day on Ioribaiwa Ridge.[34]

* * *

Back in Port Moresby, Captain Gordon Darling, the ADC to Lieutenant General Sydney Rowell, recorded in his diary late that night: '21 Brigade are cut off all day, but bits and pieces regain touch with our Lines of Communications about 2300. They must be absolutely exhausted, having fought hard all day. Heavy Casualties to both sides'.[35]

18

'WE WERE ALL IN TATTERS, HALF-STARVED, MUDDY AND BLOODSTAINED'

Further south at Nauro, Major Hugh Challen had placed a standing patrol of militiamen from 15 Platoon, 'C' Company, 3rd Battalion to cover the Naoro River crossing halfway between Menari and Nauro. It was now that this battalion got word that Lieutenant Colonel Allan Cameron, who had commanded the 53rd Battalion, was their new CO.[1] Sergeant Colin Kennedy with 15 Platoon recalled his OC of the patrol heading towards Menari had a chance meeting with Brigadier Arnold Potts, who was now heading for Ioribaiwa to meet with Brigadier Selwyn Porter:

> The village was alive with activity but not so congested as at Uberi. The villagers were nowhere in evidence. There would have been no more than 7 or 8 huts not including one built for troops in transit. The design of the huts was very good, unlike the heavily thatched ones in the lowlands. Built well above ground level they also afforded open shelter against the heavy rains, which were almost a daily occurrence.... 15 Platoon were sent forward on the track which leads to Menari the village over the next intervening range. A breakdown in communication occurred at this time. The platoon commander decided to keep his orders to himself.
>
> When two returning soldiers coming back along the track inquired of him what his task was, the response was 'I've got

CHAPTER 18

my orders'. One of the two then announced that the other was Brigadier Potts. Had he been wearing his badges of rank no doubt the response to the question would have been different. Because of the closeness of the fighting all badges of rank were required to be removed. The day before he had experienced enemy fire raking his headquarters and most of his force being cut off …. The track along this stretch was in a low-lying area not far from a river draining this deep valley. The fluid mud was about 15 centimetres deep. The platoon commander ordered the men to fix bayonets in case of ambush, a thought which would not have occurred to me, his platoon sergeant. A rifle was heavy enough to aim without a bayonet attached. In the jungle, bullets were quickest ….

When 15 Platoon reached the creek (it not being a river) we found a log bridge over a shallow stream about 5 metres wide. There was but a limited field of fire because of the jungle undergrowth. A defence position was prepared more or less in a line parallel with the creek and about 20–30 metres from it. There was no time nor means to thin out the vegetation. Rocks from the stream were used to make above-ground defences, but they left much to be desired. There was only the bayonet and steel helmet to dig a hole, and this was virtually impossible. Over the creek the track commenced its climb to Menari.[2]

With the continued withdraw of Potts' men, orders were now sent for 'B' Company, 3rd Battalion, who were conducting patrols from Edwalla, to return to Ioribaiwa Ridge to join the men of 'A' Company who had taken up a defensive position there. Lance Corporal Patrick 'Nance' Hogan, a 21-year-old grazier from Crookwell in New South Wales, with 'B' Company, 3rd Battalion recalled: 'That evening an urgent message (brought through by runner) was received and with about half an hour's daylight left we set off on a return journey. We made terrific pace and reached the river where we had dinner … After crossing the river with the aid of light from a torch, we began to ascend the terrific mountain on the opposite side and after going a short distance one chap fell down a small cliff and dislocated his shoulder .... After this it was decided to abandon the impossible task and we bivouacked for the night, just tumbling down where we stood and making the best of it until dawn, when once again we were on the trail'.[3]

Meanwhile, their mates of 'D' Company had also begun their advance into the Owen Stanleys, having been transported to Ilolo the day before. At 7.15 pm, however, Cameron issued revised orders; the battalion was to move up the Nauro Track at 6.45 am the next morning (10 September). The men of 'A' Company would head forward of the Battalion Headquarters with the Headquarters Company just behind. The battalion 2i/c and staff would then move forward with 'B' Company when these men arrived.[4]

* * *

Still behind the southern slopes of Ioribaiwa Ridge at their Red Shield post was Major Albert Moore and Corporal Jock Inglis. The salvationist officer recalled years later: 'I was watching anxiously for some coffee, tea and biscuits to arrive. Sugar was almost out. Surely no Red Shield post had ever catered for so many with so little. That day we served 35 gallons [160 litres] and supplied about 200 of the 53rd Battalion, who had been withdrawn. News from the fighting front was not good. We were still being forced back and the brigade retired to Nauro'.[5]

* * *

It was now that the bulk of the survivors of the 39th Militia Battalion began to arrive at the end of the Kokoda Track, just north of Port Moresby, as recalled by Lieutenant Harry Mortimore, a 28-year-old bread carter from Red Cliff in Victoria. He was a veteran of the Middle East Campaign, having served with the 2/14th Battalion; however, since then, he had been assigned to 'B' Company, 39th Battalion and had been in the thick of the fighting since the Japanese invasion.[6] The

CHAPTER 18

young lieutenant recalled the final day's trek to get out of the Owen Stanleys: 'There were supposed to be trucks waiting for us at Ilolo to take us to our camp site on the Laloki River. But there weren't any – they were being used to bring up fresh troops from Moresby wharf. So, we marched. I remember passing through one battalion area .... They had arrived from Australia only the day before and they just stood and stared at us. We were all in tatters, half-starved, muddy and bloodstained. Along with lots of others I had had dysentery for a week, and most of us hadn't had our clothes off for several weeks. We eventually reached our camp site where we were issued with clean clothing, soap, razorblades, etc. Then we piled all our old rags in a heap to be burned and had a glorious wash in the river'.[7] The battalion war diary merely records for 9 September: '39 Bn moved to 30 [Brigade] I.T.B. area'.[8] Mortimore and the others of 'B' Company had set out to cross the Owen Stanleys almost two months to the day, 8 August 1942, to conduct patrols around the Kokoda–Buna area, arriving there just one week before the Japanese invasion at Gona and Buna.

Also arriving in Port Moresby were the veterans of the 2nd AIF, 25th Brigade, as recorded in the 7th Division war diary: 'HQ 25 Aust Inf Bde, together with 2/25, 2/31, 2/33 Aust Inf Bns and det 2/4 Aust Fd Amb, arrived PORT MORSBY on morning of 9 Sep 42. Brig [Kenneth] EATHER and staff reported to HQ 7 Aust Div at 1100 hrs same day .... The main body of 2/4 Aust Fd Amb did not arrive until the second flight and some ten days after 25 Aust Inf Bde went into action'.[9] These men were the first Australian troops to be issued with 'jungle greens', unlike those now fighting the Japanese who were still mostly dressed in khaki. They were also well supplied with 'emergency rations in any quantity, also ammunition in any quantity'.[10] Now attached to the 25th Brigade were the men of the 2/1st Pioneer Battalion (less on company), commanded by 48-year-old Lieutenant Colonel Arnold Brown from Sydney; he and his men would go into battle with the brigade.[11]

SAVING PORT MORESBY

* * *

To the north and still west of the Kokoda Track, Lieutenant Robert McIlroy and his lost patrol of the 2/14th Battalion were trying to get across what was almost certainly Efogi River, located at the bottom of a steep and treacherous ravine west of the track. They were soon helped by local Papuan Seventh Day Adventist parishioners at Hailo: 'During the morning came on a native store hut and obtained sufficient yams and sweet potatoes from it for a meal, while we were cooking these, some natives arrived and were very upset to find their store hut open, and it turned out that we had been eating their seed yams. We therefore returned what we hadn't eaten, and the natives became very friendly and led us to their village (Hailo) about ½ an hour away. For the rest of the day, they kept us feeding continuously on yams, taro, bananas, pawpaw and sugar cane and finally gave us a hut to sleep in'.[12]

Not far away to the east was Lieutenant Maurice Treacy and his two men, also with the 2/14th Battalion, who had been sent south by Captain Sydney Buckler. The young lieutenant recorded later in his report:

> After travelling some distance, the party arrived at an old village. A search for food was being made in one of the huts when it was noticed that the village was occupied by JAPS. The party retired to a vantage point to observe and a few minutes later a detachment of 33 enemy inf. joined those in the village.
>
> The party encircled the village and in arriving back on the track almost ran into the same detachment which was moving on after a rest. Giving the enemy fifteen minutes start the party followed across a ford in the nearby creek.
>
> On reaching [the] WEST bank of the stream, the party was surprised by about fifteen native bearers returning empty handed. These men were not escorted by JAPS, they appeared to be quite friendly, saluting as they passed. They did not speak.

CHAPTER 18

Continuing on, another enemy bivouac area was reached. Enemy stragglers here and the party's rice supply increased to about 11 pounds. Previously the rice obtained was of poor quality and slightly mouldy, one of the men [Japanese] discovered here had newer polished rice, he was evidently a recent reinforcement. It was apparent that the enemy force as a whole was suffering from an epidemic of dysentery. That evening a halt was called near [another] JAP bivouac ….

Sounds of mortar fire had not been heard since the previous day and it was a bitter disappointment to know that no assistance had been obtained for Capt. Buckler's party which by now would have moved.[13]

* * *

Meanwhile, the War Cabinet and the Advisory War Council were increasingly concerned about the situation in New Guinea, and on 9 September, the Australian Minister for the Army, Frank Forde, had asked General Blamey to go to Port Moresby to confer with Lieutenant General Sydney Rowell and report back to them.[14]

19

'NAURO WAS A REPETITION OF MENARI'

At 6.45 am, 10 September, the advancing elements of the Japanese *III/144th Regiment* encountered the forward position of the Australian 2/16th Battalion, which was located just south of Menari. The Australians broke contact and began to retire further south down the Kokoda Track. An ambush position had been prepared by Corporal William 'Bill' Speed, a 31-year-old labourer from Claremont in Western Australia, with 'D' Company. The corporal and his section were a few hundred metres forward of the Australian perimeter, and when the Japanese advanced, they inflicted at least ten casualties against the Japanese before withdrawing.

By 7 am, the remainder of the battalion was withdrawing, having made a clean break from the enemy. Thirty minutes later, Major Hugh Challen, the temporarily CO, along with Captain Phillip Rhoden and Lieutenant Stan Bisset and his intelligence section had moved back to the top of Maguli Range between Nauro and Ioribaiwa. Here, they reconnoitred a fallback position for the small number of troops who were to hold back the Japanese while the remainder of the battalion continued to withdraw reaching Nauro at around 11 am.[1]

By 8.15 am, a standing patrol from 'A' Company, 3rd Battalion forward of Nauro reported that the men of the 2/16th Battalion were passing through their position. Earlier, at 6.45 am, the other companies of the 3rd Battalion had left Ioribaiwa to move towards Nauro on the main track, while Corporal Christie and one of his men

CHAPTER 19

from the intelligence section were conducting a reconnaissance of the police track from Ioribaiwa to Nauro, which runs north-east from the main track from Ioribaiwa before turning directly north to Nauro.

By 10 am, 'A' Company of the 3rd Battalion was arriving at the position held by 'C' Company and the Battalion Headquarters' position located just south of Maguli Range. The battalion CO, Lieutenant Colonel Allan Cameron, as well as Adjutant Captain John Jeffrey, a 32-year-old bank clerk from Merewether in New South Wales, and the battalion intelligence officer, Captain John McCracken, a 34-year-old banking officer from Canberra, went forward to conduct a reconnaissance of the main track to Maguli Range. On returning about an hour later, Cameron deployed along the track with 'A' Company forward and the Headquarters Company behind them.

By 2 pm, the 21st Brigade Headquarters had arrived at their location and took up a position in the centre of the 3rd Battalion's sector. Later that afternoon, at around 5.30 pm, Cameron's men of 'B' Company arrived at is position, which was soon followed by the mortar team that had been located north at Nauro.[2]

By the early afternoon, Caro and his men of the 2/16th Battalion had fallen back to take up a position along Maguli Range. Not far behind them were the men of the 2/14th Battalion (now represented mostly by the men of 'D' Company and the Headquarters Company with surviving remnants of 'A', 'B', and 'C' companies), which was now led by Brigade Major Hugh Challen.

The 2/14th Battalion history records that the Japanese were now moving up slowly and seemed to have lost some of their momentum. The Australians rested on the peak of Maguli Range and dried their clothing in the sun and prepared their bivouac for the night. Within hours orders came through that both battalions, with a combined strength of just 307 men, were to form into a single composite battalion under the command of Lieutenant Colonel Caro, with Major Challen as his 2i/c.

Additionally, word came through from Major General Arthur Allen to Brigadier Arnold Potts that he was to be replaced by Brigadier Selwyn Porter.[3] Allen sent his last message to Potts as commander of Maroubra Force: 'Porter going forward as your personal relief. Would like you to return to me when you have handed over to Porter. My plans will be prejudiced if we cannot hold Nauro as a supply dropping ground. Otherwise, all supplies must be carried from Uberi. Can you cover Nauro until Porter can put his plan into operation? …. Advise immediately Cooper is contacted and repeat to Porter. Advise urgently strengths your units'.[4]

Meanwhile, the newly formed composite battalion formed a weak perimeter astride the track on Maguli Range with 'D' Company, 2/14th Battalion located in the forward defence line right of the track, while 'A' Company, 2/16th Battalion covered the forward position left of the track and the cobbers of 'C' Company would act as a counterattacking force. The men of 'B' and 'D' companies, 2/16th Battalion completed the perimeter by covering the rear of the men of the Headquarters Company. The remnants of 'A', 'B', 'C' companies, 2/14th Battalion, who were under the command of Lieutenant Linday Schwind, would support the forward units.

At this time, Captain Thomas Boag's militiamen of 15 Platoon, 'C' Company, 3rd Battalion remained in their forward position to cover the Naoro River crossing forward of Maguli Range. A 12-man patrol led by 23-year-old Corporal James Jeffrey from Caulfield in Victoria with 'C' Company, 2/14th Battalion was also sent to conduct reconnaissance along a track leading back to Itiki.[5]

Also now moving up was 'B' Patrol, 2/6th Independent Company. Commanding the 2/6th Independent Company, which was basically a commando unit, was Major Harry Harcourt, a 50-year-old civil servant from Hobart.[6] Harcourt had sent out 'B' Patrol commanded by Lieutenant Richard Egan, a 39-year-old accountant from Melbourne, to cover the left flank of the Australians along the Kokoda Track. Egan and his 50-man patrol were to search out

CHAPTER 19

Japanese infiltration parties. Harcourt recorded the situation of his company as the Japanese continued their advance across the Owen Stanley Range:

> On 30 August, the Coy came under command of 7 Aust Div and on 6 September was moved to the left flank of OLVRERLY in the Laloki area to prevent Jap infiltration round this flank. From this base, between 6 September and 12 October, protective and offensive patrols were sent out on both the right and left flanks of the main Moresby – Kokoda Trail. These patrols varied in size from a subsection to 150 men. The duration of the patrols varied from 5-6 days up to four months. In some cases, these patrols acted as independent, long-range patrols, in others they were in close contact with regular formations. Some patrols covered several hundred miles over mountainous country and crossed ranges of between 10,000 to 12,000 feet [3,000 to 4,000 metres]. On some patrols up to 10 days rations were carried [by each] man. Supply to these patrols was either by lines of communication organised by Coy HQ or by plane dropping. These patrols covered a great deal of country and carried out much useful work on track reconnaissance in their areas and in most cases sent back the only reliable information of tracks which was ever in possession of the General Staff.[7]

* * *

Having evacuated Menari the day before for Nauro, Captain 'Doc' Vernon was now ordered to withdraw towards Ioribaiwa Ridge; he recorded in his report to ANGAU for 10 September: 'Nauro was evacuated by 1100, stores and buildings being set on fire. I left just before the rearguard came in and we stayed the night on the ridge overlooking the Ioribaiwa ravine'.[8] In his personal diary, he was more expansive:

Nauro was a repetition of Menari. I sent all the patients and practically all the stores of any value to Ioribaiwa, and then we waited through the inevitable Nauro post-noon downfall for the next move ... with all the remainder of my staff we started back with hundreds of soldiers, the firing in the valley behind us being then very close. However, the Japs did not enter Nauro till the next day. At the top of the ridge, I joined the PIB for lunch [and] we sat in a row on rocks on each side of the track and were waited on by the regimental cooks just as if we were at a peacetime picnic. In fact, the whole withdrawal or at least our part of it, was no more exciting than a picnic, yet but for our air force, how mercilessly we might have been pasted from the skies. We spent that night on the ridge close to WO Davies' bivouac and pushed onto Ioribaiwa early on the 11th, taking a much-needed dip in the creek on the way.[9]

Not far away on Ioribaiwa Ridge was Captain Kienzle with ANGAU who had established his camp – as recorded in his report: 'I made a new camp on the ridge defended by our troops. I met Brig Porter and Lt Col Cameron .... After seeing that all the carriers were in, I made a reconnaissance of another track – the old Kokoda trail leading over [the] ridge to [the] west of our posns. I sent Cpl SAUBO with 1 Koiari to reconnoitre as far as possible along the old Kokoda track towards NAURO and to observe enemy movements and report back. Col. Cameron CO of 3 Bn took up defensive posns ... enabled the carriers to have more rest and better food. We could not hold up the enemy on the ridge south of NAURO and were forced to withdraw through IORIBAIWA, where I established a new camp at the creek south of IORIBAIWA to be known as Station "44" [Dump No. 44]'.[10]

* * *

CHAPTER 19

Close by, just behind the southern slopes of the ridge, was Major Albert Moore and Corporal Jock Inglis at their Red Shield post. The major recalled:

> September 10. Jock was sick most of the night but was still on his feet at the crack of dawn and together we set about to meet the needs of the day. We served a company of the 3rd Battalion moving forward, and there were many sick and wounded moving back. That day a regimental aid post retired to our area and a grass hut was built to accommodate the doctor and his staff. It was good for the wounded and sick to get comforts and refreshments to help them on their way, but we were in a little conflict with the medical officer, who preferred the men to keep moving. However, I served them and withstood the pressure.
>
> As supplies were well-nigh exhausted and there was no response from Moresby to my signals, I decided to race into Uberi and see what I could scrounge to keep the post operating. I almost ran up [*sic* down] the golden stairs in my hast to reach the native village and made it is record time .... I received a message from Warrant Officer Dick Bell, who informed me that George Woodland had despatched supplies and that he would come through on Sunday [13 September]. I wondered if we could remain in that position as long as that if the retreat continued.
>
> The ration corporal at Uberi promised to help me and later sugar and milk arrived from his store, and we were able to continue .... I almost ran up the golden stairs in my hast to [return] and made it in record time ... found Jock flat out catering for the men.
>
> News came that the brigade had fallen back to Ioribaiwa, the next stage to where we were operating, and the possibility of a hostile patrol, particularly at night, was very real. Forty-seven gallons [215 litres] of coffee and tea were served to the passers-by on that day, and the task was a thrill.[11]

SAVING PORT MORESBY

Major Moore had good reason to be concerned about still having his Red Shield post operating just below Ioribaiwa on Sunday 13 September.

* * *

Earlier, Brigade Major Brian Beresford, a 32-year-old Barrister from Portland in Victoria, with the 30th Militia Brigade reported to Potts that Brigadier Porter was coming forward to Ioribaiwa; Potts moved back to meet him there, by now aware he was to be replaced by Porter. On the way, he came across the badly wounded Private Clarrie Maskiell, who he had earlier goaded into moving back from Menari:

> 1000 steps from the bottom and 200 from the top of Ioribaiwa Ridge, Potts paused where a man lay on the track. It was Maskiell. Two tall police-boys stood by. Potts knelt to examine him, noticed a flicker of his eyelids, and for the second time eluded a weak right hook. He went on, chuckling, to the village, and ... was summoned to the telephone, and at the earliest opportunity, he asked [Colonel Frank] Norris if he could spare a stretcher party to carry Maskiell up.
>
> 'Is this the man you mean?' Norris asked, and Potts saw Maskiell lurch over the last step, followed by the two police-boys. Whenever he seemed about to fall, they seized his tattered shirt, and as often as they touched him, Maskiell lashed out angrily with his sound arm. Three feet from Potts, he swung another blow. It carried him off his feet, and he fell with his arm around Norris' neck.[12]

* * *

After conferring with Potts, Porter took over command at 2.30 pm while Potts hurried back to Port Moresby to report to Major General Allen. Porter's orders were to command all troops from Uberi forward – he was to stop the Japanese advance and stabilise the position, and to

CHAPTER 19

gain what ground he could. In addition to the 21st Brigade, he also now commanded the militiamen of the 3rd Battalion (less one company), the 2/1st Pioneer Battalion (less one company), and the 2/6th Independent Company, which were now conducting patrols north-west of Ioribaiwa. He was also informed that the 25th Brigade would soon arrive to spearhead offensive operations against the Japanese.[13]

It was now that Brigadier Porter considered launching offensive operations against the Japanese. He had ordered Lieutenant Colonel Brown of the 2/1st Pioneer Battalion forward, and when these men arrived, his force would consist of the 3rd Militia Battalion, as well as remnants of the 2/14th and 2/16th battalions, and the 2/1st Pioneer Battalion. As he was planning his offensive, however, he got word that the 2/1st Pioneer Battalion had not even left Ilolo. So, Porter was forced to cancel his planned attack. He then ordered the men of these 2/14th and 2/16th battalions to hold the position until 8.30 am the next day when they were to fall back to Ioribaiwa Ridge.[14]

The *Report into Operations, 21 Brigade – Owen Stanley Campaign* records: 'At this stage it was decided to combine 2/14th and 2/16th bns into one bn under Lt Col. Caro. The strength [was] approx. 100 and 200 [men] respectively. The dispn. at nightfall were 2/14th Bn. [as well as] elements holding F.D.L's, Bde HQ and two coys, [and] 3rd Bn approx. ¼ hr along the track towards Ioribaiwa. The plan was to hold this position for as long as possible to allow 25th Bde, which was moving up, to commence offensive ops. from a suitable jumping off place. In view of the incidents [that occurred on] 8/9 Sept. Bde Comd decided it would be impossible to counterattack along the track in the event of the Jap getting behind 2/14th, 2/16th Bn posn. He therefore decided to withdraw to the high ground at Ioribaiwa which offered far more tactical possibilities'.[15]

Porter would withdraw his entire force to Ioribaiwa Ridge. Not far from his rear, however, was Imita Ridge, and just south of that was Port Moresby.[16]

SAVING PORT MORESBY

* * *

Sometime that day, 1st Lieutenant Horibe, *No. 6 Company, II/144th Regiment*, recorded in his diary for 10 September just one word: 'WAMAI' [Nauro].[17]

20

'CAPTURED LIEUT. COLONEL KEY AND FOUR OTHERS'

During the morning of 10 September, the ridges just north of Port Moresby were being heavily reinforced. The men of the 25th Brigade moved out towards the Owen Stanleys having just landed in Papua the day before. The first to do so during the early morning hours that day were the men of 'D' Company, 2/33rd Battalion. The battalion was led by Lieutenant Colonel Alfred Buttrose, a 30-year-old wool expert from Woodville in South Australia. In all, 29 officers and 550 other ranks of the battalion would move forward. And only a small element of the battalion's Headquarters Company, its mortar platoon (less one detachment), and carrier platoon (less one Vickers machine gun) would remain in the Port Moresby area.[1]

The historian of the 2/33rd Battalion, William Crooks, a 23-year-old apprentice printer from Sydney, now a platoon sergeant with 'D' Company, 2/33rd Battalion recalled within 24 hours of arriving in Port Moresby that the men of the 25th Brigade were heading straight into the mountains to support the hard-pressed men of the 21st Brigade:

> In an atmosphere of urgency, we hurried about sorting stores, issuing ammunition and changing our khaki drill for green-dyed drill, the first A.I.F troops to do so. These greens were being dyed in Sawyers stoves which had been set up by 21 Brigade men, supervised by Australian Army Ordinance Corps [AAOC] men. One simply stripped naked, moved over to a Sawyer issue point, dumped one's own khakis, picked up equivalent sizes,

SAVING PORT MORESBY

wet or dry, and put them on. The same AAOC people issued us with long knee-length American gaiters in place of our own web ankles.

By then it was dark, and our evening meal of bully beef over, [so] the rifle companies all tried to get some sleep, while BHQ, the Signal Platoon and the RAP checked the 108 wireless sets and sorted out the cable and carriers, medical panniers and packs. All realising by now that the job ahead of us would be tough, and whatever we fought with would be carried on our backs, as this was the end of any form of roads.

For Battalion Headquarters the night was a busy one. On the return of the CO an immediate 'Orders' Group was held with the company commanders. The situation was far from pleasant. 21st Brigade was withdrawing and had been for over a week. The 2/27th Battalion was missing completely, having just disappeared off the track during an engagement the day before. Our 25th Brigade was to move up immediately and assist 21st Brigade to stabilise the situation and then counterattack with a view to pushing the Jap back to where he came from. The CO was unable to learn just where in fact 21st Brigade was. Only two maps were issued to Lieutenant Bobbie Howland, the intelligence officer. Each consisted of a printed green expanse, blue-printed rivers and black-printed tracks and village names. Contours, grids, vertical intervals, heights, except for the main range, were just not shown, nor was there any information about 21st Brigade or Jap depositions. Twenty-four one-colour sketches showing an area from beyond Uberi to the top of the main range were issued the following day. On these, company and platoon commanders were to fill in information as we went along ....

The 'bombshell' order that if wounded, the men would just have to fend for themselves was a sickening one to the troops,

CHAPTER 20

coupled as it was with the information the Japs were known to take no prisoners on the battlefield, and invariably bayoneted [the] wounded .... The hurried orders to D Company – to be ready to move at 0400 hours – were not issued at section level until midnight. They were received in silence, and with a sense of foreboding. The platoon commanders gathered the men together under tent flys in the light of single torches and gave out the grim orders. At this distance more than twenty-seven years later, that night was the low point of the war in D Company. Whatever lay ahead we were prepared [to] face. But the piece of information that if wounded they would have to fend for themselves, received in the darkness of a tent at midnight, to men just awakened from their sleep, was shattering .... With this grim thought almost certainly stuck in the darkened recesses of the mind, though not one single word was spoken about it by anybody, the company boarded the same 2/27th Battalion trucks and moved down the three miles or so to the road junction outside McDonald's plantation homestead. There the men debussed in the darkness and in single file, led by the OC, Captain [Trevor] Clowes, moved off up the fifteen-feet-wide dirt road that led to Owers' Corner and the start of the Kokoda Trail.[2]

Each man carried five day's rations of bully beef, his share of the Bren gun magazines, two bandoliers of ammunition – each containing 50 rounds – two four-second grenades, emergency ration and field dressing, and for those not carrying any Bren magazines, they carried a two-inch mortar-bomb stuffed into their hip pocket. In five hours of marching along the track, the company was strung out for one kilometre, and as they pushed forward, the wounded militiamen of the 39th and 53rd battalions and the 2nd AIF of the 21st Brigade were still coming down in the opposite direction. That night, the men of 'D' Company, 2/33rd Battalion camped at Uberi. The next day, they were to push onto Imita Ridge; the remainder of the battalion was a day's march behind them.[3]

* * *

Also moving up the line towards Ioribaiwa that day was the CO of the 25th Brigade, Brigadier Kenneth Eather, who passed Brigadier Arnold Potts coming down in the opposite direction. Potts also again came across the wounded Private Maskiell as he pushed south to Uberi. The private was now on a stretcher and Potts 'lifted the banana leaf shielding Maskiell's face from the sun, and saw a contented grin spread through the stubby beard. "You Old Bastard," said Maskiell'.[4]

Eather was an energetic leader who had commanded militia battalions before the war, and he had sailed with the Australian 6th Infantry Division, commanding the 2/1st Battalion. Allen, who was then commanding the 16th Brigade, was quick to recognise Eather's abilities and considered him the best of his three battalion commanders.[5] Eather was now ordered by Major General Allen to use offensive action to halt the Japanese advance towards Port Moresby as far forward as possible. He was to regain control of the Kokoda Track and push forward through the Isurava-Deniki area, with a view of retaking Kokoda Plateau and its airstrip. He was warned that the Nauro, Menari, Efogi, and especially Myola air-drop zones were critical for supply, and he needed to control them before advancing towards Isurava and beyond. He was informed that the 21st Brigade had also been instructed to hold a defensive position north of Ioribaiwa. Eather and his small reconnaissance group would soon reach Uberi.[6]

American historian Walter Borneman concluded in his assessment of the Kokoda campaign at this point: 'The Australians fought bitterly during their retreat …. In part swayed by Willoughby's [MacArthur's intelligence chief] continuing reassurances, MacArthur remained convinced that the Japanese strength on the trail was light and that there was little prospect of their advancing on Port Moresby. He could not understand the repeated Australian withdrawals. In fact, Horii

CHAPTER 20

had pushed the maximum number of troops he could reasonably supply over the harsh ground along the trail, and it was not until one of the Australian brigades [25th Brigade] hurriedly disembarked at Port Moresby ... [and was rushed forward] that the Allies achieved some measure of parity'.[7]

* * *

Further north was Captain Charles Sims and his men of 'A' Company, along with Lieutenant Colin Sandison leading his two platoons of 'C' Company, 2/27th Battalion. The previous day they had lost contact with Lieutenant Colonel Cooper and the bulk of the 2/27th Battalion. Most of the day was spent cutting a track through thick entanglements of lawyer vine, bamboo and secondary growth of all kinds. Just on dusk, his men found themselves overlooking Nauro. A patrol then went forward, only to report that the enemy were in possession of the village.[8]

Sims briefly considered attacking the Japanese, but just as quickly dismissed the idea, recalling years later: 'Well this would have been no good; we couldn't allow an attack on the L of C. We didn't have the ammunition, and we only had rifles, no other support at all. We also had people who were pretty well [fatigued] and some of these walking wounded from other battalions'.[9] He conferred with his officers and it was decided they would have a better chance if 'A' Company split into platoon groups with each to independently make their own way back to the Australian lines. Sims recalled: 'I got all the officers and said: "Well my thoughts on this now would be to divide into small groups, tactical groups" .... So, it was decided that we'd move off in these groups of eight or ten men. I gave them an indication with the points of the compass where Moresby was, what to look for; aircraft; the rivers running at right angles to the track and so on, and finally we said, "Best of luck see you there"'.[10]

Close by, Sandison and his men of 'C' Company Headquarters, 13 Platoon and two sections of 14 Platoon (minus Caddy's Section),

2/27th Battalion had halted for the night; they intended first thing the next morning to move east for one day and then south, believing it would bring them to Port Moresby.[11]

A few kilometres north of their position, Lieutenant Colonel Cooper and the bulk of his men were also still lost in the jungle. Food was also scarce, as most since leaving the Kokoda Track three days before had with them just one tin of bully beef and their emergency rations. Having heard nothing from Sims or Sandison, Cooper decided to move out; however, just after leaving, they heard firing from the direction of the track between Menari and Nauro. Because of the stretcher-cases, Cooper and his men were forced to remain in the cover provided by the jungle, and they moved across the high ground along a route that was roughly parallel and east of the main track. The men of the Headquarters Company were now forced to help carry the stretchers, as many of the Papuan carriers disappeared into the jungle that night. In darkness, Cooper and his men bivouacked near a native garden, having encountered no Japanese.[12]

Corporal Clive Edwards recalled the day's events: 'We didn't get a very early start despite being up at first-light and we chafed with impatience, but this horrible feeling of frustration was to increase as the day wore on, and a tortuous track had to be carved down towards a fairly large stream (the same which flows past Menari) and as the pangs of hunger began to hit us in earnest we chafed even more at the inactivity .... Dysentery in a mild form ... hit the lads and some of them, particularly Dave Sheppard, are in a bad way. The track on this side of the stream as it followed the ridge along, up and down and precipitous, was even worse than on the other, and we finally had to halt for the night at a place where I had to use my string to tie my gear to a bush to prevent it sliding downhill. I have only the tiniest of hip-holes in which to sleep'.[13]

* * *

Heading north, Captain Sydney Buckler and his small party of the 2/14th Battalion were continuing their journey since becoming lost

CHAPTER 20

two weeks before during the fighting around Isurava Rest House. He was trying to find an easier way over the mountains for the walking wounded and stretcher-cases. Their immediate destination was the village of Sengai below the northern slopes of the Owen Stanleys, just south-west of the village of Gorari. They had been living mostly on boiled sweet potatoes, and during the cold nights, they were forced to try and sleep in the dank jungle and bush in their soaked uniforms. Corporal John Metson, a 22-year-old salesman from Richmond in Victoria, was still on his hands and knees crawling, half-starved but unbeaten; they would not reach Sengai for another ten days.[14]

Still heading in the opposite direction was Lieutenant Maurice Treacy and privates Fredrick Rockliffe and William Avent, who had been sent south by Captain Buckler to get urgent help from Myola. With the Japanese well south of the dry lakes, Treacy knew there was no way he could get help for his OC, as each day they got further apart:

> The party commenced movement again at an early hour and at about 0900 hrs encountered an enemy party of about 8 almost face to face on the track. Unfortunately owing to the extremely wet conditions which had prevailed during [the] preceding week, the weapons carried were seriously in need of oil, being in a rusty condition. The T.S.M.G. [Thompson submachine gun] failed to fire, the rifle carried by one member of the party fired one round after which the trigger didn't function, and a grenade dropped failed to explode. However, the pistol was effective and some of the enemy were accounted for.

> During this encounter Pte Rockliffe was wounded and the party then withdraw and dressed Rockliffe's wound which was a bullet wound in the upper leg. He showed great fortitude and after a short rest was able to return to the track and continue. After travelling a short distance an enemy solder was seen lying on the track about 50 yards ahead. He was seen to move slightly and

may have been wounded. This appeared to be a trap and it was then decided to quit the track and travel S.E. through the jungle for two days then S.W., in our attempt to [reach] the main track behind our own lines.[15]

Indeed, the Japanese reported this fight two days later and sent out an order to remind their men to be vigilante regarding the remnants of the Australians thought to be still lost behind their lines: 'On 10 Sep, at 7.40, fighting took place at a point in the KAGI area between enemy remnants and our troops at the rear, resulting in a number of wounded. I view of this occurrence; every TAI must exercise still stricter guard precautions. In case runners, etc., have to act alone, they must be especially warned'.[16]

* * *

Meanwhile, Lieutenant Robert McIlroy with the 2/14th Battalion, was south-west of Treacy's party. He and his men had been lost in the jungle since 2 September after leaving Templeton's Crossing to patrol the old pre-war track. They were now still at the Seventh Day Adventists village at Hailo. The villagers provided two guides to help McIlroy and his men along a track that would take them to Menari. In the early afternoon, Corporal Lewis Waller, a 22-year-old printer from Bairnsdale, along with privates Albert Bell, a 29-year-old labourer from Geelong, Alexander Roberts, a 21-year-old farm labourer from Werribee, and Lionel Smith, a 24-year-old clerk from North Balwyn, all from Victoria, having no groundsheets or blankets, moved forward to reach Menari ahead of McIlroy before the afternoon rains set in. As they swung onto the main Efogi–Menari Track, they were targeted by a Japanese machine gun covering the junction. Corporal Waller and Private Bell were killed as they attempted to move back to warn McIlroy, while privates Alexander Roberts and Smith were cut off from their officer and the rest of the lost patrol.[17] Private Lionel Smith recalled their lucky escape years later:

CHAPTER 20

I went forward, and I saw these Tolai boys. Now they're the boys from Rabaul that the Japanese brought across as carriers and they were a group of about ten or twelve Tolai standing, I just came out of … the jungle … I said, 'Where's the Australian soldiers?' and I had already walked past two Japanese weapon pits. They were 15 feet behind me and I looked, he said, 'Japanese soldier over there' and if he hadn't talked as loud as he bloody well did, well the Jap wouldn't have heard anything, but the Jap looked up as I looked across and he had, both of them had Australian groundsheets, they loved the Australian groundsheets because it rained every afternoon at 2 o'clock or thereabouts, [so] you put the groundsheet up and get under it. They had these groundsheets rigged up over their weapon pits, tied to the saplings around them … that Jap tried to pull the gun over and he couldn't because of the saplings and having the groundsheet tied he couldn't pull the thing over … and so I just took off to the right and I saw tracks and I started to go down the tracks.

Down the track were more Tolai boys [carrying] … mountain gun ammunition … and they all dived into the jungle on the left and I dived into the jungle on the right … I struck a little rivulet, only say 6 feet wide, very dense it was … and there was a river, it was the Menari River. I got onto the side of the Menari River [likely the Fagume River], and I thought, 'If I walk down here for a while and cut back up to the right, straight up the hill again I'll probably get back to our group of blokes' …. I heard gunfire and any rate I did go back into the jungle from the creek or river …. I heard this crashing and bashing about 20 feet to my left and out onto the riverbank came … Alex Roberts, and you know I said, 'What's happened' and he said, 'They let fly at all our blokes up there …. I said, 'Well look there's no point in going back up there, it will be mayhem, so we might as well try and make it on our own.' So, we then started to go …. I said, 'If we go in this direction we'll be going back towards our troops at any rate, even if we keep parallel to the Japanese'.[18]

Privates Smith and Roberts eventually made their way to Ioribaiwa, where they rejoined the battalion on 14 September.[19]

Meanwhile, McIlroy was advancing with one of his sections down the track, and on hearing machinegun fire, he rushed forward to see what was happening, concerned that his four men had fallen into a Japanese ambush; he recorded in his reported:

> At about 1400 hrs we halted for 10 minutes, permission was given to 4 men who at this stage had with them [no] blankets or groundsheets, to go on ahead so that they would have a chance of reaching MENARI before the afternoon rain. When the patrol moved on the native guide disappeared and about ½ an hr later we passed through an unoccupied Jap bivouac area. At this stage I took a section forward of the patrol and shortly afterwards reached the main EFOGI and MENARI track and was greeted by several rifle shots from the opposite side of the track, followed by movement in the scrub towards MENARI.
>
> I then moved the advance section down the main track to investigate, leaving the main [body] on the side-track, a few minutes later was fired on again this time with LMGs and rifles, one of my men being wounded. I then decided to return, withdraw to the mission village to patch up the wounded man. On the way back it was discovered that the main body had disappeared.[20]

* * *

Meanwhile, Lieutenant Colonel Arthur Key, commanding the 2/14th Battalion, along with his small party were captured by the Japanese. With elements of his staff, Key had become separated as they were

CHAPTER 20

swept into Eora Creek during the fighting of 30 August at the Isurava Rest House. Key and his men had been stumbling through the valley for two days when they fell victim to a Japanese ambush, forcing his party to split into two smaller groups. Key led one consisting of: Captain Thomas Hall, a 32-year-old bank manager from McKinnon; privates John O'Sullivan, a 24-year-old pump hand from Northcote, Walter Greenwood, a 23-year-old railway porter from Maryborough; Gordon Etty, a 28-year-old bank clerk from Armadale; and Lawrence Veale, a 27-year-old sign writer from Bendigo – all in Victoria.[21]

Japanese 1st Lieutenant Hirano, leading *No. 3 Platoon, No. 1 Company, I/144th Regiment*, recorded in his diary for 10 September: 'Captured Lieut. Colonel Key and four others. Though questioned the prisoners stubbornly refused to speak. Tied them securely for the night and decided to send them to the battalion commander tomorrow morning'.[22] The fate of Key's men was likely a swift beheading; however, it appears that Key managed to escape soon after interrogation and spent more days wandering in the jungle before being recaptured. The final mention of Lieutenant Colonel Arthur Key is from a captured Japanese document dated 13 October, in which it states he was interrogated at the Kokoda Headquarters of the Nankai Shitai.[23] Soon after, he was at a Japanese field hospital in Sanananda, as recorded by Father James Benson, who was the sole survivor of the Gona and Sangara missionary parties.[24] From here Lieutenant Colonel Arthur Key disappears from the historical record. Within less than two months, the Kokoda Campaign had claimed the deaths of three Australian battalion commanders.

SAVING PORT MORESBY

PART 3
THE BATTLES FOR IORIBAIWA
RIDGE

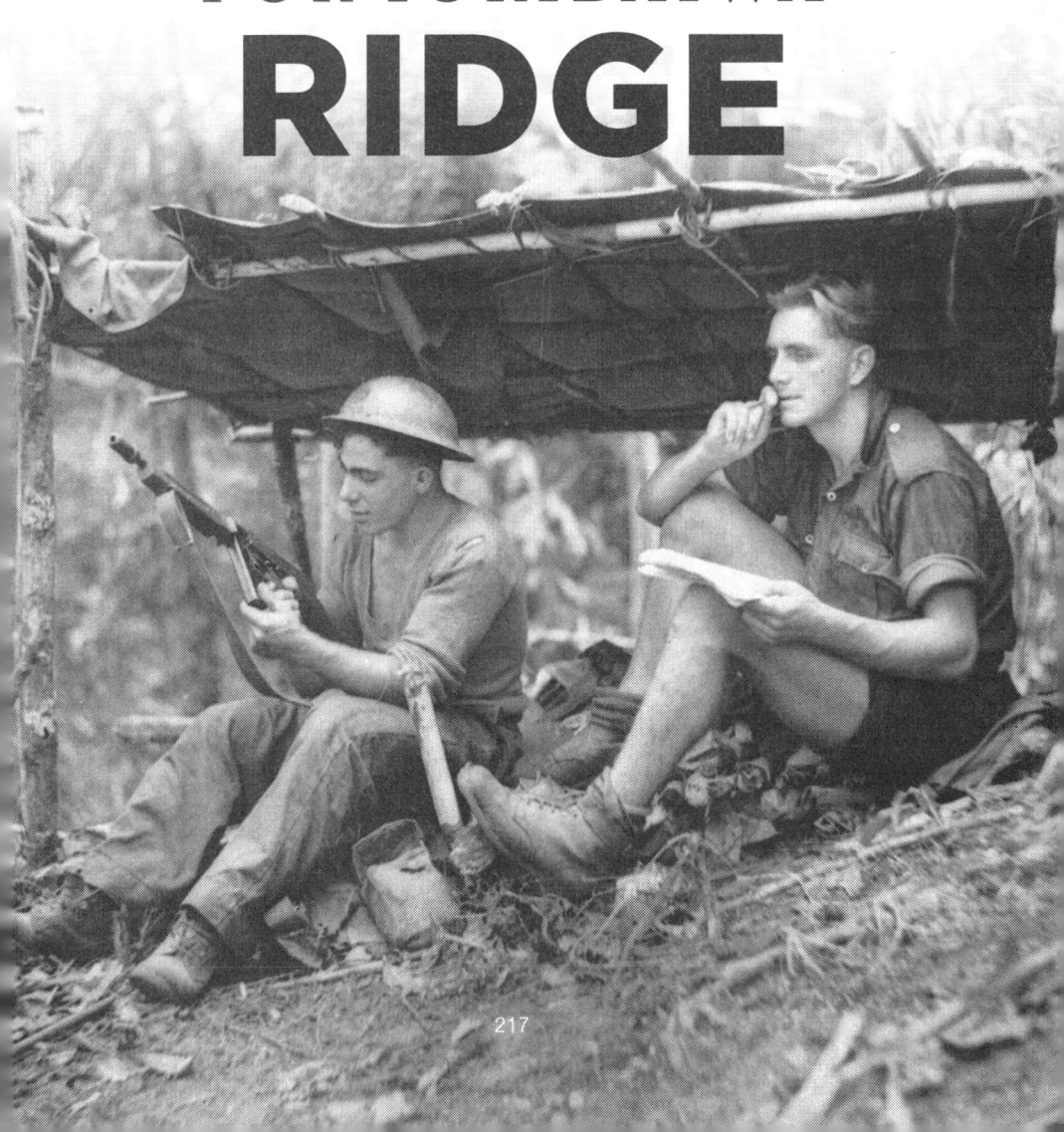

21

'ONE ENEMY BOMBER BOMBED US THIS MORNING'

At 6.30 am, 11 September, Captain Phillip Rhoden along with the battalion's intelligence officer Lieutenant Stan Bisset and several men from the 2/16th Battalion moved south from Maguli Range to find the next fallback position. Just as Caro and his men of the composite battalion were preparing to withdraw towards Ioribaiwa Ridge, they were attacked by the Japanese at 8 am. Just south of these men, 'B' Company, 3rd Battalion had withdrawn approximately 200 metres down the track to take up a holding position for the brigade, reporting that by 8.45 am they were in position. It was now that Captain Boag and his men from 'C' Company, 3rd Battalion, passed through from their forward position to the battalion's sector now located behind Engineers Ridge.[1]

Up front, the men of 'D' Company, 2/14th Battalion were able to hold back advanced elements of the Japanese troops of the *III/144th Regiment* who were trying to flank their position. Among the men of 'D' Company, 2/14th Battalion was Private John Shimmin, a 24-year-old labourer from Mildura in Victoria, who was occupying a forward foxhole. To his astonishment, just before the fighting broke out, he saw a 'six-foot Japanese fatalist obviously despatched to draw fire. The Jap was strolling along the track muttering to himself, with his rifle in the slung position'.[2] Shimmin shot and killed the soldier. It was during this fighting that Warrant Officer Wilford Noble, who had broken through

CHAPTER 21

the Japanese lines as he tried to reach Brigadier Potts' position on Brigade Hill, was killed while several others were wounded.[3]

The bulk of the 2/16th Battalion was ordered to withdraw further down the track while the men of 'C' Company, 2/16th Battalion were to remain behind to set up an ambush position. The track here was in a terrible condition. The battalion's historian recalled: 'The mire was never less than six inches. In places the track was knee deep in mud and filthy stinking water. Fit men in peaceful times would not have attempted to go on. Men, all desperately tired and some wounded and deliriously ill, had to fight a rearguard action in those conditions'.[4]

It was at 8.30 am that word reached the 2/14th Battalion that they were now to fall back.[5] Sergeant Albert Sargent, a 24-year-old truck driver from Wangaratta in Victoria, described by the official historian of the Kokoda Campaign as a brave and resourceful young leader was with two others cut off during the withdraw.[xvi] They protected a wounded mate, 33-year-old Private John Hall from Bendigo in Victoria, and all eventually made their way back to the Australian lines on Ioribaiwa. Another who brought in a wounded mate was 25-year-old Canadian-born Private George Bennett from Brisbane, who while under fire carried Private John Nelson, a 20-year-old labourer from Adelaide; however, the private died of his wounds three days later.[6]

As the 2/14th Battalion fell back, the Japanese of the *III/144th Regiment* were close behind, but they fell into the ambush south at the former engineers' camp that was set up by the men of 'C' Company, 2/16th Battalion. It was assessed that the Japanese suffered 22 casualties before the men of 'C' Company fell back.[7] The fighting and ambush brought the Japanese advance to a holt as they apparently licked their wounds. The 3rd Battalion reported that by 11 am, the last of the 21st Brigade had retired through their lines. The 2/16th Battalion diarist recorded that an hour later, 'B' and 'D' companies were in a position

---

xvi   The later Lieutenant Albert Sargent would join Australian Special Forces (Z-Force) on the attack against Singapore Harbour as part of Operation Rimau in 1944; he was captured and later executed by the Japanese just weeks before the Japanese surrender.

north of Ofi Creek crossing, which they were to hold until the forward elements had passed through.[8] At 1.30 pm, Lieutenant Colonel Allan Cameron ordered his militiamen of 'A' and 'C' companies, 3rd Battalion to retire to Ioribaiwa Ridge, where they would take up a position on the high ground east of Ioribaiwa village. By 3 pm, the bulk of the composite battalion were in position on the ridge half an hour north of Ioribaiwa located along the track, positioned in depth with 'A' Company, 2/16th Battalion forward, followed by 'D' Company, 2/14th Battalion, and then 'B', 'D', Headquarters, and 'C' companies, 2/16th Battalion.[9]

Captain McGee, commanding 'A' Company, 2/16th Battalion that was now representing the most forward element of the brigade, placed an ambush party led by Corporal Brian Moloney, a 27-year-old miner from Kalgoorlie in Western Australia, just south of Ofi Creek crossing. He and his men watched as 30 to 40 Japanese moved down the creek searching abandoned Australian shelters for food. Moloney observed in amazement as the Japanese began fighting among themselves over tins of bully beef. Within minutes, the ambush was sprung, and McGee and his men withdrew to the brigade area. Half an hour later, a report came in from 'A' Company that the Japanese were on the right flank forward position.[10]

By 4 pm, the militiamen of 'B' Company, 3rd Battalion who had been conducting a holding action forward of the brigade arrived on Ioribaiwa Ridge. Cameron placed this company on the right forward slope near the village, while 'D' Company, who had just arrived from the rear, occupied the position on the high ground right of the village. The remainder of the battalion was placed in a perimeter defence along the track on the reverse slope. However, 'D' Company soon advised that they could not reach the high ground, so Cameron placed them to the right of 'B' Company.

Just before 7 pm, Cameron received orders from Brigade Headquarters that he was to send a patrol along the eastern subsidiary track to Ponoon, and once there, establish a position on the high ground

CHAPTER 21

behind the village. An officer and eight men from 'C' Company were accompanied by a guided from the intelligence section to occupy the position.[11] Meanwhile, patrols of the 2/6th Independent Company were working their way forward to the west, looking for Japanese infiltration parties.[12]

The war diary of the 21st Brigade records at 6.45 pm the 2/14th Battalion reported a Japanese advance guard 'patrol of 5 all killed. At the moment the Japs are working up the right flank calling out, "Withdraw". 1840 hrs 3 Bn were asked to put out a patrol if possible. But the coys were not yet in position'.[13] Sometime during the withdraw from Menari to Ioribaiwa, the acting CO of the 2/14th Battalion, Captain Rhoden, felt compelled to give his exhausted and battle-weary survivors a short speech: 'Don't give in now, we've got this far, and we've done well even though we've lost a lot of people. Australia depends upon us … don't give in now'. Years later, he reflected on this speech: 'I believe it had some effect. But it was no rabble rouser, it was only a sensible, I hope, exposition of where we were and what we'd done and how well we'd done it. It was a reminder call. It wasn't a wakeup call, they didn't need waking up … I wasn't that type of commander that talked at people, I talked to them'.[14]

The 21st Brigade's *General Narrative of Events* records: 'Booby traps were laid just off the track and during the night four out of nine laid were sprung with apparently good results, a badly damaged helmet and a bloody sleeve being found in the vicinity next day. Telephone comm with Brig EATHER at UBERI was established that evening and track traces and marked maps were prepared and sent back to HQ 25 Bde'.[15] Regarding maps, the historian of the Australian Independent Companies, Greg Blake, records: 'Maps which were available were generally found to be hopelessly inaccurate, little more than glorified sketches and useless from a military point of view. Typical of these maps were examples which had large blank spaces on them stamped with such useful information as, "no information available", and more disturbingly, "suspected cannibals"'.[16]

SAVING PORT MORESBY

* * *

Finally arriving on the ridge that day was Captain 'Doc' Geoffrey Vernon, who quickly moved further south to Dump No. 44, about two kilometres south to establish his new aid post. He recorded in his report to ANGAU for 11 September: 'Arrived Ioribaiwa at 1100 and found the patients and stores except large quantities of Epsom salts and a few bottles of acriflavine solution had already been got back. All medical personnel moved down to Camp 44 where we established an aid post. I sent surplus N.M.Ords. back to await instructions at Uberi'.[17] In this personal diary, the former Gallipoli and Australian Light Horseman veteran recorded having earlier retired to what was likely Ofi Creek, just forward of Ioribaiwa Ridge. Here he documented that the former racial distinctions between white and black were crumbling:

> This bathing pool was typical to everyone on the track now that the Army was on the move and had invaded the solitudes of the Owen Stanley Ranges; the water was full of diggers and the banks dotted with parties resting or boiling their billies, but my boys [Papuan carriers] jumped in quite unconcernedly among the white troops. This showed that the old barriers between white and black were breaking down as mixed bathing in this sense was most unusual in pre-war Papua.

> Our hospital at Ioribaiwa had been emptied in advance, and NMO Nanadai was hastily packing our drugs and equipment. Nanadai is the only orderly I need to mention by name in this diary; for pluck and loyalty he stood high above the rest. Months later at Popondetta, he was given the opportunity of returning to Moresby, but he would not take it; he said, I'd sooner wait here with you till we finish the Japs. By all working feverishly we cleared the hospital at Ioribaiwa of everything of value except innumerable packs of salts; Ioribaiwa was certainly a dysentery

CHAPTER 21

centre, but the stock of Epson Salts exceeded its needs; it looked very much as if the entire supply for the line had been held up there, a proceeding which in the early days of short supplies had happened much oftener that we admitted. In the end we had to leave them, and we found ourselves short at Camp 44, I would have liked to return to get them, but by that time Ioribaiwa was too risky a place to be poking about in.[18]

* * *

Still at the base of the southern slopes of the ridge was Major Albert Moore and Corporal Jock Inglis at their Red Shield post; he recalled:

> September 11 saw a constant stream of sick and wounded, many of whom were pathetic cases. The men were shadows of their former selves.
>
> One fellow, weary, sick, all spirit sapped from his broken body, took the cup of coffee I offered him and sat on the track with his back against the embankment. His head drooped and I saw the tears course down his cheeks. He was only one of many men in that broken condition. The track was as near to hell as anything I could imagine, and something I would never hope to see again. That day the doctor and his staff fell back to Uberi, and I was glad of the use of the grass hut which was left behind. Brigadier Potts and Colonel Norris also called on their way out. I was glad to serve the brigadier and he was very appreciative of the hot drink. I felt that he had been willing to let me serve on the Trail.
>
> I received a warning to be ready to evacuate the post at a moment's notice, so we packed our surplus gear. Seven natives were left with me to help me move if the danger signal arrived. I found it hard to keep the natives from clearing out as they were fearful of the invader over-running the area.[19]

SAVING PORT MORESBY

* * *

Just after midday, Major General Horii, who was located near Nauro, issued Nankai Shitai Operational Order 'A' No. 115:

> The SHITAI has completed the pursuit of the enemy as scheduled, in accordance with 'HORII' Operation Order No.109, and at 1000 hrs today reached the western [southern] end of the Range, which formed the objective of our action. The enemy in this sector are Australian forces, consisting of 39 Bn and 53 Bn of 22 Bde [*sic* 30th Brigade], together with 21 Bde (14 Bn, 16 Bn, 27 Bn). Following their defeat by our SHITAI, and the subsequent close pursuit, it is evident that these troops are in a state of utter confusion, and to a great degree have lost the will to fight ....
>
> The SHITAI is firmly holding the key points in the vicinity of our present line and is about to prepare for a future advance. The commencement of the advance [attack against Port Moresby] depends upon the situation but is expected to be about the 20th .... The KUSUNOSE Regt [144th Regiment] will form the first line and holding the key positions of the heights west of WAMAI [Nauro] will reconnoitrer enemy conditions and terrain of the sector with a view to the advance on MORESBY. The YAZAWA Regt [41st Regiment] will take up an proximate position in the area adjoining the main road of the left bank of the Nauro River [*sic* Naoro] .... When the medical TAI has disposed of the patients now under its care, its main strength will be stationed near the KUSUNOSE BUTAI, and a part near the SHITAI HQ for relief of casualties. After the attack, FL will be established in the vicinity of WAMAI.[20]

That day, Horii's headquarters also distributed Intelligence Record No. 18 to the senior commanders of the *South Seas Force*:

CHAPTER 21

The enemy forces with which we were engaged in the ISURAVA Area consisted of the main strength of the Australian 22 Bde [*sic*] (39 Bn 53 Bn) (49 Bn of said is thought to be on guard at MORESBY) and the main strength of 21 Bde (14 Bn and 16 Bn). 27 Bn of this Bde came up later than the other forces and joined the engagement near EFOGI ....

The enemy attempted to hold the Japanese forces at ISURAVA, but were unable to carry out their plan, due to the SHITAI's energetic attack. The enemy attempted resistance during their withdrawal and in the early stages (EORA and EFOGI) fought with a certain amount of determination. However, after being thrown back by the SHITAI's rapid pursuit and energetic offensive, the enemy have, to a large degree, lost the will to fight, and have only been putting up resistance on a small scale during their retreat. They are thought to have abandoned their last key position, the heights to the west [south] of WAMAI [Nauro] in order to defend MORESBY.

According to statements by prisoners, etc. [all would be executed soon after] the enemy suffered enormous losses in their successive resistance and appear to have lost their fighting spirit.

The following are some instances:

(1) Col OWEN Comdr 39 Bn was killed at KOKODA.

(2) Col KEY, Comdr 14 Bn is missing at ISURAVA.

(3) Comdr of A Coy, 14 Bn is missing; of the 3 pl comdrs, 1 was killed, 1 wounded and 1 missing. Coy Sgt took over command but was taken prisoner. This Coy, whose normal strength is 120, numbered only 50 as on 9 Sep.

(4) Comdr of B Coy, 14 Bn was killed; of the 4 pl comdrs, 2 were killed, I wounded, 1 missing. A Sgt became acting Coy comdr but was taken prisoner on 9 Sep.[21]

SAVING PORT MORESBY

Later that night, Captain Takase, a member of Horii's staff, completed his situational report, recording available rations and supplies to the *South Seas Force* then on the Kokoda Track:

> In regards to the use of provisions and forage, the daily ration (of Polish Rice) was provisionally fixed at 3 GO [about 450 grams] and later, due to changes in the situation, 3 GO was made the absolute maximum. However, while troops are waiting for action, the basic ration will be 2.5 GO per man per day, and the fullest use must be made of taro, wheat and other local produce and captured food in order to maintain fighting strength.
>
> Captured provisions will be dealt with similarly to regular provisions and the eating of captured food between meals is strictly forbidden. Every TAI comdr must exercise rigid supervision and guidance in this matter.
>
> Each TAI will report in writing its present position as regard provisions and the prospect of making them last until the 22nd or 23rd. Each Regt and Bn – this also applies to other BUTAI(s) – will send an … officer or NCO in charge of rations to the SHITAI HQ at 1100 hrs on the 21st to discuss investigation concerning provisions.
>
> In spite of reported warnings, cutting away of wiring previously laid down by the enemy (red wire) continues. Now there is a tendency to cut away our own wiring (yellow wire). This makes wired communications between TAS(s) almost … [impossible] and affects adversely the operations of the SHITAI. All TAI comdrs will make all troops down to the last man thoroughly understand the foregoing and take strict precautions against further removals of wire.
>
> As a substitute for candles, a minimum quantity of enemy telephone wire will be issued. A fatigue party consisting of 12 2 Bn, 10 of the KAWASHIMA TAI [*No. 3 Company, 55th*

CHAPTER 21

Cavalry Regiment] and 30 men from the YAZAWA BUTAI, carrying their launchers, will assemble at the NAURO [Naoro] river Bridge by 1000 hrs tomorrow, the 12th, and report to Cpl KONDO of the SHITAI's Sig HAN (cutting tools to be taken).[22]

Sometime that day, 1st Lieutenant Horibe, *No. 6 Company, II/144th Regiment*, recorded in his diary: 'On hill south of "MAWAI" [Nauro]'.[23] Close by would have been 1st Lieutenant Hirano, leading *No. 3 Platoon, No. 1 Company, I/144th Regiment*, who recorded in his diary: 'One enemy light bomber bombed us this morning'.[24]

22

'TAKE AS MANY JAPS WITH US AS WE CAN'

Now coming up the track, Brigadier Kenneth Eather, commanding the 25th Brigade, had already reached Uberi just south of Imita Ridge. There he telephoned Porter at Ioribaiwa Ridge about eight kilometres north from Uberi, as the crow flies. Eather was tasked with conducting offensive operations against the Japanese and decided to adopt the following plan as reported in the 7th Division war diary for 11 September:

(a) 2/33 Aust Inf Bn to make an outflanking movement to the right through PONOON village thence along the police track to NAURO.

(b) 2/31 Aust Inf Bn to move along a track on left flank to converge upon NAURO.

(c) 2/25 Aust Inf Bn to push through IORABAIWA on the main track towards NAURO

Whichever of the two outflanking bns reached NAURO first was to capture the village and then to retire along the line of adv to a posn overlooking NAURO. The bn not engaged to adopt a similar defensive posn to overlook NAURO. The whole of this plan depended upon freedom of movement.[1]

Meanwhile, the men of 'D' Company, 2/33rd Battalion, led by Captain Trevour Clowes, a 24-year-old regular soldier from Warwick in Queensland, moved out of Uberi at 6 am. Within an hour, they

CHAPTER 22

were climbing the heart-breaking 1800 steps cut into the ridge already known as the 'Golden Stairs', leading to the top of Imita Ridge. It took each man around four hours to climb. As recalled by Sergeant William Crooks: 'At each half-hourly halt, and later, when they were spaced at quarter-hour intervals, the men would lie gasping, some retching, as knees shook and trembled. Brens were passed from one to another at about five-minute intervals. By 1100 hours the top of the first spur had been reached, and we were faced with a narrow slippery drop of nearly 1000 feet [330 metres]. The afternoon clouds came over and the rain beat down turning the already muddy track into a nightmare slide. At 1300 hours we passed through the 2/1st Pioneer Battalion'.[2] Eather had earlier ordered the men of the pioneers to advance and conduct a reconnaissance of the tracks north of Uberi.[3] By the time the men of 'D' Company reached the summit of Imita Ridge, they could not locate the main track, given its poor condition and the mist that quickly closed in around them. The men of 'D' Company bivouacked that night on what they hoped was the top of Imita Ridge.

Clowes was all too aware that there were no Australian troops between 'D' Company and the 21st Brigade up ahead on Ioribaiwa Ridge and ordered 28-year-old Corporal Harold 'Brick' Bradford from Sydney with 1 Section, 16 Platoon to occupy a position about 15 minutes forward of the ridgeline. Meanwhile, the young captain had placed the rest of the 16 Platoon astride what was thought to be the track on the crest, with 18 Platoon on its left and 17 Platoon to its right.[4] Sergeant Crooks recalled: 'Here the country was thick with great trees rising eighty and a hundred feet into the sky, with little scrub and brush about. As dark descended D Company took up its allocated positions, unable to light fires, or even to dig-in, as we had no tools – not even entrenching tools. The night was dark and foreboding, as the jungle was lashed with heavy bitterly cold rain. The wind blew in gale force, first from the north, then from the east, from which directions there were no close features to shield the ridge.[5]

SAVING PORT MORESBY

Just days before, as the 2nd AIF men of the 25th Brigade landed in Port Moresby, they were told stories mostly by militiamen who had not themselves fought the Japanese – even so, there was some grain of truth to the stories. Now as these men took up their positions for the night on Imita Ridge, with no knowledge of where the Japanese were, they likely recalled these stores:

> Japs attack at night. They crawl silently up to your position and bayonet you while you sleep. They call out your name in the night and you don't know if it is your mate or not. They scream in the night yelling in English to 'come and help me Jack'. They all carry lights and let off crackers to draw our fire. They creep all around the flank and come charging in on HQs and RAPs. They have a small deadly mortar fired from the knee. They all wear green uniforms and surround themselves in natural camouflage. You can't see them until a bayonet is in your guts. Nowhere has one of our companies or battalions been able to hold them. He has horses to carry his stuff and the men carry nothing but arms and grenades. He has been torturing and bayoneting all wounded. He has mountain guns that blast you out of existence, and we have nothing beyond a few three-inch mortars. The jungle is a bloody nightmare, everything that moves turn out to be Japs trying to surround you and sneak up and cut your throat.[6]

Few of these men likely got a good night's sleep. Sergeant Crook recalled at the time: 'If this is to be, then we'll take as many Japs with us as we can when the time comes for us to go'.[7] Corporal Bradford and the men of 1 Section, at the forward listening post, must have especially experienced a rough and sleepless night. Meanwhile, further south, the remainder of the battalion had by 4 pm reached Uberi.[8]

* * *

CHAPTER 22

Eather that morning had also ordered the 2/31st Battalion – commanded by Lieutenant Colonel Colin Dunbar, a 40-year-old insurance representative from Warwick in Queensland – into the Owen Stanleys. They were to bivouac just behind the men of the 2/33rd Battalion near Uberi. Following behind them the next day would be the men of the 2/25th Battalion, commanded by Lieutenant Colonel Charles Withy, a 52-year-old company manager from Ascot in Queensland.[9]

* * *

Further north, Lieutenant Colonel Cooper and his lost party of the 2/27th Battalion were moving south at first light; however, the path they followed was exceedingly difficult, and in places it was said it 'would test even a mountain goat'.[10] Great credit was given to the stretcher-bearers who had the worst of it, and who like all were weakened by the scarcity of food. The diarist of the battalion recorded: 'HQ A Coy, Bn HQ and 18 Platoon detailed to carry stretchers. Relay system used although in the more difficult places a human chain had to be formed to move the stretchers up or down steep gradients or re-entrants. Slowness of progress may be gauged from fact that an ANGAU party passed through this Unit after having covered in approx. a half hour a distance which this unit traversed in a full day'.[11] That night, Cooper and his men slept on the high ground north-east of Nauro, not having seen or heard any Japanese troops that day.

* * *

Still heading north with his men of the 2/14th Battalion was Captain Sydney Buckler. The officer was attempting to find an easier way across the mountains as he had four stretcher-cases and several walking wounded. At the end of each day, the last to come in was Corporal John Metson, who was crawling on his hands and knees, refusing all help to carry him. That day, Buckler spotted the tell-tale

Japanese web-toed boot prints in a native garden, and he ordered his men out of the patch; they then continued north. His party now totalled 47 men: two officers and 45 other ranks, including eight of who were wounded.[12]

Also, north of the Australian lines but heading south was Buckler's junior officer, Lieutenant Maurice Treacy, who he had sent the week before to get help from Myola. Treacy and his men were still lost in the jungles in the heart of the Owen Stanleys, east of the main track. Treacy later wrote in his official report: 'On Friday 11th a dry lake was reached, where that evening a wild pig was shot and the party built a fire and enjoyed some grilled pig's liver'.[13]

Meanwhile, Lieutenant Robert McIlroy and his seven-man section – who had the day before lost contact with the main body of his patrol and had reached one of several Seventh Day Adventist mission stations in the area; this one was west of the main track – stayed here for two days to provide medical attention and rest for the soldiers who had been wounded in the previous day's ambush.[14]

* * *

Late that night, Brigadier Arnold Potts arrived at the headquarters of the 7th Division in Port Moresby to report to Major General Arthur Allen. His CO was initially far from cordial to his brigadier. Their discussion, however, did much to placate Allen because Potts outlined the realities of fighting on the track. General Staff Officer 1, Colonel Charles Spry recalled Potts' arrival at the divisional headquarter late that night: '[He] looked as if he had had a very gruelling time, but he was bright and vigorous. As far as I could judge, he had saved Moresby from invasion by [the] Kokoda route by forces superior in numbers to his own and had prevented the catastrophe nearly brought about by the neglect of New Guinea by authorities before the AIF came. He was, I think, just in time'.[15] A decade later, Potts recalled to the original historian of the Kokoda Campaign, Raymond Paull, his discussion with Major General Allen:

CHAPTER 22

Potts quickly sensed the attitude towards him ... the hollow, simulated cordiality and the more obvious avoidance which told him plainly that, excepting for a few staunch friends, he was regarded as a failure, and his troops shared his ignominy. Naturally this coolness towards him put a sharp edge on Potts' characteristic pugnacity. Allen's ADC wrote the next morning: 'Brigadier Potts arrived and reported. He looked well but had lost a lot of weight. Given a warm welcome by GOC and given a good clean change by ADC. Told me it was the first night's real sleep he had had since the action began'.

Allen had already spoken to many of the troops. By this time, he had become convinced that the Japanese would have reached Port Moresby had it not been for the brigadier's tactics and the stubbornness and endurance of the troops who had blocked their path. Allen, however, received his friend with a friendly warming: 'You're going to run into a lot of criticism here, old man.'

'You and I discussed all manner of contingencies before I went up the track,' said Potts. 'We understood each other, so you've no cause to be ashamed of your troops.'

'What are they like, and how is their morale?' Allen inquired.

'Bloody sore feet, and savage as hell at giving up ground. But as soon as they get some good tucker into them, and have a chance to recover, they'll be right on their toes again'.[16]

Clearly, Lieutenant General Sydney Rowell and Major General Arthur Allen had no real idea what the fighting at the sharp end was all about – it was not just the Japanese that the brigadier and his men had to contended with, but multiple vectors, including the terrain, weather, jungle, the lack of supplies, and substantive support from the Australian High Command.

23

'... PREPARED FOR AN ATTACK ON MORESBY'

On 12 September, the position held by Lieutenant Colonel Albert Caro's composite battalion was reorganised, while elements of the 3rd Battalion moved south-east along the ridge to cover Caro's right flank. The militia patrol from 'C' Company that had been sent out earlier to reconnoitre east of Ponoon reported back at around 3 am that they were unable to find the track to the village and their guide from the intelligence section had become lost. At 5.45 am, Lieutenant Colonel Allan Cameron ordered 'C' Company to move east to find the track and occupy the village.[1]

Sergeant Colin Kennedy with 15 Platoon 'C' Company recalled: 'Captain Boag sent 15 Platoon to scale the very high ground on what then became the extreme right flank of the position held by the Australians. The "distance" wasn't more than a 30-minute climb through the kunai grass and up a small cliff face. Being visible through binoculars from across the valley no one was permitted to descend except at dusk and early morning in the days following. No. 14 Platoon held the saddle itself and 13 Platoon took up a position on the lower slope of the pinnacle. Corporal Alan Smith, of Goulburn, was in charge at the time. At no time were any Australians placed in a defensive position between 13 Platoon and our 'D' Company the distance between the two perhaps being as much as four kilometres in a direct line over the watershed of the pinnacle'.[2]

Orders were issued that Ioribaiwa Ridge was to be held until relieved by Brigadier Kenneth Eather's 25th Brigade. By 7 am, the men of 'A' Company, 2/16th Battalion were passing through the Headquarters

CHAPTER 23

Company of the 2/14th Battalion in Ioribaiwa village. These men occupied the ground left of the village, taking over the position held by 'A' Company, 3rd Battalion, who now moved out to establish a forward ambush position. Only now did the Headquarters Company, 2/14th Battalion move back to take up a position to the rear of 'D' Company, 2/14th Battalion, who were located behind the village. The men of 'C' Company, 2/16th Battalion took up a position on the right, while the men of 'B' Company, 2/16th Battalion remained forward of the composite battalion along the track. At this point, the 3rd Battalion sent out a patrol further down the track towards the creek crossing, while the men of the 2/6th Independent Company patrolled the ridge to the west.[3]

The 2/14th Battalion war diary records: 'By 0800 hrs our coys were in their new positions and a BHQ had been established just South of the village [Ioribaiwa]. Except for a slight interchange of shots, no contact was made with the enemy. The combined 'HQ' people of 2/14 and 2/16 bns cooperated in providing tea, water, and rations to all troops. IO and patrol from D Coy contacted 2/6 Independent Coy some 45 yards along the ridge West of EUROBIRA [*sic* Ioribaiwa]. Forward B Coy 2/16 Bn erected booby traps astride track leading up from creek, and towards dusk several loud explosions indicated that some of these had been sprung. Enemy was contacted then, and two rifle posts and one LMG post (enemy) were knocked out'.[4]

At around 10.30 am, the 21st Brigade Headquarters received the welcomed news that the battalions of the 25th Brigade were moving up the line to begin offensive operations against the Japanese. Indeed, Captain Kienzle of ANGAU was still organising supplies and carriers at Dump No. 44; he recorded that day: 'Bns of 25 Bde passed through our camp to go to the front at IORIBAIWA. The Japs were using their mountain guns and heavy mortars'.[5]

* * *

Lieutenant Colonel Alfred Buttrose, CO of the 2/33rd Battalion, had already been instructed that he and his men were to conduct an eastern flanking movement from Imita Ridge in a wide sweep towards Ioribaiwa Ridge to retake Nauro. Leading the advance would be the men of 'D' Company who were still on Imita Ridge. Throughout the early morning hours of darkness, these men were experiencing the height of the thunderstorm that had broken out the night before. Sergeant William Crooks recalled how the wind was 'shrieking and moaning through the trees, and streaks of lighting preceding mighty claps of thunder'.[6] Just before dawn, the storm finally abated, and by 10 am, the rest of the battalion had finally arrived on Imita Ridge to join Captain Trevor Clowes and 'D' Company. Clowes was now informed by Buttrose that he and his men were to lead the battalion's flanking manoeuvrer.

A guide from the PIB accompanied the battalion, and his job was to guide these men to Nauro in their eastern flanking movement. Within hours, however, Clowes and the rest of the battalion were strung out, lost in the jungle in another thunderstorm. These men eventually came across what Sergeant Crooks described as: 'A crest that fell away on all three sides in sheer drops of hundreds of feet. The area was strewn with gigantic volcanic boulders. Six-inch thick vines intertwined with each other in the close low scrub covering the jungle floor, the track could not be found'.[7] Buttrose sent out several patrols to find and cut their way through the jungle, each to use the compass bearing for Nauro, but to no avail. Buttrose and his men were forced to admit defeat, and they took up a defensive perimeter just before darkness, somewhere between Imita and Ioribaiwa ridges.[8]

* * *

Meanwhile, the 2/31st Battalion had also arrived on Imita Ridge. While recovering from malaria in 1943, the stretcher-bearer with the battalion, 21-year-old Private Geoffrey Hamlyn-Harris from Cairns in Queensland, recalled the journey and the first night spent in the jungle:

CHAPTER 23

Pulling ourselves up the face of praecipes by the poles of trees, we staggered along native footpaths, tramped all day on the edge of arching razorbacks with pitfalls yawning on either side, lunged up false crests and around spurs which led to yawing gulfs, and slid down into river gorges to climb out again well nigh exhausted on the opposite side. On some of the steeper ascents the fuzzy-wuzzies had cut steps; and these made a lot of differences where we otherwise could not have gained foothold .... Leg weary and sore all over, and in my case with a headache, we eventually arrived at what seemed to be the top of the range, only to see still more ranges ahead even higher and far into the interior, clothed all over with the same slushy denseness of jungle ....

The jungle noises, pressing in from every side, from the trees above and the gullies below, were as weird and strange as they were varied. There were all manner of crickets, shrilling in sharp contrast with the sound of running streams, with the wind in the treetops and the swish-swish of continuous showers of rain. One cricket sounded like an air raid siren, and there was another that recommenced its shrill paeaning at seven o'clock pm with such punctuality that one of the chaps who knew about it set his watch to it.

Other sounds of jungle life we heard for the first time that night were squeaking, scuttling and grunting's of prowling beasts and reptiles, together with the croakings and pantings of a thousand different species of frogs, toads, and other things of an aquatic nature. Night birds came out and wooed their mates with strange flutings or screamed or squawked in the tooth and claw of their natural enemies. A thought crossed my mind that perhaps this was all we humans, too, amounted to in the long run, animals still in a wild state, fighting one another for food or mates or whatever our needs might be.[9]

An unknown soldier with the battalion also recalled the journey: 'The men had believed they were fit but nobody was tough enough to live through an hour's march without exhaustion. Passing the grass-hut village of Uberi, they pulled themselves up the face of precipices, trod apprehensively along razorback tracks, slid down twisting muddy hills and struggled to climb the next hillside – and the next … and each one became steeper'.[10] The battalion diarist recorded that 'D' Company led the advance and soon took up a position astride the track on the forward slope of Imita Ridge, which had a field of fire of only a few metres due to the steep slope and thick vegetation. Behind them were their mates of 'A' Company holding a narrow 'gap' through which the tracked led. Taking up a position on the right was 'B' Company, with the Headquarters Company on the left, and 'C' Company to the rear. The diarist concluded: 'Quiet night'.[11]

* * *

During the early morning hours, the 2/25th Battalion were moving up to Uberi. By 4 am, 'A' Company was on the move, followed two hours later by 'B' Company with the Battalion Headquarters' staff and Headquarters Company. They were followed by the men of 'C' Company at 7 am, and finally by 'D' Company at 8 am. The first men began to arrive at Uberi at around 1 pm. They would continue the advance at 6.30 am the next morning to reach Imita Ridge.[12]

* * *

Meanwhile, around Ioribaiwa Ridge, Lieutenant Colonel Allan Cameron had by 1 pm ordered another patrol from 'A' Company, 3rd Battalion to move further forward. Sergeant Donaldson Hill, a 26-year-old dairy farmer from Mittagong in New South Wales, and a nine-man section crossed Ofi Creek on the right front and took up a position on the high ground right of the track, returning after dark reporting they had not observed any enemy troops.[13]

Many of the 21st Brigade veterans strung out along the ridgeline

CHAPTER 23

realised that the Japanese appeared to be losing some of their aggressiveness – no doubt feeling the strain of the long-drawn out fighting and handicapped by their extended supply lines, just as the Australians had suffered in the earlier days of the campaign. However, unlike the Australians, the Japanese had no capacity to be supplied by transport aircraft – they were in a dire situation. Late in the afternoon, the Japanese engaged the Australians with fire, but a quiet night followed as the exhausted men from both sides took stock of their position and the tactical situation.[14] The war diary of the 2/16th Battalion concludes for the day: '1630 Enemy opened fire on B Coy with LMG – apparently attempting to draw our fire. B Coy did not return fire'.[15]

At 6.30 pm, a 21st Brigade conference was held with Brigadier Porter issuing the following orders: (a) Be careful of 2/33 Bn on the right; (b) Prepare better defensive positions; (c) Insist on better camouflage; and (d) 3 Bn to send patrol on left flank to attack creek crossing from rear.[16] The 25th Brigade war diary records for 12 September:

> 1040 hrs: 2/31 Bn to take over from 2/33 Bn approx. 1700 hrs. 2/33 Bn to move to right of 21 Bde (situated on ridge running east and west through IORIBAIWA). Close contact being maintained between 21 and 25 Bdes and 21 Bde passing all infm possible about the enemy to 25 Bde. 2/25 Aust Inf Bn arrived UBERI. Object of 2/33 Bn to move to NAURO.

> 1345 hrs: Orders by Big EATHER. 2/31 Bn to move fwd at first light and contact Brig PORTER, 21 Bde and seek his advice as to most suitable routes, east, centre and west to NAURO and try to take and hold NAURO. 2/25 Bn to move from UBERI ½ hr after first light and take over 2/31 Bn posns at IMITA RIDGE.

> 1810 hrs: Infm received from 21 Bde, that the JAPS killed on 11 Sep were wearing new issue uniforms of darker green than formerly and wearing haversacks at sides instead of on back.[17]

SAVING PORT MORESBY

* * *

That day, Major General Horii, who was located near Nauro, issued Nankai Shitai Operational Order 'A' No. 116:

> According to reports by special patrols ... the main strength of the enemy appears to have retreated to an area west of IORIBAIWA, while a small force of the enemy has been found on the western [southern] bank of the stream to the east [north] of IORIBAIWA. From recce by YAZAWA BUTAI, there would appear to be a native road leading NW [southwest] from the vicinity of the crossing point on the Nauro [*sic* Naoro] River.
>
> A part of the SHITAI will attempt to capture the key position(s) near IORIBAIWA.
>
> The KUSUNOSE Regt will commence operations tomorrow the 13th and capture the key position(s) near IORIBAIWA.
>
> Part of both the SHITAI's Sig HAN and Medical TAI will cooperate in this action.
>
> All other TAI will remain in their present positions.
>
> I will remain in my present position and, upon the advance of the frontline to IORIBAIWA (expected to be about the 14th), I will proceed to an area near the summit of the highlands to the west of WAMAI [Nauro].
>
> HORII Tomitarô
SHITAI Comdr
>
> Method of issuing:
>
> To be telephoned to Comdr of KUSUNOSE BUTAI. To be communicated verbally to persons appointed to receive orders and take down by them in writing.[18]

CHAPTER 23

Later that day, after issuing his Operational Order No. 116, Horii finally received his orders from the Imperial Headquarters that had been issued two weeks earlier – 28 August. On reading the message, he must have been concerned: 'Advance to the southern slopes of the Owen Stanley Range and destroy the enemy troops there. Use a portion of your strength to secure the front but amass your main strength on the north side of the range in preparation for future operations'.[19] While of concern, it appears that Horii had already come to this conclusion if we are to believe a conversation he reportedly had with Major Koiwai Mitsuo, commander of the *II/41st Regiment*. Koiwai recalled Horii admitting:

> 'I wanted to capture Port Moresby at any cost, but a little while ago when I washed my hands at the stream and saw a soldiers' mess tin, I changed my mind. With the rice the soldiers have, it would be difficult even to capture the enemy base in front of us. There won't be any soldiers who have two gōs [about 300 grams] rice for tomorrow's lunch. Most of the soldiers said that was the last batch of rice. I was well aware of the lack of food but did not realise the situation was this bad. If we continue to advance, we will quicken the suicidal act.' Koiwai replied: 'I was wondering in the beginning if the two weeks' ration would be enough for the campaign. Why did you come to this conclusion?' Horii was generally upset by the question, replying: 'I thought this campaign would not be feasible without the prospect of faultless supply. So, I took the staff officer, Tanaka, with me from Rabaul to the 17th Army [in Davao, Philippines] to submit my opinion. But the 17th Army said it would do whatever was necessary for the matters involved in the frontline and that the South Seas Force should only carry out the order. We were told so sharply, as if the South Seas Force were cowards. So, I had nothing more to say, but accept the order to attack unreservedly'.[20]

Meanwhile, advanced elements of Colonel Kusunose's *144th Regiment* had approached the northern slopes of Ioribaiwa. The colonel was still ill, suffering from malaria, but he continued to function and had been carried forward on a stretcher for the last few weeks with each advance.[21] With this force were the guns of the *55 Mountain Artillery Regiment*, and the regimental gun company was placed under the command of Lieutenant Colonel Hozumi Shizuo, commanding the *55 Mountain Artillery Regiment*. Hozumi sent spotters out to locate targets for his guns, which were now located near the lower slopes of the ridge.[22]

The Japanese official history described that the Japanese combat engineers had done great work in improving the conditions of the track: 'The road south of Mawai [Nauro] had been repaired and provided easy going for the advancing troops. The pursuit party consequently reached Ioribaiwa on 12 September and made contact with enemy positions'.[23] Close by would have been Lieutenant Sakamoto Atsushi, commanding *No. 2 Machinegun Company, II/144th Regiment*, who recorded in his diary: 'Arranged weapons, stored forage ... and prepared for an attack on Moresby. Information received Enemy strength at Moresby – 20,000'.[24]

First Lieutenant Hirano, leading *No. 3 Platoon, No. 1 Company, I/144th Regiment*, who had spent much of his time at Myola on carrying duties through much of the fight, was again called to conduct this task. He recorded in his diary that he and his men were ordered back to the depot at Isurava to bring back supplies: 'Battalion Order was delivered, instructing me to take charge of 100 men of the Battalion to carry back provisions. At 1720, 1st Lt. YOKODA of Regimental Infantry Gun came to make various arrangements. Decided to leave tomorrow'.[25]

By now, Japanese war correspondent Okada Seizo had pushed south through Menari and was with the rest of Horii's force in the vicinity of Ioribaiwa Ridge. He later recorded: 'We had lost a considerable portion of the troops killed and wounded in action. Besides a growing number of men were suffering from malaria, colonitis, weakness of eyesight amounting to night-blindness, pneumonia and other fevers

CHAPTER 23

of undetermined causes, nervous breakdowns, diet deficiency diseases, fainting – illnesses brought about by insufficient [medical] equipment, especially by extreme shortage of food, exhausting march, the heat and humidity and cold of the mountains and other indescribable hardships under which we had been struggling along. All the troops now showed unmistakable signs of weakness and exhaustion. But they marched desperately on and on, with only one objective in view, asleep or awake – Port Moresby'.[26]

* * *

Still behind Ioribaiwa Ridge was the Salvation Army Red Shield post. Major Albert Moore recorded:

Saturday, September 12 was a quiet day. It seemed uncanny. There was little or no news coming from the front, and it seemed like a calm before the storm. That day the signallers installed a phone at our post but moved it out later in the day. I felt it a comfort to have some contact with the rear in case of danger. That night I had a stretcher-case, badly wounded man stay in the grass hut. He has suffered a deep chest wound from shrapnel. The natives who carried him camped alongside the wounded man and I went to him a number of times during the night. His fortitude and courage amazed me. In the morning I gave him a little nourishment and as the natives lifted his stretcher to their shoulders, I put my hand on his and uttered a prayer that God would go with him and take him safely to the head of the road and to hospital. He said to me: 'You are worried about me. I will be ok and will be at the hospital in a few days'.[27]

* * *

Early that morning, Brigadier Arnold Potts met with both Major General Arthur Allen and Lieutenant General Sydney Rowell for a more detailed debriefing. Both generals asked about Japanese tactics, supplies, the lack of reserves, the Australian supply failures, the inability of Australian and American aircraft to evacuate Australian wounded at Myola, and the weakness of the Australian 53rd Militia Battalion during the fighting for Isurava and Alola.

To this last question, Potts replied: 'I don't know the bastard who sent the 53rd up the track'. Rowell ignored this critique as he was aware of the 'Bastard'. Potts continued: 'There's no better way of killing men in large numbers than sending raw and undisciplined troops to fight a jungle war against troops of seven years' experience'. Potts also commented on the supply situation: 'It's a pity that the two planes left from the mess at 7-Mile [Japanese successful bombing of the airfield on 17 August] didn't drop some rations at Myola before they pushed off to Australia. After all, Myola is only thirty-five minutes [flying time] from here'.[28]

Overall, Rowell and Allen were sympathetic, and they were both in agreement that there was no reason to remove Potts from command of the 21st Brigade, as he had done the best that could have been expected under the conditions. That would not stop others, however, from treating the brigadier and his men of Maroubra Force as ignominious failures – generals MacArthur and Blamey among them.[29]

Later that day, General Blamey arrived in Port Moresby for the first time and held discussions that would result in an ill-conceived and prepared plan to outflank the Japanese using the route north from Abau to the east. This doctrine was very much in line with MacArthur's ongoing demands to outflank the Japanese along the Kokoda Track. Clearly, the American warlord had no idea of the type of terrain and vegetation the Australians were fighting in. A small advance party, under American Brigadier Hanford MacNider, had arrived that morning to make preliminary arrangements for American operations in Papua. After consulting with Rowell, MacNider decided he would go by way of Rigo, a coastal village about 70 kilometres east of Port Moresby. He now planned for the US 126th Regiment, 32nd Division to move up the valley of the Kemp Welsh River, which flowed

CHAPTER 23

almost due south from the mountains. They were to make their way to Dorobisolo on the southern slopes of the Owen Stanley Range. They would then strike north-east across the mountains to Jawerere located at the headwaters of the Kumusi River – about 60 kilometres east of Nauro, as the crow flies. From Jawerere, they would follow the line of the Kumusi, flowing north-west to Wairopi on the Buna side of the Kokoda Track. A small Australian reconnaissance patrol would be sent out on 17 September, led by Lieutenant Sydney Smith of ANGAU, tasked with looking over parts of the proposed route.[30]

Rowell later denied Blamey's claim that he had been directed to reconnoitrer this route, writing: 'I had talked with MacNider and expressed my views as to sideshows. I said that, as he was acting on GHQ orders, I couldn't veto his proposals, but I made it clear that there would be no diversion of facilities from the main task. I discussed this with Blamey … and he was in complete agreement'.[31]

Rowell had planned a much smaller and more focused attempt at flanking the Japanese to cut their supply lines near Menari and Nauro – he did not attempt a grandiose manoeuvre being planned by the Americans of crossing the Owen Stanleys with a whole regiment of troops (equivalent to an Australian brigade) to attack the Japanese just south of Buna. Indeed, the day before Blamey's arrival, Rowell had placed Lieutenant Colonel Ralph Honner in command of what was designated 'Honner Force'. Honner's command was a composite company made up mostly of militiamen from the 36th, 49th and 55th battalions along with a detachment from the 2/6th Independent Company. Within days, these men were pushing up the Laloki–Goldie River to the west of Port Moresby; they were tasked with cutting Japanese lines of communications.[32]

With Honner Force was 20-year-old commando Private Allen 'Ossie' Osborne, who was with the 2/6th Independent Company and from Sydney. He recalled: 'A composite militia force headed by Lieutenant Colonel Honner was formed and it was our job to blaze the trail and lead them to the point of intersection with the Owen

Stanley Track. These troops, who a few months later did very well over in the Sanananda area, were certainly not up to this trip, neither physically or mentally .... We were fairly used to moving through the bush and, anyhow we didn't move in very big groups and we, generally speaking, didn't stir the tracks up much, but once a group of infantry started moving over them, the tracks started to suffer .... I remember one place where there was a brown strip of mud up over this mountain and the only things standing in it were the big trees that wouldn't come out when they pulled themselves up. It was nearly impossible to move over it'.[33] The war diary of the 2/6th Independent Company also recalls that: 'On their passage through our camp many of the infantry were supplied with long trousers in place of shorts and 50 Thompson submachine guns were loaned to them'.[34]

Brigadier John Rogers, a 47-year-old professional soldier from Penguin in Tasmania, and Blamey's Director of Military Intelligence was in Port Moresby with the Australian commander of SWPA. He and several American officers now witnessed some of the survivors of Maroubra Force making their way back from the front and were truly dismayed to see the condition of these men of the 2nd AIF. The brigadier soon after was quoted in the Melbourne *Argus*: 'I remember the fighting at Gallipoli and Passchendaele in the last war and in Greece and Crete in this war, but I have never before seen solders so fatigued so grizzled and so hammered as these men back from the Kokoda front. They had been through hell, and no tribute is too great for what they have done'.[35]

* * *

Meanwhile, the CO and the bulk of the 2/27th Battalion since the battle of Mission Ridge were still north and east of the main track, with the walking wounded and stretcher-cases. They moved out at first light but the last four days of strenuous climbing up and down ridges and gullies with little food had taken its toll. Cooper decided to rest the men within hours of first moving south. The remaining

Lieutenant General Sydney Rowell, GOC New Guinea Force. During the Kokoda campaign his feud with General Blamey would cost him his job and create long-lasting controversy. By the 1950s he had returned to favour and would serve as Chief of the General Staff (modern-day Chief of Army) (AWM026582).

Major General Arthur 'Tubby' Allen, GOC 7th Division, seated next to a portrait painted by Official War Artist William Dargie. Allen had been decorated for service during the Great War and had remained active in the Australian militia during the inter-war years (AWM023271).

Brigadier Arnold Potts, commander of the 21st Brigade. The Kokoda campaign was Potts' first assignment as a brigade commander. It would prove a stern test of his leadership and mettle (AWM099103).

Lieutenant Colonel Ralph Honner, CO of the 39th Battalion. A schoolteacher in civilian life, Honner was the calm and experienced leader the 39th Battalion needed during the Battle of Isurava-Abuari (AWM005638).

Lieutenant Colonel Albert Caro, CO of the 2/16th Battalion, with Captain Frank Sublet on his left. Sublet was Caro's most trusted company commander and later commanded the 2/16th in his own right (AWM026752).

Lieutenant Colonel Charles Withy, commander of the 2/25th Battalion

Lieutenant Colonel Geoff Cooper, commander of the 2/27th Battalion.

Officers of the 2/31st Battalion, left to right: Lieutenant Colonel Colin Dunbar, Major Ewan Robson, Lieutenant Charles Birnie, and Captain Edward Beazley.

Lieutenant Colonel Alfred Buttrose, commanding the 2/33rd Battalion with Brigade Major George Larkin just behind on the Kokoda Track.

Gallipoli veteran, Medical Officer Captain Geoffrey 'Doc' Vernon, ANGAU who cared for the sick and wounded of Maroubra Force and the Indigenous carriers.

Papuan carriers bringing Australian wounded back to Port Morseby, without the dedication of these carriers many Australian troops would not have survived the campaign.

Captain Bert Kienzle ANGAU, who commanded the carrier force on the Kokoda Purchasing vegetable from Papuans during the Kokoda Campaign.

Papuan carriers a load of around 40 kilograms to supply Australian troops on the Kokoda Track.

Salvation Army Captain Albert Moore at his Red Shield Post at Myola.

Captain Albert Moore at one of his Red Shield Posts on the Kokoda Track.

Salvationist Corporal Jock Inglis (foreground) was Captain Albert Moore's (centre) most trusted assistant who like Moore traverse the track in both Directions.

One of the most iconic photographs taken of the Kokoda Campaign – the parade of the survivors of the gallant 39th Battalion at Menari who had alone for many weeks withstood the Japanese advance (a still taken from Damien Parer's *Kokoda Frontline*).

Father James Benson, head of the Gona Mission, the sole survivor of the Gona and Sangara Mission parties who were held prisoners by the Japanese.

Lieutenant George Pearce leading, 16 Platoon, D Company, 2/14th Battalion was last seen going out with some of his men to take out a Japanese machinegun just outside the Australian northern perimeter at Isurava, weeks later he and his surviving men would make their way back to the Australian lines.

Corporal James Metson, 2/14th Battalion, who with others of Maroubra Force were forced into Eora Creek during the Japanese attack against the Isurava Rest House on 30 August, although wounded in the ankle, he refused all help and crawled on his hands across the Owen Stanleys only to be murdered by the Japanese along with other wounded mates.

Lieutenant Maurice 'Mokka' Treacy, 2i/c of 'B' Company, 2/14th Battalion would spend many weeks lost in the Jungle with his men before finally making his way to the Australian lines much further south; he would soon after write a fascinating account of his journey which is preserved in the battalion war diary.

Another iconic image of the Kokoda Campaign, Lieutenant Colonel Ralph Honner proudly addresses his men of the 39th Battalion; those 'Ragged Bloody Heroes' at Menari on 6 September 1941 (another still taken from Damien Parer's *Kokoda Frontline*).

Ay Myola before the fighting for Mission Ridge and Brigade Hill, men of Lieutenant William Grayden's 15 Platoon, 'C' Company, 2/16th Battalion; left to right Private Daniel O'Connell, Lance Corporal Francis Cunningham, Sergeant Reginald Burges, and Private Harry Kelly.

An exhausted and half-starved, Private Jack Scally, 2/27th Battalion, who like many of the 2/27th Battalion was trapped behind Japanese lines after the fighting for Mission Ridge; he would spend 16 days lost in the jungle before returning to the Australian lines.

Corporal John Burns, who with Private Alfred Zanker stayed behind to help the wounded and sick of the 2/27th Battalion, they and the wounded would eventually be found by Australian troops, both men were Mentioned in Dispatches for their dedication.

Lieutenant William Grayden, 2/16th Battalion later wrote a remarkable detailed account of his experiences during the Kokoda Campaign; he was lucky to survive the campaign.

Three men from the 2/14th Battalion who had become separated from their unit in the Myola area, finally arrived behind the Australian lines after 42 days lost in the jungle and mountains of the Owen Stanelys.

Hungry members of the 2/27th Battalion line up for a meal at Itiki in October 1942. The troops were lost in the wilderness of the Owen Stanleys after being cut off from Maroubra Force at the conclusion of the Battle of Brigade Hill (AWM027018).

Men of the 2/14th Battalion enjoy a rare moment of rest under a makeshift shelter close to their positions on Ioribaiwa Ridge. Lance Corporal Harold Atkinson (left) cleans his Thompson sub machine-gun while Sergeant Harold Phefley writes a letter home (AWM026727).

Captain Claude Nye, leading 'B' Company, 2/14th Battalion would with several of his men be killed in the vicious fighting to retake Brigade Hill.

Sergeant William Crooks, with Clowe's Company was involved in constant fighting and would survive the war to write the most detailed Australian battalion history of the Second World War.

An Australian 25-pounder of the 14th Field Regiment in position shelling Japanese positions on Ioribaiwa Ridge, this was the first time that Australian artillery was used during the Kokoda Campaign, and was a great morale booster for the Australian infantrymen.

The 14th Field Regiment's 25-pounder guns being hauled through the jungle in the vicinity of Uberi on the Kokoda Trail. Members of the regiment are assisted by the 2/1st Pioneer Battalion (AWM026855).

Major General Arthur Allen and Brigadier Kenneth Eather, commanding the 25th Brigade discuss the operation to retake Ioribaiwa Ridge in late September; to the right is Lieutenant Dalrymple Fayle, Allen's personal assistant.

Captain Trevor Clowes, who with his men of 'D' Company, 2/33rd Battalion was involved in constant fighting patrols around Ioribaiwa and Imita Ridges; here he holds Japanese stores that are clearly marked AIF Malaya.

After months of remaining in Australian general MacArthur finally makes an appearance in Port Moresby as the Australian offensive is launched, General Blamey had visited Port Moresby rarely and for just a few days at a time, and with the Australian advance he would now be based at the township.

Major General Horii Tomitarô, Commander of the Nankai Shitai.

Kusunose Masao commander of the 144th Regiment.

Sergeant Imanishi Sadashige, I/144th Regiment, one of the few Japanese to survive the Papua Campaign.

Private Nishimura Kokichi, II/144 Regiment, another who was lucky to leave Papua alive and would manage to survive the war.

Japanese troops at Ioribaiwa Ridge, looking towards the Australian positions on Imita Ridge.

Colonel Tazawa Kiyomi, commander of the Japanese 41st Infantry Regiment.

Japanese troops slashing their way through the jungle with bayonets, likely during the Kokoda Campaign.

Japanese mountain gunners in action somewhere on the Kokoda Track.

Japanese officers and troops of the 144th Regiment blazing a trail through the jungles of the Owen Stanelys.

CHAPTER 23

Papuan bearers were also beginning to complain about the lack of food; they had even less to eat then the troops. At 11 am, the battalion moved slowly on its way and on reaching a suitable spot at around 3 pm, Cooper decided to halt for the night and allowed the Papuan carriers to cook some food they had collected the previous day. The men of 'D' Company were now detailed to return to a native garden that they had passed through the day before to collect enough yams and taro to feed the whole party.[36]

Earlier, Cooper had sent out 24-year-old Sergeant Robert Paine, from Adelaide, who led a patrol towards Nauro, to ascertain how many Japanese occupied the village. Paine returned with news that there were few enemy troops in the village, also reporting he had contacted some men from 'A' Company and informed Cooper that after leaving Menari, both companies had struck the Japanese and it had become necessary to split both companies into platoons. The war diary records: 'Consequently A Coy was formed into parties of five, each party with a compass and orders to make [their] own way back to Base. C Coy decided to march EAST for one day and then SOUTH to base'.[37] That night, Cooper and his men bivouacked and awaited the return of 'D' Company with the yams and taro.

The adjutant of the battalion, Captain Harry Katekar, also recorded the plight of the stretcher-cases: 'The wounded, God only knows, were in purgatory, hungry and in great pain. Some of our natives began to desert, meaning that our men had to replace them as bearers, "Doc" Viner-Smith allowed the maggots to remain on the wounds in order to eat the rotting flesh and so prevent gangrene. That night we were still short of Nauro. I found it a great mental strain and so did the CO and other officers, with that great responsibility of not only saving our wounded but of saving ourselves from starvation'.[38]

24

'THE MEN SLITHER DOWN A CLAY SLOPE, AND THEN, MIRACLE OF MIRACLES ...'

During the morning of 13 September, Colonel Kusunose ordered scouts be sent out to reconnoitre the enemy positions along the ridge, while also sending out the men of the *II/144th Regiment* to the west to swing around the Australian flank and then attack from the rear; a task these men were now all too familiar with.[1]

Lieutenant Colonel Hozumi and his gunners were also up early that day, and at first light they opened fire. In addition, several Japanese mortars bombed the positions occupied by Caro's men strung out along Ioribaiwa Ridge, killing and wounding several from 'A' Company, 3rd Battalion, representing the units first casualties. In retaliation, a mortar team from the battalion bombed just north of the Ofi Creek crossing, while a patrol from 8 Platoon, 'A' Company, 3rd Battalion was sent out as ordered the previous night at 6 am on the left flank as a standing patrol, with 1 Section to patrol forward of the creek where it crossed the main track. At around 10 am, Lieutenant Colonel Allan Cameron sent out another patrol, led by 22-year-old Sergeant Keith Sackley, from Undanderra in New South Wales, to search the ridge to the right front of the battalion's position. At dusk, both patrols would return reporting no contact with the Japanese, but Sackley was able to report that enemy machinegun fire appeared to originate from the track south of Nauro.[2]

CHAPTER 24

Still approaching Ioribaiwa Ridge from the south were the Australians of the 2/33rd Battalion who remained unsure of their position. At 9.45 am, 25-year-old Brigade Major George Larkin from Sydney, with the 25th Brigade, rang the 21st Brigade stating that the 2/33rd Battalion had failed to make any progress and requested the 21st Brigade send additional guides to lead the battalion up to their position. By 10 am, Major Hugh Challen had sent guides to Brigadier Kenneth Eather to bring up the battalion.[3]

Meanwhile, the Japanese continued their probing attacks, against the approach to Ioribaiwa Ridge, but some blundered into booby traps set by 'B' Company, 2/16th Battalion, positioned in the forward area. In this sector was Lance Corporal William Cousens (now an acting corporal) with 'B' Company who recalled years later:

> Our unit was ordered to dig in about half way up Ioribaiwa Ridge in the forward position. Claude Turnbull and his No. 2 on the Bren gun Joe Paull dug in behind a big tree on the left of the track while I was detailed to dig and camouflage a fox hole on the right of the track opposite the tree, and while this was going on, Lieutenant Ross Watts and a detail of men went down the track to set up some booby traps out of 4 second grenades, before returning to await the advance of the Jap troops. They were really going to cop it this time ….

> Then it all happened, with Claude's Bren exploding into life, but I couldn't see anything as I peered out of my fox hole, as there was a slight bend in the track about 45 or 50 yards down which cut off my view, until two Japs jumped across Claude's fire, to set up the prefect target for my Tommy gun, they went down. All hell broke loose then, with the most intense and prolonged fire power I had ever heard before or since. They knew Claude and Joe were behind the tree, but didn't know where I was. It seemed that their intention was to blow that tree to pieces to expose Claude and Joe, but we just calmly let them go on wasting their

ammo while waiting further results, which came about when a huge branch extending out past my fox hole come down with a big clunk just in front of my nose.

That was a lucky escape, because had it gone over the opening of my fox hole there would have been one soldier less on our side. Soon after that their fire began to slacken, causing me to pull out a grenade and glance across to Claude while pointing at the grenade, Claude and Joe cottoned on and we let them have a salvo of 4 second grenades, which put an end to that battle there and then. They just off with their dead and wounded without bothering to pick up a dead young sergeant in front of my fox hole, and that's the last of them we saw ever on that particular front, they never tried again. While Claude savoured that victory with gusto, pointing out that only three of us faced the whole Jap army that day and sent them reeling in no uncertain manner ....

There was a very ominous and quiet couple of hours after that set back to the Japs, resulting in a patrol being sent down the track led by Lt. Ross Watts with me being one of the party ....to find out what the Japs were up to. Our booby traps had proved to be very successful, judging by the blood and gore and fragmentary clothing lying around the grenades that were set off. The Jap had really got a caning up that time all round, and it was beginning to hurt him badly. Without superiority in numbers and arms he was a very ordinary foe indeed. But he still had a big pull over us in heavy equipment such as his mountain guns and four-inch mortars which out ranged our three-inch mortars by a considerable distance.

However, we carried on with our patrol almost to the heavy water course, where the track took a bend about fifty yards up from the water. Lt. Watts halted us here, saying that he would reconnoitrer the last little bit to see what was going on down

CHAPTER 24

there, while at the same time borrowing a Tommy gun from one of our patrol for the occasion, before taking off. We were only there about five minutes before being alerted by several bursts of fire from the Tommy gun and wondered what the heck was going on. There was quiet for about three minutes after that with us on the point of going down after Wattsy, dead or alive. When we were diverted by the sound of heavy rustling of the thick foliage coming up from the river directly below us. It was either Wattsy or Japs, so lets wait and see. The noise was getting closer by the second, and still we couldn't see who it was. When about two seconds off trigger pulling time Wattsy poked his head out of the scrub, much to our relief, then we hauled him onto the track to hear all about it.

Wattsy was in high glee explaining with loud guffaws how he spotted a line of Japs strutting down the track like turkeys to the water, when he just lined them up with a sitting shot and got the lot of them with the Tommy gun, and then made his way back to us .... He was content to sit there in loud voice and laughter, pointing out different aspects of the incident, when a shot rang out down near the river, reminding us that there was a war on and that it was time to move back to our own lines.[4][xvii]

The battalion diarist recorded several patrols went out that day: '0900 B Coy set ten booby traps. 10.30 Report received from B Coy that eight of the ten booby traps had exploded, a pool of blood alongside each trap. Also, blood near MG post and 500 empty cartridge cases. Patrols ordered to move further down track to river crossing if possible'.[5]

Around noon, at least one Japanese mountain gun opened fire against Ioribaiwa village, with shells exploding beyond in the sector held by the 21st Brigade Headquarters. Here, 31-year-old Corporal

---

[xvii] Cousens also states that the much quoted episode of Watts then taking a leisurely bath in the river after having killed the Japanese never happened.

Roland Lamb, from Northam in Western Australian, formerly with the 2/16th Battalion Intelligence Section, was killed, while five others were wounded.[6] The 2/16th Battalion diarist concludes for 13 September that intermittent fire was heard coming from south of their position until 1 am. The next morning, it was found that some men from the 2/33rd Battalion who had finally made their way onto the ridge had mistaken some wild pigs for an enemy infiltration party in the darkness. One of their officers, 27-year-old Lieutenant Graham Barclay from Launceston in Tasmania, commanding 11 Platoon, 'B' Company, was killed in the confusion by friendly fire.[7] The diarist of the 2/14th Battalion recorded the day's events including Watt's patrol:

> Patrol from our forward Coy investigated booby trap area and discovered all except two booby traps had been sprung; large pools of blood showing that the enemy had suffered several casualties. Two patrols, both 3 Bn CMF [Citizen Military Force], set out in morning, one to left and one to right, the left with object of harassing enemy near creek crossing, the right mainly for recce purposes. Our patrols connecting with 2/6 Independent Coy on left, continued all day. At 1215 hrs approx., enemy shelled village and beyond with light field pieces, inflicting casualties on Bde staff. Word was received that 2/33 and 2/31 bns were almost up to our position ....
>
> Captains RHODEN, [John] O'NEILL and [Medical Officer Donald] DUFFY were sent back a short distance for a rest. Still no word of 2/27Bn and C Coy and portion of HQ Coy 2/14Bn. A patrol of forward Coy 2/16 Bn encountered 7 Japs at Creek crossing; 5 were killed the others escaping. Patrol then withdrew to forward positions. At 1730 hrs, forward Coy was reinforced by the men from C Coy. 2/33 and 2/31 bns arrived and were in position some distance behind our position. During the night, an interchange of shots and grenade explosions took place. In

CHAPTER 24

the morning it was discovered that the nights disturbance was caused by 2/33 Bn.[8]

Meanwhile, Cameron at around 5 pm had sent instructions to the rear echelon of the 3rd Battalion, ordering all available riflemen to advance and bring with them all Thompson submachine guns from 'E' Company (machinegun company).[9]

* * *

Just south, the men from the 2/33rd Battalion, led by Lieutenant Colonel Alfred Buttrose – who with his men had become lost the previous day as they tried to conduct a right flanking movement from Imita Ridge – were still located somewhere behind Ioribaiwa Ridge. He now sent out the battalion's intelligence officer, Lieutenant Robert Howland, a 28-year-old bank official from Newcastle in New South Wales, and the signal officer, 23-year-old Lieutenant William Weale from Sydney, to find the main track, which was somewhere west of their position. They were to tap into the telephone line to inform Brigadier Eather that they had become lost, and he was withdrawing the battalion to the main track; from there he would get his bearing and continue their flanking manoeuvre. By 10 am, the battalion had found the track, where the two officers informed their CO that Eather had told the young officers that his original orders were cancelled, and the 2/33rd Battalion was to move directly up the track onto Ioribaiwa Ridge and once there await further orders.[10]

As the men advanced, the shadow of Ioribaiwa Ridge loomed up ahead, and they soon arrived at the Salvation Army Post manned by Major Albert Moore and his assistant, Corporal William Inglis. Sergeant William Crooks recalled at the first creek crossing, 'we saw the Salvation Army post, where Albert Moore, the Salvation Army representative with the 21st Brigade, had set up a shelter to give tea to returning wounded, some thirteen of whom were then enjoying a "cuppa" as we crossed the rushing stream of Ua-ale [*sic* Ua-Ule]

Creek'.[11] The men of the battalion pushed on and were soon passing Dump No. 44, which was an Army Service Corps food dump for the men of the 21st Brigade. Here, around 400 Papuans were taking a well-earned break from their carrying duties. By 3 pm, the men were approaching the slopes and heights of Ioribaiwa Ridge and its village located along the track. Here the brigade major informed Buttrose he was to peel off to the right, move up the ridge and adjacent high ground and occupy the area immediately to the right of the militiamen of the 3rd Battalion.[12]

The 530 officers and men of the battalion who had so far survived the track moved out – around 50 men (representing half a company in strength) – had already fallen out due to ill-health or injury. As recalled by Sergeant Crooks: 'We moved along a rising narrow track through thickly covered jungle slopes, over little crests, down ravines and across rock strewn outcrops'.[13] Now the men of 'B' Company took the lead followed by the Battalion Headquarters, then the men of 'A', 'C', and 'D' companies. In the darkness, the men moved along the ridge as a steady drizzle set in. It was now for the first time that these men heard the fighting with the distinctive sound of light and medium Japanese machinegun fire, followed by the quick bursts of Bren and Tommy gun bursts; all hearing and occasionally seeing the exploding shells and bombs of the Japanese mountain guns and mortars.[14]

It would be hours yet before these men would finally bivouac for the night. Sergeant Crooks recalled the journey along the ridgeline: 'The inky blackness of the night was now and then lighted up with phosphorescent fungi and fireflies. Exhausted by our two days of climbing and descending, we moved clutching the bayonet scabbard of the man in front. At times we tripped, and blindly clawed at the thick and dripping jungle undergrowth, while the sound of mortar and gunfire could be heard echoing through the valleys. We didn't feel at all happy'.[15] It was not until around midnight that the men were told to rest for the night, all were unsure of their positions, as the

CHAPTER 24

battalion was strung out along the ridgeline well east of the track and Ioribaiwa village.[16]

*　*　*

Further south, the men of the 2/31st Battalion had moved down from Imita Ridge, and like those before them, these men had no detailed maps, so what lay ahead remained largely unknown. They merely followed the Kokoda Track with their scouts up front. The men of the battalion soon also came across Major Albert Moore's Salvation Army Hut close to Ioribaiwa Ridge where Moore had hot tea waiting for them. Here were several more exhausted wounded men of the 21st Brigade who were making their own way back to Port Moresby. The war diary of the battalion recorded that the men on reaching Ua-Ule Creek, just short of Dump No. 44, 'bathed in creek and everyone's spirit soared. All spoke in highest terms of Salvation Army Padre. More tough going followed. MG and Mortar fire heard at intervals. Night was spent in defensive role on high ground in front of IORIBAIWA. During night MG, rifle and mortar fire sounded very close – yet we made no contact with enemy'.[17] As the men of the 2/31st advanced towards Ioribaiwa Ridge, their mates of the 2/25th Battalion had moved forward from Uberi with 'A' Company and the Headquarters Company arriving at Imita Ridge just before 11 am. When the rest of the battalion arrived on the ridge, they bivouacked there for the night.[18] No doubt, Major Moore and Corporal Inglis would have a cup of tea and biscuits ready as these men passed through their Red Shield post.

Another who commented on the Salvation Army post in the jungles near Ioribaiwa was Australian journalist Geoffrey Hutton, who was reporting from the front: 'The men slither down a clay slope, and then, miracle of miracles, they see round a bend in the track a native grass shelter, a tent, and a Salvation Army banner. Cauldrons of water are boiling, and each man gets a mess-tin of steaming hot tea. They mumble their thanks and sprawl on the ground, giving themselves up to the enjoyment of it. This post has become famous, because every man going up to the frontline or coming back sick or wounded is sure of a cup of tea or a packet of chewing-gum when he passes through. It is run by brisk and cheerful Adjutant Albert Moore, who made a name in Syria by setting up his canteen under the [Vichy] French guns. A direct hit once messed up his truck and peppered him with shrapnel, but he laughed it off. "They were only small pieces", he says'.[19]

Major Moore recalled the men of the 2/31st Battalion, and soon after those of the 2/33rd Battalion, as they passed through his Red Shield post just behind Ioribaiwa Ridge:

> I learned that the 31st Battalion was on its way [up] the Golden Stairs. We stirred the fires into life and had tea ready for them as they arrived. Drinking vessels were a problem, as we only had the tins that the coffee and milk had been emptied from, and with such a crowd of men we were many tins short. Men searched in the jungle for empty tins discarded by other troops, gave them a rinse in the creek and came for their issue. I even saw several men come to the urn and take their coffee in their tin hats. I hated to think just what the flavour was like.
>
> We had just met the needs of the 31st when I heard that the 33rd were close behind. I have no idea just how we made our supplies stretch out, but we catered for every man and, of course, they were ushered along the track as there was just no room for men to congregate in numbers.
>
> The signallers again installed a phone at my post, mainly so the officers going forward and falling back could ring through to headquarters for instructions. Word had gone right over the hills that the Salvos were on the Trail, and weary men would be enquiring for hours as to how far they had to go before they reached us. I longed in the early stages to have a Red Shield post between each village, but we did not have the men to fill the need.

CHAPTER 24

As the 33rd passed through, a lad came for a cuppa who had been with the Red Shield in the Middle East. I asked him if there was any chance of him joining us and he said he would be happy if the commanding officer would agree …. Running along the Trail I caught up with the colonel and asked him if there was any chance of the man, Arthur Hunter, joining me. 'Did Hunter ask you to get him or is this from you?' asked the colonel. I assured him that it was my request, as I badly needed someone to push supplies to out post. The colonel did not hesitate further and said I could have Hunter. Hurrying back, I immediately despatched my new helper into Moresby to stir things up there and personally get supplies to us (Arthur was still with the Red Shield when I left New Guinea about 18 months later). During the night I kept my ear close to the phone in case we might be warned to get out.[20]

* * *

Just before 6 pm, Brigadier Eather issued orders that the 2/25th Battalion was to move forward from Imita Ridge to Ioribaiwa at 6 am the next morning. The 2/33rd Battalion was ordered to be in position to launch an attack against Nauro on the morning of 15 September, while the 2/31st Battalion was to occupy a position just west of Nauro on the same day. The men of the 2/25th Battalion would advance along the main track in support of the operation.[21]

The Australian 7th Division Headquarters recorded at the time that due to the 21st Brigade encountering superior enemy numbers, they had been forced to fall back from the crest of the Owen Stanley Range. Although the scene of the fighting was only 50 kilometres from the base area of Port Moresby, it was noted that there was a lot of difficult country between the enemy and the township, and that the 25th Brigade was moving forward to attack the Japanese. Patrols had been pushed out on the tracks leading into the Port Moresby area. It was also noted that many Japanese spoke English and were in the habit of giving orders to the Australians to fall back during darkness.[22]

SAVING PORT MORESBY

* * *

Japanese scouting patrols were already active on the northern lower slopes of Ioribaiwa Ridge. Horii now decided that Ioribaiwa Ridge would be his final objective for the *South Seas Force*, until they received the final order to renew the assault against Port Moresby. He decided that Colonel Kusunose and his veterans of the *144th Regiment* would take and hold Ioribaiwa Ridge, while Colonel Yazawa and his men of the *41st Regiment* would be held in reserve around the Nauro. These men were also to protect the supply line back to Kokoda as well as conduct patrols south to identify other approaches to Port Moresby in preparation for a future advance to take the township. Even so, once the advance was renewed, he expected Colonel Yazawa and the men of the *41st Regiment* to capture Port Moresby. Yazawa's first objective was for either the *I/41st Regiment* or *III/41st Regiment* to capture Kila-Kila airfield southeast of 7 Mile Drome near the township itself, while the other battalion would capture the township and occupy its peninsula. The *II/41st Regiment* would remain in reserve to support the *144th Regiment* in its advance and occupation of the base area, including 7 Mile Drone.[23]

First Lieutenant Horibe, *No. 6 Company, II/144th Regiment*, recorded in his diary for 13 September: 'On a hill north of IORIBAIWA – Bn decided to camp here for further reconnaissance'.[24] Close by would have been the unknown leader of the signals unit with *No. 2 Infantry Section, No. 1 Company, I/144th Regiment*, who had a busy day organising and maintaining communications between several headquarters. Some time that night, he recorded in his diary: 'Bde HQ, Regt HQ, FA Bn HQ, No. 2 Bn HQ and No. 3 Bn HQ – connect on a single line. Bde HQ, Regt HQ and Bn HQ – connect on one line. Regt HQ, No. 2 Bn HQ, FA No. 3 Bn HQ – connect on one line. Two telephones were installed in Regt HQ. One spare reel of wire was removed from Regt HQ. Left at 0600'.[25]

CHAPTER 24

First Lieutenant Hirano, leading *No. 3 Platoon, No. 1 Company, I/144th Regiment*, was now heading north as ordered; it was critical that he and his party bring forward much-needed supplies. He recorded in his diary how his proposed party of 100 was now reduced to fewer than half that number: 'At 0515, assembled 35 men from Battalion Infantry Gun [Company] and reached the Field Hospital. Men from machinegun and other BUTAI did not assemble as desired. Carrying sick and wounded men, departed at 1000. Halted for the night at an old artillery position'.[26]

* * *

Earlier that morning, Cooper and his lost men of the 2/27th Battalion were still waiting on the return of the men of 'D' Company with the yams and taro. The leading elements came in at around 8.30 am, and soon the food was being cooked. The battalion CO rested his men, and he and his officers went about checking their weapons, ammunition and numbers. With him were 18 officers and 293 men, including a dozen stretcher-cases (most of these critically wounded were men of the 2/14th and 2/16th battalions) and three men of the 2/16th Battalion who had joined his party after having already spent two weeks lost on the jungle. The wounded, the stretcher-bearers of the Headquarters Company, and the men from 'D' Company were fed by noon, and by 2 pm, the remaining men had received their share of yams and taro, with a good quantity of yams remaining. At 3 pm, the battalion was on the move again and did not stop until dusk, when all slept on the narrow track. That night, one of the stretcher-cases died.[27]

* * *

The lost party led by Lieutenant Maurice Treacy of the 2/14th Battalion east of the main track had spent the previous day resting, probably feasting on the wild pig they had managed to kill the previous day. Treacy now decided they would attempt to push south-west, believing that by now they had flanked the Japanese. They moved out along a native track carrying about nine pounds of rice and 12 pounds of pork.[28]

Meanwhile, having stayed at the Seventh Day Adventist Mission Station west of the main track for two days, Lieutenant Robert McIlroy and his seven-man section, with the 2/14th Battalion, moved out to try and reach the Australian lines beyond Menari; he later recorded: 'As the wounded man had recovered sufficiently to walk, we started off again travelling the main track. After ½ a day we reached a small garden, and as the wounded man was not fit to travel further, [we] halted there for the night. In the evening one of the men injured his knee and became a [stretcher] case'.[29]

Further north, Treacy's OC, Captain Sydney Buckler, and his small party of the 2/14th Battalion were continuing their journey north with the stretcher-cases and walking wounded in their attempt to reach Sengai Village below the northern slopes of the Owen Stanleys. From there they would attempt to cross the Owen Stanleys from an easier approach to the south coast. They were surviving on eating mostly sweet potatoes. Still with them was Corporal Metson on his hands and knees crawling, half-starved but unbeaten; he and his mates would not reach Sengai for another week.[30]

* * *

That day, still in Port Moresby, General Blamey remained completely ignorant of jungle warfare. Australian journalist Chester Wilmot had just days before returned from reporting the fighting along the Kokoda Track. Unlike Blamey, he, along with fellow journalists Geoffrey Hutton and Osmar White, had firsthand experience of what fighting in the Owen Stanleys was all about. Wilmot was appalled by Blamey's flippancy in regard to the very basics demonstrated when he asked Blamey about the urgency in supplying his troops with jungle green uniforms. Wilmot recorded Blamey's refusal to consider the sensible suggestion: 'Khaki had been designed in India as the ideal camouflage

CHAPTER 24

for the jungle; and that he had no evidence that this jungle was different from that in India. I offered to provide him with several thousand witnesses who had fought in the country and who thought otherwise. General Blamey returned to Australia and apparently told the War Council that green uniforms were unnecessary in New Guinea, even though, at that time all our troops were changing into them'.[31] Maybe if Blamey had bothered to take a short trip to the extreme southern slopes of the Owen Stanleys while he was in Port Moresby, he may have come to a different conclusion. Chester Wilmot, like anyone who dared to question him, was now on Blamey's hit list.

25

'... THE "WOODPECKER", WHICH SLOWLY SPITS OUT LEADEN DEATH'

During the morning of 14 September, the 25th Brigade had concentrated in the Ioribaiwa–Uberi area and Brigadier Porter now handed over control to Brigadier Kenneth Eather. The brigade had been slated to conduct offensive operations against Nauro the next day. No doubt the battle-weary and sick men of the 21st Brigade were relieved to see the fresh and fit troops of the 25th Brigade arrive. Now standing on Ioribaiwa Ridge for the first time was Private Julian Waters, a 28-year-old labourer from Brisbane, Queensland, with the 2/25th Battalion, who recalled: 'I was in A Company HQ .... Our HQ position was on the safe side of the ridge. My first impression was of Drover Dick [Captain Richard Marson] standing right on top of the feature with Colonel Withy and Brigadier Eather. These three officers were evidently evaluating the whole position. Their calm demeanour would leave one to think that it was no more than taking part in manoeuvres. It must have been an inspiration to all though to see their leaders so boldly placing themselves in full view of the enemy. Almost from the moment we reached the top of the ridge, casualties were occurring'.[1]

Hours earlier, Lieutenant Colonel Alfred Buttrose and his men of the 2/33rd Battalion had awaken to find themselves strung out along the ridgeline. Each platoon leader and company commander gathered and organised their men. They were close to the militiamen of the 3rd Battalion, and soon the battalion was

CHAPTER 25

on the move further to the right to take up their position on the high ground of Ioribaiwa Ridge. Buttrose's orders were now to occupy forward defensive positions along the ridgeline in front of the 3rd Battalion, who were then covering the extreme right flank; offensive operations against Nauro, for now were cancelled. The 3rd Battalion war diary records that earlier that morning, Boag and his men reported lights on the ridge opposite their position with two Japanese flares fired from the vicinity of a banana patch; however, by mid-morning, the men of the 2/33rd Battalion had gone into defence on the forward slope below and right of Boag's position covering the extreme right flank.[2]

By 10 am, the 2/33rd Battalion had taken up their positions. The men of 'B' Company were located to the right of the militiamen along the ridgeline on the track. The men of 'D' Company had moved further east along a track that led from the small village of Ponoon, over a small crest to the north. They then moved down the forward slopes of the ridgeline for 30 minutes taking up a position in an open native garden. Captain Trevor Clowes positioned 16 Platoon on the left, 17 Platoon on the right, with 18 Platoon covering the left rear, while a company listening post was located 15 minutes forward. Now covering the extreme eastern flank of the Australian position on the ridge was 'C' Company, who occupied the highest ground with 13 Platoon forward, 14 Platoon left front, and 15 Platoon in reserve to the right rear. Just behind them in support were the men of 'A' Company.[3] It was not long before Lieutenant William Weale and his signallers had established a line to brigade, but little information came through, only that the 'Japs were probing about the forward slopes and the 2/31st Battalion was heavily engaged'.[4]

Sergeant William Crooks concluded for 14 September: 'With no further movement that day, other than local clearing patrols, the battalion again just lay on the ground in the now familiar wet and tried to get some rest, but few slept. We were listening to the chatter of fire over towards the unenvied and unfortunate composite battalion, which was steadily being whittled down'.[5] These men, while hearing the fighting, had yet to see any Japanese troops – they would not have long to wait.

*　*　*

Early that morning, the sick commander of the *144th Regiment*, Colonel Kusunose Masao, had gathered his officers to explain the situation as he saw it: 'It appears that over 200 of the enemy have occupied a position in the Ioribaiwa area. 2nd Battalion's main force is making a flanking attack to the W[est] ... and will be the right frontline. 3rd Battalion [less detachment] will be the left frontline and will attack the enemy position at Ioribaiwa parallel with the main road [Kokoda Track]. The mountain artillery will stay at its present position and cooperate with the regiment and with the 3rd Battalion. The regimental gun company will establish itself with the mountain artillery and cooperate with 3rd Battalion. The communication unit [signals] will liaise between the frontline battalions and Regimental Headquarters. I will advance at the rear of 3rd Battalion'.[6]

Meanwhile, his commanding officer Major General Horii, still located near Nauro, was issued Operational Order 'A' No. 117 at 4 pm. Horii now decided to take advantage of the Australian's continuing withdrawal by pushing further south to occupy Imita Ridge. For the first time, there was mention that his force was to be supplied by Japanese aircraft – nothing, however, would come of this:

> The KUSUNOSE Regt, having defeated a small force of the enemy, are now advancing towards IORIBAIWA. Observation shows smoke rising in the direction of UBERI. This may indicate that the enemy is destroying munitions. Apparently, there is no branch road leading north [west] from WAMAI [Naruo].
>
> The SHITAI will take advantage of the period of waiting to defeat the enemy in this sector and advance towards ERURU [Imita Ridge?].

CHAPTER 25

The KUSUNOSE Regt will defeat the enemy to the heights on the eastern [northern] side of ERURU, while with part of the main force will capture ERURU and hold. Advances to be made thereafter will be the subject of a separate order.

The YAZAWA Regt will set out on the morning of the 16th. With the advance of the forces in the first line, the vanguard will advance to the valley and on the western [southern] side of IORIBAIWA. Personnel engaged in clearing casualties as well as those collecting resources may move up progressively as the situation develops.

To deal with the dropping of provisions and medical supplies from the air, 20 men under the command of an officer will be left at WAMAI. This detail will also be responsible for precautionary guard duty for the casualty clearing station in this area .... I will remain at my present location tonight.[7]

* * *

Just before 7 am, the Japanese opened fire on the forward positions of the 21st Brigade. The Japanese of the *II/144th Regiment*, which had attempted to flank the Australian left flank, were now facing the newly arrived men of the 2/31st Battalion. To their left, the Japanese of *No. 7 Company* and half of *No. 9 Company, III/144th Regiment*, supported by elements of the engineers, were attacking the centre of the Australian position held by the composite battalion. The Japanese attacks were supported with artillery and mortar fire. The diarist of the 2/16th Battalion recorded: '0945 Mountain gun still worrying our tps – the slightest movement in the open would result in shots from [this] gun immediately'.[8] By 10.30 am, reports came into brigade that a patrol by a section from 9 Platoon on the Australian left flank had observed Japanese troops on the track between 'A' Company, 2/16th Battalion, and further west near Spotter's Hut.[9]

A patrol from 'D' Company, 2/14th Battalion, led by 24-year-old Corporal Arthur Morris from Geelong in Victoria, had returned with news that the Japanese were close to Spotter's Hut, around one kilometre north-west of Ioribaiwa village along the ridgeline. They had gone out in the morning and contacted the men of 'C' Patrol, 2/6th Independent Company, led by 29-year-old Lieutenant Arthur Tregarthen, from Brisbane, Queensland.[10] Twenty-year-old Private Gordan Green from Queensland, a member of 'C' Patrol, recalled: 'We were known as Tregarthen Force and we were the Brig's personal patrol, he would select the job for the day. The range was over-run with native pads and the maps were well out of calculation. Old *Trig* could walk like a kangaroo hops and a very good bushman. He would have been one of our best men in the bush – he was that much older than us 20-year olds [sic]'.[11] That night, Brigadier Eathen asked Lieutenant Tregarthen if he and his men could go out the next day and silence a Japanese mountain gun that had been targeting them all that day.

Meanwhile, on returning to their position, Morris and his men were fired on by a Japanese patrol that had managed to infiltrate through parts of the slopes near the hut. Morris' force was too small to attack these troops, and by 11 am, the young officer had made his way back to report to headquarters the location of the enemy position.[12]

The 2/14th Battalion war diary concludes with the tactical situation along with the welcomed news that Captain Russell and his men who had conducted the rearguard action at Menari had made their way back to the Australian lines: 'During the night enemy activities were confined to a few scattered shots and an attempt to creep up on our FDLs. Grenades accurately thrown by our troops discouraged the enemy and their attack was abandoned. The position of 2/25 Bn at nightfall was 200 yds to our rear. Word was received that Capt. RUSSELL, Lieut. RAINEY and small party had reported in at a rear camp'.[13] The war diary of the 2/16th Battalion also concludes that at 5 pm several booby traps were again heard to explode.[14]

CHAPTER 25

The 21st Brigade held the main track, but in doing so, they lost several more men killed and wounded by Japanese gunfire. Night found the survivors still there with the 2/25th Battalion to their rear, who had arrived at their position by 4 pm. Lieutenant Colonel Charles Withy placed 'A' and 'C' companies forward, 'B' Company on the right, and 'D' Company on the left, while the Battalion Headquarters and Headquarters Company were located to the rear. By now, the Japanese were creeping about the Australian forward positions on the ridge.[15]

* * *

The men of the 2/31st Battalion had by 10 am swung off the track, moving through 'A' Company, 2/16th Battalion. These men were happy to see their former battalion commanding officer during the days of the Syria Campaign, now brigade commander. Porter reviewed the men as they came onto the ridgeline, and the men noticed that the brigadier still wore his Black over Red colour patch of the 2/31st Battalion on his slouch hat. Here, 'Uncle Bill' had a word with his veterans of the Middle East Campaign and the entire battalion felt better for the unexpected presence of Brigadier Porter.[16]

These men moved out using a side track running north-west from Ioribaiwa along the ridgeline, which brought them towards the Spotter's Hut area. The top of the ridge was narrow with several bottlenecks, making manoeuvring extremely difficult. Leading the way were the men of 'D' Company who were now targeted with fire from nearby Japanese. Up front was one of their scouts, Private Oscar Vidler, a 24-year-old labourer from Murwillumbah in New South Wales. He was among the first to be hit and wounded with Japanese machinegun fire, but he stayed on his feet directing return fire against the enemy position before he was forced to fall back for treatment. It was also here that the men of the battalion suffered their first man killed – Lance Corporal Hugh Dutton, a 42-year-old labourer from Northam in Western Australia, who charged with his Bren gun to clear the way forward. The rest of 'D' Company were ordered not to follow, as they were then in a holding position. The Great War veteran's body was later recovered having suffered multiple bullet wounds.[17]

It was now approaching 2 pm, and Lieutenant Colonel Colin Dunbar ordered a flanking movement by 'B' and 'A' companies. Captain Lloyd Hurrell, a 26-year-old patrol officer from Wingham in New South Wales, who had signed up while in New Guinea, commanded 'B' Company. He and his men were to circle around the same track that Private Dutton had earlier charged. They pushed along the upper slopes of the ridge for about a kilometre with two platoons – the leading platoon was led by Lieutenant Desmond MacDonnell, a 28-year-old clerk from Cairns in Queensland, and behind him was their OC and the reserve platoon, led by 34-year-old Sergeant Bullingham from Subiaco in Western Australia. They were within 20 metres of the ridgeline when they were hit by intense machinegun and rifle fire, quickly followed by grenades.

It was now that a runner from the Battalion Headquarters reached Hurrell with orders for he and his men to fall back. As they began to fall back, MacDonnell was hit in the groin. The bulk of the company held their position while the wounded officer was carried to the rear, but he soon died from his wounds. Lieutenant Desmond MacDonnell was the first officer of the battalion to be killed in the Kokoda campaign.[18] To make matters worse, heavy rain now set in, turning the slopes to thick mud, and as both companies withdrew to the battalion's holding position, the Japanese followed them in with a strong thrust which raked the ridge with fire. Dunbar's men settled in a tight perimeter for the night.[19]

The stretcher-bearer with the battalion, Private Geoffrey Hamlyn-Harris, recalled the 'blooding' of the battalion during its first engagement with the Japanese: 'Accounts there were … of the action of the "woodpecker", which slowly spits out leaden death. It was doubly sad to find that so many familiar faces with their accustomed cheery smiles were missing …. Missing someone I remembered, I heard that

CHAPTER 25

he had been blinded for life. He was throwing his hand grenade when it hit a tree near him, bounced back and exploded at his feet, covering his body and face with shrapnel and mud'.[20] The 2/31st battalion war diary records the day's events:

> After walking up the tough slopes to IORABAIWA ... we were met by Brig PORTER. After a short rest we moved out into enemy territory through the 2/16th Aust Inf Bn lines in the following order: D, A, Dunbar Gp (Bn HQ), B, HQ Coy, and C Coy. We had barely cleared the 2/16 Aust Inf Bn along the track on the ridge running NW from IORABAIWA when the Jap opened fire on leading elements of D Coy ... (at this point ... the ridge was on the average 20 yards wide on top, and at some points it had 'bottle necks' five yards wide. So that an extremely narrow field of advance was possible). B and A coys tried flanking movements, but the steep sides of the ridge and the presence of well-camouflaged snipers in the treetops prevented any more advance than the bottom of the ridge. To add to the difficulties of the initial entry into this campaign, rain began to fall in torrents.[21]

The battalion's intelligence report adds: 'We lined the edges of the ridge making a tight perimeter. A quiet night followed interrupted by the Jap calling out "Aussie" and "Hey Aussie" at regular intervals – apparently a small team of callers worked singly'.[22]

* * *

Corporal William Cousens, 'B' Company, 2/16th Battalion recalled the Japanese shelling throughout the day and an anxious night: 'It was just a matter of waiting in your position with the hope and prayer that one of those missiles was not going to land in your lap. This went on for most of the day, with us ever on the alert for the presence of the enemy .... We had been issued with a password throughout the day with the words "Leaping Lulu," which was just about impossible for the Japs to say, as the letter L was hard for them to get their tongue around. This was in case they tried to infiltrate our positions throughout the night. We were all desperate for sleep and either one of us would rather have died than let our mates down by going to sleep. Our duties were two hours off and two hours on guard throughout the night with those two-hour intervals coming up much too frequently with only three men to execute them through those long hours of darkness'.[23]

* * *

First Lieutenant Horibe, *No. 6 Company, II/144th Regiment*, had fought that day against the men of the 2/31st Battalion near the Spotter's Hut area. He recorded in his diary: 'Marched up the road. Commenced attack at 1145. Met stiff resistance. The battle did not progress as desired. At 1230 reported to the coy comd the casualties of one killed and four wounded. Sighted the enemy slowly withdrawing from the right. The enemy, which was facing the Coy, however, continued to hold its position. Their strength became stronger'.[24] He also recorded that his position was targeted with a barrage of mortar-bombs, wounding five more of his men. The men of his platoon immediately divided into their trenches and foxholes. Horibe told more men to stand-to that night, expecting an attack, which did not materialise. He also recorded that one of his wounded, Private Nakao, died during the night. Orders soon arrived that they were to continue their attack the next morning.[25]

Somewhere nearby would have been the unknown leader of the signals unit with *No. 2 Infantry Section, No. 1 Company, I/144th Regiment*, who recorded in his diary that day: 'SOGABE and NYUDO were assigned to the No. 3 Bn to maintain communications between the No. 3 Bn and Regt HQ. The sec. leader and Pfc SHIRAKI cooperated with Regt HQ, No. 2 and No. 3 bns unsuccessfully attacked IORIBAIWA. Bivouacked on the riverbank near IORIBAIWA'.[26] It was also now that Horii's final reserves finally arrived in the Nauro

CHAPTER 25

area; the men of the *I/41st Regiment* had days before been ordered from Kokoda Plateau to join the rest of their regiment.[27]

Still heading north with his mixed party to bring back suppliers to the frontline troops was 1st Lieutenant Hirano, leading *No. 3 Platoon, No. 1 Company, I/144th Regiment*. He recorded that night, clearly in frustration, the day's events: 'Departed at 0430. Attacked by enemy planes. Men were exhausted and our progress was slow. Men from M.G. [unit] were not cooperative. Battalion infantry Gun [men] camped along the creek of KAGI'.[28]

Still at Myola was Private Teruoka Akira with the *Tanaka Unit, 55th Division Medical Unit*. That morning, Allied airmen also bombed Myola. Soon after, Teruoka saw the remaining elements of the *I/144th Regiment* move out to rejoin the rest of the regiment. The week before, Teruoka had recorded in his diary of wanting to regain his strength, but he was still far from it, recording: 'We sick and weak were left behind'.[29]

* * *

Still just behind Ioribaiwa Ridge, near Ua-Ule Creek, was Salvationist Major Moore. That night, he was waiting near the telephone for any word that the Japanese had broken through the Australian lines of the ridge to his front. He soon got a call, requesting that he deliver an urgent message to Lieutenant Colonel Arnold Brown, commanding the 2/1st Pioneer Battalion on Imita Ridge:

> September 14 found us again busy all day. Further ground had been lost to the invaders and we were now within easy earshot of the mountain gun in use by them. We could also hear the small arms fire. That meant that the distance between us and our forward line was very short indeed. It was a quiet day and Lieut. Austin, from the Red Shield in Moresby, arrived at our post to give a helping hand. He was the only one of our team to get through to the post.

Late that night the phone rang. It was division calling the Salvation Army post. I was asked if it might be possible [for me] to take a message to Colonel Brown who was holding with his unit, the next line of defence on Imita Ridge. This ridge was ... between us and Uberi. I agreed to try to reach the colonel with the message. It was a perilous task on that track in the pitch dark and teeming rain. As I climbed towards the top of the ridge I was challenged by a sentry. I identified myself and asked if he could guide me to Colonel Brown. I found the colonel sheltering from the rain in a sitting position under a groundsheet. He thanked me for conveying this official instruction to him. I then cautiously returned down the ridge to our post. It was a sleepless night as we felt now that we could expect anything at any time and could not afford to sleep.[30]

* * *

Lieutenant General Sydney Rowell designated the forward area that would be the responsibility of the Australian 7th Division. This was defined by the area around the Kokoda Track north and north-east of the Laloki River which ran roughly parallel with the coast and some 15 kilometres north of it, east of an imaginary line through Hombrom Bluff, which lay halfway between Port Moresby and Owers' Corner (Hombrom Bluff was the last ridge before Port Moresby itself). The units allocated to the division were those of the 21st and 25th brigades, the militia of the 3rd and 55th battalions, along with the 2/1st Pioneer Battalion, 2/6th Field Company, 2/14th Army Field Company, and the 2/4th and 2/6th field ambulances.[31]

The 30th Militia Brigade was tasked with stopping any penetration into the Port Moresby base area through the country from Hombrom Bluff west to the Brown River, which flowed roughly south from the mountains to join the Laloki some 14 kilometres north-north-west of Port Moresby. Within this area, the Goldie River ran south-west to the Laloki, close to the Kokoda Track. Additionally, the brigade was

CHAPTER 25

to guard against any unlikely Japanese landings in the Port Moresby area. To carry out these tasks, the brigade could call on the militiamen of the 39th, 49th and 53rd battalions, although the men of the 53rd Battalion were again largely employed unloading ships. Also attached was Honner Force, and 'D' Patrol 2/6th Independent Company, less some detachments then on long range patrols.[32]

The rest of Rowell's force was split into three principle groups: the 14th Militia Brigade was used in a coastal defence role at Port Moresby, with only the 36th Battalion assigned to this task; an artillery group – including the 13th and 14th Field Regiments – were tasked with helping to defend Port Moresby from both the land and the sea, while the NGF Composite Carrier Group, made up of the bulk of the Bren gun carrier and mortar platoons of the 21st and 25th brigades, were to be used as a reserve and to act as a mobile striking force.[33]

In addition, Major General 'Tubby' Allen had established a company strength force east of the track made up from the rear details of the 21st Brigade, commanded by Major Alfred Robinson, a 36-year-old hotel owner from Geraldton in Western Australia, with the 2/16th Battalion to be based at Jawerere. He now led what was designated 'Jawforce' and was tasked with preventing any Japanese advance south from Nauro by using several subsidiary tracks that led through many kilometres of broken country.[34] Meanwhile, to the west was Lieutenant Honner and his men of Honner Force.[35]

* * *

The journey for the two men from Lieutenant McIlroy's lost patrol, Corporal Lionel Smith, and Private Alexander Roberts, finally came to an end, making their way back to the Australian lines after being separated from McIlroy's party. For a short time, they were back with their mates of the 2/14th Battalion before making their way to Port Moresby for medical treatment and rest. The battalion recorded: 'About midday Cpl SMITH and Pts ROBERTS, 2 members of a patrol sent out from EURO CREEK, under Lieut. McILROY, on

2 Sep, reported in and gave an account of their experiences during their 13 days in the bush. They gave a verbal report to Brig PORTER and passed on valuable information regarding tracks and Jap activities along the enemy L of C .... At 1300 hrs Capt. [Peter] SMITH and one OR of 21 Bde [also] reported in through our lines, after having been missing same time as 2/27 Bn and stretcher parties from 2/14 and 2/16 bns (Wed 9 September)'.[36]

* * *

South of Lieutenant Colonel Geoffrey Cooper and the bulk of the lost men of the 2/27th Battalion, Captain Sims and his now small party were able to shoot a pig. As they enjoyed the feast, more lost men, mostly of the 2/14th Battalion, joined their party. Nothing was left, all was consumed: eyes, brain, intestines, and the bones broken up for marrow. Sims recalled years later: 'It's amazing what you do when you've got a bit of hunger! ... because we drew lots and you'd have a bit of this and a bit of that, and I had a bit of his hind quarter with a bit of knuckle'.[37]

Meanwhile, their CO, Lieutenant Colonel Cooper, and his large party east of the Kokoda Track, were carrying a dozen gravely wounded and heading south. Before setting out that morning, they buried the stretcher-case who had died the night before. By 8 am, they were on the move and made good progress with the improved track leading down a hillside. However, it was not to last. At 11 am, they spotted an Allied reconnaissance aircraft and tried to make contact but failed to do so. The country they passed during the early afternoon was by far the worst they had encountered, with the war diary recording: 'One steep pull tested courage of stretcher patients as session was rough and painful'.[38]

At some point during the afternoon, 13 men from the 2/14th Battalion, who had been lost in the jungle since the withdraw from Isurava towards Myola, passed through the battalion. They had been struggling in the jungle while making their way south for over two

CHAPTER 25

weeks, using a compass in their attempt to reach Port Moresby. They told Cooper that they had encountered no Japanese until south of Myola where they were forced to dodge three enemy patrols in two days. By 5 pm, Cooper called a halt on high ground overlooking Nauro, and in a sheltered position, they cooked their yam rations as daylight faded.[39]

* * *

Back in Port Moresby, the well-informed Australian journalist George Johnston reported: 'A good many things have been happening on the fighting fronts. Most important, of course, is the fact that the "impassable" barrier of the Owen Stanleys hasn't held up the Japs. They are through the pass and advancing through the dripping trees and muddy tracks somewhere between Menari and Ioribaiwa Ridge, well into the southern flanks of the range …. They are still employing their well-tested tactics of flank attack and infiltration synchronised with heavy frontal assaults. Their object always seems to be to try to cut large bodies of Australians into small sections and then annihilate those sections one by one. Our troops have countered many of these moves with the point of the bayonet … many of our patrols have been engaged in desperate hand-to-hand clashes in the shadowy wilderness of the range.[40]

26

'A PARTY OF ABOUT 70/80 JAPS WENT PAST'

The morning of 15 September would see General Horii and his men of the *South Seas Force* launch their major operation to shift the Australians off Ioribaiwa Ridge. The men of the *III/144th Regiment* would attack the Australian right and centre, while those of *II/144th Regiment* would continue their attack on the left.[1] The Japanese official history records: 'Major General Horii attacked the front of the enemy positions with the 144th Infantry Regiment and assembled at the ready the main strength of the 41st Infantry Regiment on the right bank area of the Nauro River [*sic* Naoro]. Since leaving Kokoda, the commander had planned to assemble both regiments for an assault on Port Moresby if Mawai [Nauro] could be reached. The 1st Battalion of the 41st Infantry Regiment, which had earlier landed at Basabua, joined the main force of the regiment on 14 September'.[2]

* * *

At first light, the centre of Brigadier Kenneth Eather's position was already under pressure from Japanese infiltration parties. The tired remnants of the 21st Brigade, holding Ioribaiwa village, were swept with machinegun bullets and targeted with mortar-bombs and artillery-shells, which killed seven and wounded 29 men of the composite battalion. Indeed, the battalion recorded that the Japanese had advanced during the cover of darkness and dug-in to within 30 metres of their forward defensive lines.[3]

The diarist with the 2/16th Battalion recorded that during the previous night and early hours, the men of 'C' Company were attacked

CHAPTER 26

by the enemy, but the Japanese were beaten off with some well-placed grenades. Even so, some Japanese troops managed to creep up and help pinpoint the Australian positions. At 6.45 am, 'C' Company's position was again targeted by Japanese mortar-bombs, and enemy troops were soon observed digging-in just forward of their perimeter.[4] Lieutenant William Grayden leading 15 Platoon, 'C' Company, 2/16th Battalion, who had been in the thick of the fighting around Mission Ridge, including breaking through to the Brigade Headquarters' position at Butcher's Corner, recalled the effects of this shelling:

> Shelling had commenced early lower down spur and wounded were passing through our position mid-way up the slope. A shirtless Corporal Charlie Lintott was one of the passing wounded. He had been struck by a piece of shrapnel across the lower back and had suffered a deep cut, 200 mm long. It had opened out so that it was fully exposed but was not bleeding heavily. When Charlie reached our position, the Japanese fired one shell from a gun below us, it missed the lip of our ridge as we had expected but burst in the canopy six metres above us.
>
> Pte John Baker and Pte Harry Whitfield were in a 600-mm deep slit trench about five metres further along the Track. Shrapnel from the shell struck Whitefield across the forehead. He was mortally wounded and convulsing in the bottom of the trench.[xviii] Lintott took shelter behind the trunk of a tree, a metre wide, on the brink of the ridge. John Baker left the slit trench with the dead or dying Harry Whitfield and joined Lintott behind the tree.
>
> Three further shells slammed into the foliage above us and Lintott was again hit in the back by shrapnel. Having been thwarted in our attempts to dig a slit trench, Sgt Pel Williams

---

xviii Private Harry Whitfield, a a 33-year-old truck driver from Maylands in Western Australia.

and I had been lying outstretched on the ground. In retrospect, in this position we probably offered a larger target for shrapnel raining down from above than we would have sitting in a slit trench or standing ….

During a lull in the shelling, I moved from my position a few metres away to put a field dressing on the fresh wound suffered by Charlie Lintott. When I reached the tree, there were now three of us behind it, me on the left, Charlie Lintott in the centre and John Baker on the right. I steadied myself against the tree, as I looked down at an angle of perhaps 40 degrees, I saw the flash of the gun. I heard no sound of an explosion. Almost simultaneously the sensation I felt was something like being in a dumper was on a surf beach.

As the tree was protruding over the brink of the ridge, I was blown sideways on to the forward slope of the ridge facing the Japanese. I must have been knocked unconscious because by the time my senses returned and I picked myself up and rejoined the reminder of the platoon on the ridge, my platoon Sergeant, Pel Williams, had already advised Battalion HQ that I had my 'head blown off'. Dutifully in accordance with standard practice, he had removed and discarded the firing bolt of the rifle I had been carrying.

I learned that a shell had hit the root of the tree that we had been sheltering behind. It had blown Charlie Lintott[xix] and John Baker[xx] to shreds. They had taken the full force of the explosion and I was covered with their human tissue. My hand had been resting against the tree which must have vibrated to such an extent that it split the skin on my thumb in a lattice pattern. In that very restricted area and brief flurry of gunfire, four 2/16th Battalion, C

[xix] Lance Corporal Charlie Lintott, a 27-year-old timber worker from Palgarup in Western Australia.

[xx] Private John Baker, a 26-year-old farmhand from Qualeup in Western Australia.

CHAPTER 26

Company men had been killed and ten wounded …. I reported to Battalion HQ to inform them that the earlier report of my death had been erroneous. Although a non-smoker, someone offered me a lit cigarette and I accepted it. To my surprise and that of HQ personnel, I found myself blowing smoke several millimetres out of both ears. Both my eardrums had been perforated. Such was the efficiency of the Japanese Mountain Gun![5]

This Japanese artillery piece was almost certainly with the gun company of the *III/144th Regiment,* which was positioned just 50 metres from the Australian frontline positions. The Japanese battalion history records the action of this gun at this time: 'In a moment the position of the enemy's Czech machine gun [Bren gun] in front was blasted away and the nearby jungle was transformed into a forest with dead trees …. The roar of the gun echoed in the valley creating an intense and heroic atmosphere. In accordance with the battalion gun, the mountain gun unit and the infantry gun unit fired at any enemy positions which perturbed them'.[6]

Also experiencing the bombing and shelling here was Corporal William Cousens, who with two others had been in an exposed forward position for days. Many years later he recalled their dire situation and the wounding of Lieutenant Grayden:

> Come daylight and the mountain gun and mortars came to life once more with a fair amount of success to the Jap gunners, but they would not attack us man to man, with it seeming pretty obvious that they were not going to attack this front position again. Ok, I will approach Lieutenant Watts and see if we could be relieved from this front position for a rest up near the top of the ridge, with the ultimate request that I contact Don Coy on top of the high ground to send three men down to relieve us explaining to them to come down on the leeside of the ridge to escape the constant shelling of the trail.

I went off on this mission with an exhausting climb up the leeside of Ioribaiwa Ridge until I reached the top, with troops dug in all around that area and hard to locate, with me very exposed at times trying to contact Don Coy, until my old mate Dave Griffith stuck his head out of a fox hole to yell, 'Come out of that you stupid b...d, you'll get blown to pieces out there.' I wasn't too happy with this prediction and used all the cover I could until I eventually come upon a Don Coy NCO, giving him the message to pass on to his superior after he had located his whereabouts, stressing the part about coming down on the lee side of the ridge to dodge the mountain gun shells. Then I retired back to our forward position to await the arrival of our relief and to supply any relevant information they may require about their new surroundings.

But what's wrong? .... We waited in vain for that relief party to show up, until the news came along that they had dead-heated with a mountain gun shell on the track, blowing two of our troops to pieces and wounding Lt. Bill Grayden in the arm. We cursed our rotten luck from that turn of events. Why in the heck couldn't they have heeded my warning and come down on the leeside of the ridge? Anyway we just had to spend another night in that forward position, using all sorts of ruses to keep ourselves alert enough for the task .... Bill [later] put the ledger straight after I had asked him why he and his two men had not sought protection to come down the other side of the track. Whereupon Bill explained that that was never the option, because it was his task to approach his two men along the edge of the track facing the mountain gun, who were entrenched in a fox hole beneath a stout tree sprouting up from the track surface. Bill was leaning on this tree with his hand while issuing orders for the lads to come with him down to our front positions, when a mountain gun shell scored a direct hit on the fox hole,

CHAPTER 26

blowing the occupants to pieces, with the tree saving Bill's life and splitting his fingers open while he had been leaning on the tree. That was the luck of the draw during those fateful days …. Anyway we were relieved next morning and took up positions around the Command Post near the top of the ridge.[7]

* * *

During the morning, the Japanese had established a mortar position to the west near Spotter's Hut, but at around 9 am, the counter-mortar fire by the 3rd Battalion mortar team succeeded in silencing the enemy mortar. Meanwhile, a 20-man patrol led by Lieutenant Stanley Steel, a 29-year-old labourer from Stanthorpe in Queensland, with 'A' Company, 2/25th Battalion, had earlier moved through the men of the 2/31st Battalion on the left to investigate the sound of felled trees coming from below Spotter's Hut. They were tasked with finding and destroying the Japanese working party that was thought to be on the forward slopes some 400 metres west of the Kokoda Track. They moved along a track about 100 metres below the ridgeline, hoping to overlook the Japanese working party, but at some point, they were observed by the Japanese and mortar fire was direct against them. Steel and his men quickly moved 200 metres down a spur for cover, and they were soon in an ambush position above a fork of a wide gully where broken ground indicated the Japanese had been passing through. However, by 2.30 pm, there had been no sign of the enemy.[8]

The battalion historian recorded: 'The men were learning the limitations of sight and hearing in the jungle. Heavy and light machinegun fire was raking the ridge directly above the patrol, mortar-bombs were exploding on the ridge and fire orders for a mountain gun could be heard, but none of these weapons could be located'.[9] Steel and his men returned to the battalion about an hour later and reported that they had observed no enemy troops; the young officer could only confirm that the Japanese were in strength on Ioribaiwa

Ridge and were making strong thrusts against the 2/31st and 2/33rd battalion positions.[10]

Meanwhile, at around 11 am, the 3rd Battalion's mortar teams, located on the ridge left of Ponoon village, now directed their fire against Japanese machinegun posts entrenched near the creek in support of the 2/31st Battalion. At 11.30 am, the Japanese retaliated against the Australian mortar fire, targeting the 2/16th Battalion's sector. As recorded by its diarist, the Japanese opened machinegun fire against their positions, with 'B' Company's sector targeted with mortar-bombs that resulted in one wounded. The diarist concluded: 'Apparently, retaliation for 3 Bn effort with mortar'.[11]

Lieutenant Colonel Colin Dunbar and his men of the 2/31st Battalion found it almost impossible to identify the Japanese positions in the thick jungle and rain. This, along with the difficult terrain, also made it difficult for patrols to make any headway. The lack of drinkable water was also becoming serious, and despite the determination of his men, it was challenging to maintain active patrolling; this was made worse by effective Japanese mortar fire. The 2/31st Battalion diary records the day's events as they unfolded:

> 0800: Standing patrols moved forward at first light and Capt. [Leslie] HURREL patrol and Lieut. [Christopher] RYLAND's patrol left Bn area to recce enemy left flank. Lieut. [Alexander] BLACK's patrol also left. Intermittent mortar and MG fire began the morning. Water became a serious problem, particularly in D Coy as movement from there to the rear was under observation. 3" Mortar was used and ranged the enemy forward of D Coy.
>
> 1115: A large patrol from 2/25 Aust Inf Bn moved through our lines to the right flank. D Coy sent out a pl to investigate the results of mortar fire. Apparently, this sortie coincided with Jap attack which was very hot – pl lost 1 killed, 2 wounded. This attack was part of a pincer movement (the right pincer) on IOBIWAIWA. Reports from D Coy indicated extreme

CHAPTER 26

difficulty in penetrating enemy camouflage [positions]. D Coy also reported the devastating effect of our mortar fire. Enemy avoids open: fires from bushes and trees .... Japs were observed to climb trees and then pull up camouflage by cord.

1600: Two Jap shells, from our forward front landed in A Coy area – 1 casualty. Slight activity till dusk and right flank and forward.

1830: Lieut. [Reginald] PHELPS' patrol returned – reported enemy digging-in 300 yds West of ridge: also, the command of cleared portion of ridge exercised by snipers in trees. Patrol lost 3 men.

1900: Japs lit fire in front of D Coy and began chatting, but we did not respond as evidently desired. NIL activity during the night.[12]

In the fighting here was 1st Lieutenant Horibe and his men of *No. 6 Company, II/144th Regiment*. Horibe reported that Australians had moved back 220 metres and Private Sasaki brought in two abandoned Bren guns. Horibe tried to press the attacker deeper into the Australian lines, but in doing so lost six wounded from his platoon. His men crept forward and fired at 'the moving enemy, and inflicted heavy loss'.[13] He concluded succinctly in his diary that night: 'IORIBAIWA – continued to attack and failed'.[14] He was ordered that night to prepare for a dawn assault against the entrenched enemy along the ridgeline.[15]

When night came, Dunbar and many of his men felt decidedly uneasy about what the morning would bring – they had good reason to be concerned.

* * *

In the centre of the Australian defence of the ridgeline, the Japanese were escalating their attacks. The diarist of the 2/14th Battalion recorded: 'The 5th day of occupation of IORIBAIWA Ridge by our unit began very badly. The enemy appeared to have been very busy increasing his strength and circling around to the flanks. He had brought in MGs and Mortars which in addition to his field piece, he used all day .... Our HQ was shelled and mortared at times'.[16] The forward company, which was manned by men of the 2/14th Battalion, was forced to give a little ground. The day's casualties were reported as seven killed and 19 wounded, while 17 were evacuated due to illness. Just before 1 pm, 9 Platoon, 'A' Company, 3rd Battalion was sent forward to strengthen the forward defence of the 21st Brigade.[17]

During the afternoon, Private Leslie 'Pappy' Ransom, a 38-year-old labourer from Regent in Victoria, with the 2/14th Battalion managed to make his way back from the previous forward position vacated earlier that day. He and his mate, Private Leonard Edwards, a 21-year-old labourer from Red Cliffs in Victoria, had been stranded. Ransom came in, stating that Private Edwards had been killed, but he had watched the Japanese setting up a couple of weapon's pits just 50 metres from his position. He managed to snipe three of the enemy before leaving his concealed position.[18] Ransom recalled:

> Len Edwards and I were both behind a couple of rails when Len said. 'This is a bad position, Pap.' So, we made a dash for a hole about three yards away and hopped in, much better there. We stayed for about an hour when Len went to check an adjoining hole fifteen paces off. He got just to the edge of the hole when he stopped a mortar round in the groin. He went over, half in, half out of the hole. Poor Blossom, as we used to call him, never knew what hit him. I did some serious thinking while I hugged the bottom of my hole, expecting a mortar round at any second. I decided to wait until light failed. It was ten in the morning, but anything I did was risky. I never shifted in the hole for an hour and a half when I heard a Jap voice. I peered over the top and, by Gawd, standing fair in the middle of the track was a Jap pointing at Blossom. He was jabbering at others I couldn't see. I thought

CHAPTER 26

for two seconds, but he was too good a target to miss. So up with the rifle. After killing him I expected all the hardware they had but nothing happened. By cripes, the stillness was getting on my nerves. I heard voices, chucked a couple of grenades. No idea what happened, but there were no more voices.

After several hours fire suddenly broke out between the Japs and our guys. I couldn't see anything, but just as I was settling back to the bottom of the hole, I saw bushes move on the next rise, and then saw a Jap's head. He seemed quite unconcerned, so I slide the rifle over the top and waited. He broke the bush fair in front, and I got a good look before I fired. He gave a hell of a scream as he fell among the bushes. They opened up with machinegun fire around Blossom's hole. I kept very quiet, thinking my chances of getting out were looking bleak. There was another lull and then more firing. I looked over the top again and noticed a green bush at the foot of a tree on the forward slope of the next rise. I thought, now that wasn't there before. I watched that bush for what seemed like a lifetime. Then it moved and disappeared behind the tree. He had leafy branches stuck in his helmet and clothing. He started having a good look-see. I waited till he gave me the best target. He spun round and fell at the foot of the tree.[19]

As it turned out, Ransom was only a couple of hundred metres forward of his battalion's position and was able to make his way back to the Australian lines without further incident. The 2/14th Battalion war diary records Ransom reporting back to the battalion and providing details of the Japanese position. Soon after, Captain Goldsmith with the 2/16th Battalion sent out a patrol to deal with these Japanese: 'GOLDSMITH … arranged with an officer of the 2/25 Bn Pl to take a patrol forward to have a crack at the enemy positions. The patrol proved successful, and both Mortar and MG crews were knocked out without any casualties to our troops. The remainder of the night was quiet except for occasional grenades and rifle explosions'.[20] This patrol was led by Lieutenant Kenneth Jefferson, a 24-year-old bank clerk from Queensland, with 'A' Company, 2/25th Battalion who was guided by a man from the 2/16th Battalion. The battalion war diary records the patrol led by Jefferson 'contacted enemy mortar and MMG emplacements and attacked with hand grenades and TSMGS. Claimed 15 enemy casualties. Unable to ascertain damage to mortar or gun owing to darkness'.[21]

The diarist with the 2/14thn Battalion recorded: 'Activity continued on our right flank most of the day, where the 2/33, 2/25 and 3 bns were endeavouring to deal with the Jap. By nightfall, our flanks were still insecure but COs of 2/25, 2/31 and 2/33 bns were confident of clearing up the situation'.[22] It was during this time that the efforts of Private David Elliott, a 24-year-old gardener from Benalla in Victoria, with the 2/14th Battalion (who had been attached to 'B' Company, 2/16th Battalion to assist with signals) repeatedly went out alone to repair the telephone lines when they were cut by enemy shelling and bombing. Finally, as recalled by the battalion historian, 'when it was beyond repair, he ran out a new line in the midst of enemy shelling, swiftly re-establishing communication at a critical time. For his valuable services under fire, he was mentioned in despatches'.[23]

Intelligence officer Stan Bisset with the 2/14th Battalion recalled the mounting casualties from the Japanese artillery and mortar fire: 'We held the main position on ... Ioribaiwa Ridge, we held that for four days. Our pits were dug, and we were being ... picked off by their mountain guns firing across the valley, over open sites, and we lost quite a few of our fellows in the last few, four days there. In the meantime, the 25th Brigade had come up behind'.[24] That day, one of the officers of the 2/14th Battalion recorded in his diary: 'This evening in the twilight I buried two Headquarters Company chaps. A very sad business as they had been terribly knocked. A shell had caught them in their slit trench. One of the chaps lending a hand fainted for a moment or two at the graveside. No one said a word – we just helped him to his feet. I noticed tears in the eyes of quite a few of the troops'.[25]

CHAPTER 26

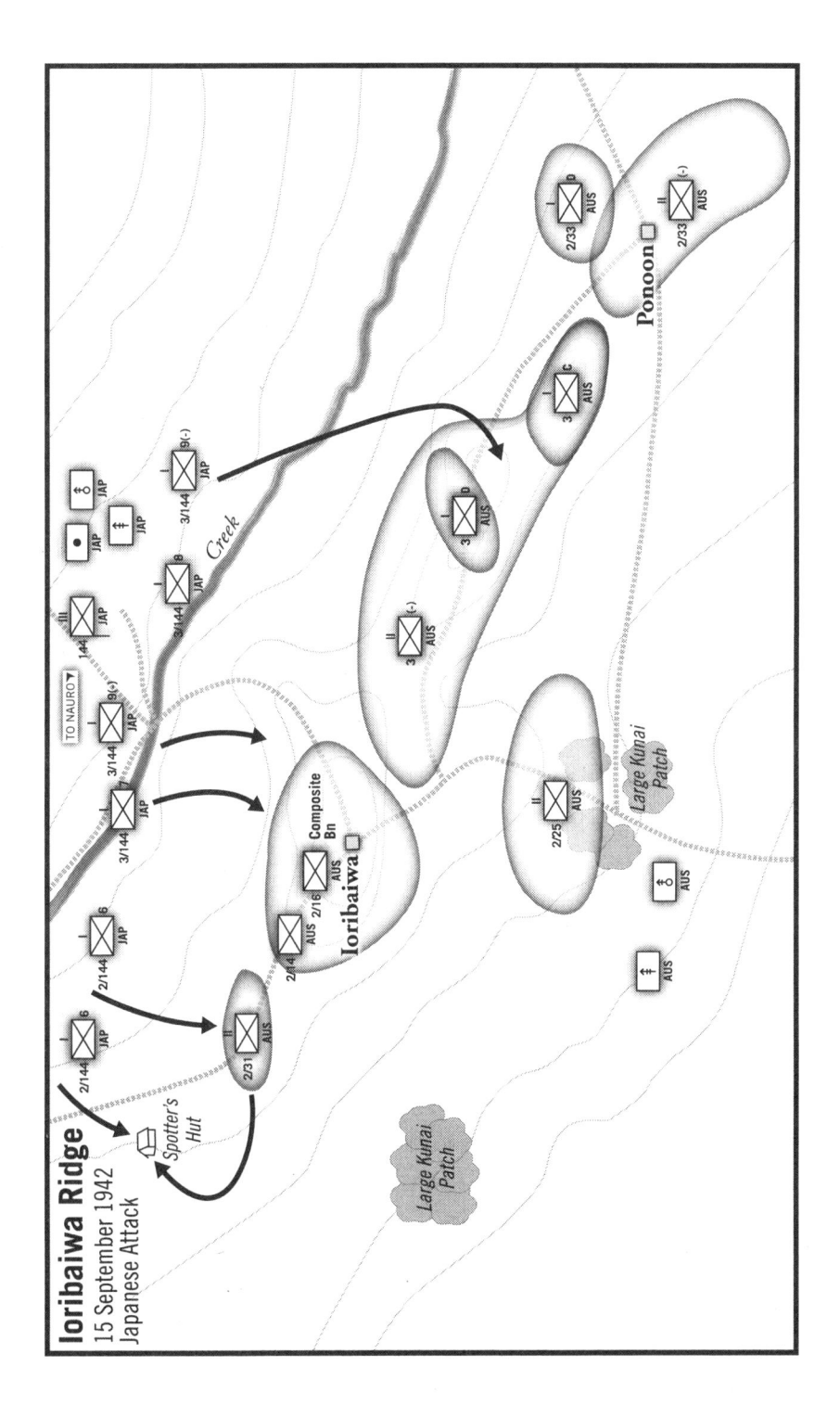

Ioribaiwa Ridge
15 September 1942
Japanese Attack

SAVING PORT MORESBY

* * *

Earlier, the commandos of 'C' Patrol, 2/6th Independent Company, led by Lieutenant Tregarthen, had left during the early morning hours to accomplish their mission of taking out the troublesome Japanese mountain gun somewhere north and west of the Australian left flank. It was estimated it would be an eight-hour hard slog to reach the Japanese position. Climbing in and out of valleys they stealthily surrounded the Japanese gunners, who considered themselves safe from Australian infantryman. Private Gordon Green recalled they quietly settled into their position and waited for their CO to give the orders, as Japanese infantry hovered around the gun: 'After lying in wait for two hours the chance came to do the deed. Everybody had crawled into position and selected their target and waited for *Trig* to give the word. Shorty and I were on the rifle grenades, Strika and his mate were on to the sentries who were pacing up and down near the gun. A party of about 70/80 Japs went past and up the hill so the word was not given until they were out of sight. Then it was on! I got two grenades away, perfect shots, I thought. The party felled the other Japs standing about and the gun crew copped it also. We then left hurriedly, led by [Private] Jim Ross, the break-away man .... The night was spent getting back.'[26]

Private 'Ossie' Osborne with the 2/6th Independent Company, who was then with Honner Force, recalled hearing of this raid: 'The Japs had lugged a damn 75-mm artillery piece right across the range with all the ammo and were harassing the infantry all the time up at Ioribaiwa. And they asked the section if they could do anything about it. So, they went across, just a small group. They couldn't do very much as they didn't have any explosives or anything available, so they went across and got themselves in position right beside the gun and just as they were about to – a bloody company of Japanese infantry arrived at the wrong moment. They were able to kill the gun crew and anyone around it which quietened it down for 2 or 3 days, so they helped them in that respect'.[27]

27
'I CONSIDER I HAVE JUST ARRIVED IN TIME'

For those covering the Australian right flank, the morning of 15 September started off relatively quiet. The men of the 2/33rd Battalion awoke to a clear and fine day and were able to 'boil up a brew'. It would, however, prove to be the 'quiet before the storm'. Clearing patrols were sent out from all companies and reported back that no enemy troops could be heard or seen. By now, the men, especially those of 'D' Company, were low on rations. These men were now digging up taro and yam from the garden position they were occupying and were forced to eat them raw. By noon, increased Japanese activity was observed in the battalion's sector.[1]

At around 2 pm, the Japanese of *No. 8 Company* and elements of *No. 9 Company III/144th Regiment*, supported by a platoon of engineers, launched their assault against the ridgeline at the junction of the militiamen of 'D' Company on the left and 'C' Company on the right (covering the 3rd Battalion's right flank), which was represented by a depression along the ridgeline. Here they were digging-in, having established a solid position. The brunt of the initial assault fell against the militiamen of 'D' Company commanded by Captain James Beckett, a 35-year-old car salesman from Queanbeyan in New South Wales. Sergeant Kennedy, who would later write the battalion's history, recalled the sector held by the militiamen covering the right flank: 'The position occupied by "D" Company 3rd Bn extended down the right incline of Ioribaiwa ridge across a depression between that ridge and a steep pinnacle further on the right. The Company HQ was located in a valley below and behind the three platoons in front. The left flank of the Company i.e., Ioribaiwa ridge was not, to my recollection, steep or heavily timbered, but the right flank was steep and heavily timbered'.[2]

The 2/33rd Battalion diarist recorded that at around 2 pm, a Japanese force infiltrated through the 3rd Battalion on their left. The men of 'A' Company, 2/33rd Battalion – led by 29-year-old Captain Kenneth Lawson from Trawool in Victoria – who were in support of 'C' Company on the extreme right, were now ordered to reach the high ground to the left, from where they could target and assault the Japanese left flank. Captain Trevor Clowes, commanding 'D' Company, 2/33rd Battalion, who were located forward from the slope, was also ordered to send out a patrol to the high ground on his left and if possible, support the attack by 'A' Company. With this, Clowes ordered 22-year-old Lieutenant Richard Cox, from Mudgee in New South Wales, and his men of 18 Platoon to conduct this patrol. Brigadier Kenneth Eather now also ordered Lieutenant Colonel Charles Withy, commanding the 2/25th Battalion then located to his rear, to advance and attack towards the high ground to the right of Ioribaiwa village where the Japanese were now entrenched; they were estimated to represent two companies at least. Withy was to attack the Japanese here with two companies. Both attacks by the men of the 2/33rd and 2/25th battalions would act as a pincer movement to cut off the Japanese. By 3 pm, the Japanese mortar and small arms fire against the 3rd Battalion had increased significantly, with the militia battalion war diary recording: 'Strong enemy patrol attacked on right flank forcing one Pl., D Coy., to withdraw'.[3]

Meanwhile, Lieutenant Cox and his fighting patrol of 18 Platoon, 'D' Company, 2/33rd Battalion had moved out as elements of the Japanese *III/144th Regiment* had wedged themselves into the Australian frontlines. Cox and his men were to advance north-west from their position on the forward and lower slopes of the ridgeline using an old police track to support the attack by the companies of the 2/33rd

CHAPTER 27

and 2/25th battalions. Sergeant 'Bill' William Crooks with 18 Platoon provides a fascinating account of this patrol in the battalion history:

> With only their weapons, the platoon was on its careful way north by 1500 hours. The platoon employed the formations such conditions called for – two point scouts leading, the section leader next, then the rest of the point section, followed by the platoon commander, the two remaining sections, then platoon HQs, the platoon sergeant [Crooks himself] bringing up the rear, with some four or five paces between men of the first section, apart from the scouts, who moved as conditions dictated, keeping each other in view. Each scout was armed with a rifle – the section leader carrying the Thompson submachine gun.
>
> From D Company position a track led across a native garden and into thick jungle, continuing on until, after some half-hour's march it came to a narrow creek with a crossing of logs. On the other side it climbed a spur. With the sound of firing still continuing on 18 Platoon's left and rear, the patrol advanced cautiously and slowly up the open spur, until about two hundred yards short of the crest, where there were two or three large native huts. Lieutenant Cox deployed the platoon astride the track. The section leader, Corporal Johnny Beck, and the two scouts, Privates John Condon and Alex Breakwell, carefully examined the village, but saw neither Japs nor natives, nor recent signs of either. In fifteen minutes, the party returned to the platoon, and then John Beck saw movement to the left, which quickly put everyone to ground. John Beck pointed out the cause of the movement – a Jap squatting in the fork of a tree, some thirty feet above ground, and on the far side of a gully that separated the spur where the platoon was located and one to the west.

The Jap was only 100 yards distant and was looking directly at the patrol. He sat perfectly still. He wore a green-coloured uniform and had leaves and ferns festooned in his netted helmet. Lieutenant Cox asked the platoon sergeant, Bill Crooks, if he could bring the Jap down with one rifle shot, to avoid the use of the L.M.G., which could indicate the strength of the patrol. Just then a runner from Corporal Wally Mackay's section, some fifty yards down the track, reported Japs on our right. Quickly everybody took cover with all peering into the thicker jungle on the eastern side of the track, where at some thirty to forty yards Wally Mackay pointed out a group of six or eight Japs spread out and advancing past our flank towards Ioribaiwa.

Lieutenant Cox, realising he could be ambushed, or caught in the middle of an actual Jap attack on his flank, decided to get out the moment the Japs had passed and report his findings to battalion. Hurriedly, but quietly the platoon fixed bayonets, sited the Japs, and waited. However, the Japs passed on, and continued away further to the east. The platoon carefully and quickly got out of the area. Moving back the way they had come, leapfrogging by sections until they began to climb up the slope towards D Company position. They arrived back at 1715 hours as D Company was beginning to move in response to an order from Lt Col Buttrose.[4]

* * *

The counterattack against the left flank of the Japanese *III/144th Regiment* force, which had infiltrated the frontlines of the 3rd Battalion, was to be led by Captain Kenneth Lawson and his men of 'A' Company, 2/33rd Battalion. Lawson placed his platoons in two single files with 7 Platoon led by 24-year-old Lieutenant John Balfour-Ogilvy, from Renmark in South Australia, on the right. To their left was 9 Platoon, which was commanded by 24-year-old Lieutenant Peter Bright from

CHAPTER 27

Adelaide. Bringing up the rear was 8 Platoon, led by Lieutenant Mervyn 'Merv' Roberts, a 23-year-old advertising professional from Sydney. These men tried to cut their way to the north-west to attack the Japanese, but it was not long before they were confronted with steep ravines to their right and visibility was extremely poor. There was just no way forward. For two hours, Lawson and his men tried to find their way, cutting through the thick vegetation with bayonets, not having any machetes, seeing neither Japanese nor the militiamen of 'C' Company. Lawson and his men had frustratingly heard over the last hour or so the sound of fighting to their left, and it appeared that the militiamen and men of the companies of the 2/25thn Battalion were attempting to push back the Japanese. Lance Corporal Nance Hogan, 'C' Company, 3rd Battalion recalled: 'Then the fireworks really did start. Bullets were flying in all directions, and many came to rest near us. The buzz of the flying shrapnel could be heard quite clearly and in one position where I was sharing a dug-out with a mate a piece of shrapnel landed nearly between us on a rifle butt, burning a very black mark on it'.[5] At around 4.30 pm, Lawson and his men were forced to retrace their steps back to their position in support of 'C' Company, 2/33rd Battalion covering the extreme right flank of the Australian position along Ioribaiwa Ridge.[6]

* * *

To the west, the two-company attack by the men of the 2/25th Battalion against the Japanese *III/144th Regiment's* right flank went forward. Leading the attack was Lieutenant Colonel Withy and his men of 'C' and 'D' companies. Unknown to these men, however, there would be no pincer attack, as Lawson and his men of 'A' Company were lost in the rugged terrain and thick jungle to the east. The attack went forward at 3.30 pm, with Withy's men of 'C' Company leading the assault and supported by 'D' Company. However, they could not dislodge the Japanese who had dug-in on the high ground even before Withy and his men could launch their assault.[7]

* * *

Meanwhile, Caro's men of the composite battalion holding Ioribaiwa village, just west of the fighting, had observed elements of the Japanese *III/144th Regiment* launch their surprise attack against the 3rd Battalion, forcing the militiamen of 'D' Company to fall back from a high point, which worryingly overlooked Ioribaiwa village. Sergeant Kennedy recalled the fighting here as recalled to him by Private William Brown, from Canberra, with 'D' Company: 'The attack, frontal, commenced … with mountaingun, machinegun and rifle fire. How they got there I do not know but snipers were in treetops. The mountaingun shells – some of them anyhow – landed very close to Company HQ. I collected two minor shrapnel scratches, one on the head and one on the shoulder. The rifle and machinegun onslaught came high on the right flank and was very intensive inflicting some casualties …. Beckett decided to withdraw …. The withdrawal was along the depression … towards Battalion HQ which if I remember rightly was on the main track leading up to Ioribaiwa. This of course meant that a wedge had been established allowing enemy movement through the depression and behind troops on Ioribaiwa itself. Fortunately, the Japanese didn't follow up their advantage'.[8]

Also with 'D' Company, 3rd Battalion, was 24-year-old Sergeant Vincent Gallagher, from Grafton in New South Wales, who recalled the fighting here: 'George Edwards was wounded in the groin and me in both thighs. I don't know how the others got back but I couldn't stand up so rolled over and over. I thought of lying "doggo" but kept going. I got to a big log and couldn't get over it. Billy Woodger reached over, dragged me across and told me to get going to the bottom of the hill as I was defenceless. I rolled my way down a steep slope till I got to where I could get some sticks to help me along'.[9] Close by would have been Private Noel Geraghty, a 21-year-old bus driver from Bega in New South Wales, who did not waste words with his first description of events on Ioribaiwa:

CHAPTER 27

When the Japs opened fire, I had just put my haversack on the ground and was chewing a biscuit when bullets started flying through the bushes just above my head. I moved very quick to get behind a small rock. It was just after this that Lieut. Woodger jumped across the track and landed beside me. As he hit the ground the bullet hit him.[xxi] The platoon that the Japs fired on were about 30 metres up the hill from where I was. The chaps who came past me wounded … were 'D' Company chaps. I was in 16 Platoon. WJ Thompson (killed) was, I think, in 17 Platoon.[xxii] One chap had a bullet furrow right across his stomach about one cm. deep, another had both legs wounded above the knees, another appeared to have all fingers missing from one hand …. It was not long after that the 2/25th Bn moved up through the scrub towards the Japs. About this time another message was relayed to me from Beckett who was well back down the hill, to go forward around the left and bring back all that were out there otherwise they would be cut off. At this time, I did not know we were pulling out.[10]

Indeed, Brigadier Eather now ordered Captain James Beckett and his men of 'D' Company to withdraw from the right flank to strengthen the centre of his position along the ridge. Thirty minutes later, the men of 'D' Company, 2/16th Battalion observed the enemy with medium machine guns 'moving over open patch NORTH of high feature previously held by D Coy 3 Bn. B Coy opened fire inflicting between 30 & 40 casualties, range 450 yds. Jap MG opened fire on village from right flank bullets landing in BHQ, no casualties'.[11] The recently promoted Captain Bert Madigan, now leading 'D' Company, led this action, forcing elements of this Japanese attacking force into

---

xxi Twenty-nine-year-old Lieutenant William Woodger, from Queanbeyan in New South Wales, survived his wounds.

xxii Lance Corporal William Thompson, a 19-year-old station hand from Hillston in New South Wales.

the open and targeting them with heavy fire that resulted in significant casualties. However, in leading this action, Madigan was wounded. Just weeks before the then, Lieutenant Madigan had bravely led 12 Platoon, 'B' Company in a holding action against the Japanese force near Myola. Also conspicuous in the fighting here on 15 September was 26-year-old Lance Corporal Benjamin Roberts, from Albany in Western Australia, who stood his ground while exposed to enemy fire and continued to bombard the Japanese with grenades. For his actions, Lance Corporal Roberts was mentioned in despatches.[12]

Hours earlier, Corporal William Cousens and his two mates had finally been relieved from their forward position half-way down the forward slopes of Ioribaiwa Ridge and had taken up a position near 'D' Company. It was not long before Captain Madigan approached the corporal with a task:

> Captain Madigan detailed me to take three men and set up an observation post about 200 yards out on our right flank below the 3rd Militia Battalion lines, to detect any Jap movement out there, and to report back to him if we detected any significant numbers of them. He stressed that we were not a fighting patrol and only to engage the enemy if attacked. Fair enough, and off we went on our appointed task, to locate a dense thicket of cover about 200 yards out, to await any results that may show up in the future.
>
> There was no immediate result to our mission as time went by out there, then after one and a half hours vigil our patience was rewarded by the movement of Japs about forty-five yards in front of us. They were advancing cautiously up the not too steep rise towards the 3rd Militia liners, where I was able to get a reasonable count of about forty of them. They were not aware of our presence, and we hung on for about another fifteen minutes to see if any more Japs were following up behind them, with negative results, I reckoned we should pull out and report

CHAPTER 27

[to] … Captain Madigan who could issue a warning to the 3rd Militia command if possible.

We made a quiet exit from the cover of our thicket and made it quietly back to our own lines, where our troubles really began, when we were challenged by our own men, who apparently didn't know we were out there and us not knowing what the new password was, if any. It was only after an explosive unmistakable stream of Aussie invective that convinced our challengers who we were, with their confession of how close we were to be blown out of existence, which showed that the boys were really on the job. I made my report to Captain Madigan, which must have been of some consequence, as a close scrutiny over towards the 3rd Militia lines revealed the presence of about seventy Japs in an exposed position, this was thirty more than the number I reported, so there must have been some ahead of the lot we spotted.

Time for action as our Bren gunners let go at that mob of Japs with never a let up, while the Bren gun barrels run red hot, making it very difficult to change barrels. The Japs suffered very badly in that exchange, with very few survivors and another shattering blow to their morale. They were well and truly on the receiving end of things despite their mountain gun and four-inch mortars, which was about the only punishment they could mete out to us now.

Darkness was almost upon us again, with me detailed to take up a forward listening post, overlooking a steep down grade out from our command post. As I was crawling out to my position in the gloom, I heard a noise behind me, which upon inspection revealed the presence of one of my mates, Vic Miles. He didn't have to come with me, and I ordered him back, as this could ne a dangerous mission if the Japs started probing round that area. But no, Vic said he was going to come anyway. Thanks Vic, and

I'm glad he did, because all night at different intervals we could hear movement down there and the clink of tins occasionally which really had us on tense alert, but no appearance of the enemy. We were hoping he would show up to break the tension to give us some relief. It didn't happen though, and to this day Vic and I often wonder who was down there that night, Japs or a stray native pig. We'll never know. [13]

* * *

Trying to maintain communications between the two attacking elements of the *III/144th Regiment* was the unknown leader of the signals unit with *No. 2 Infantry Section, No. 1 Company, I/144th Regiment* who recorded in his diary later that day: 'The battle situation for the No. 3 Bn is not favourable. The communication network between the bns is satisfactory. The Wire Sec attached to the No. 3 Bn was aided by 4 men from the Wireless Sec. Superior Pvt YAMAMOTO, Yasuhide, Pfc TAKAMASA Tamotsu, Superior Pvt KALODA Teyoki, and Pfc TAKEUCHI Yoshio'.[14]

* * *

Meanwhile, just as Lieutenant Richard Cox and 18 Platoon, 'D' Company, 2/33rd Battalion arrived back at the company's position in the native garden, Lieutenant Colonel Alfred Buttrose, having heard no word from Captain Kenneth Lawson and 'A' Company, now ordered Captain Trevor Clowes and his 80 or so men of 'D' Company to move out to attack the Japanese left flank. The men were on the move by 5.30 pm, heading south-west from the native garden to flank the Japanese who had infiltrated the perimeter of the 3rd Battalion. Sergeant Crooks recalled:

> Captain Clowes, after ordering D Company to get ready to move was very much relieved to see his 18 Platoon coming up the hill .... Lieutenant Cox was hurriedly given his orders to follow on

CHAPTER 27

in rear of D Company, which was going to push west below the rising cliffs of Ioribaiwa until they struck the Japs. Lieutenant Cox had not had time to pass on his patrol report, nor reflect in the glory of seeing the battalion's first Jap. At 1730 hours with the late afternoon mists now rising up from the creek far below, the company left the open garden patch of pawpaw and banana, crossed a low pig fence at its edge, and moved into mixed grass, palm and bamboo bush to the west ....

The going was slow. After clearing the immediate bush outside the garden area [we were] following the native track that passed below sheer bamboo-covered cliffs and rocks. After about half-an-hour – it being then a little after 1800 hours – Peter Barr, the leading scout, turned an elbow of the track that looked down into an area part kunai and part garden and was reasonably open. Cautiously he moved to approach a ten-foot-high rock that was just to the left of the now open track, and at fifteen or twenty yards came in sight of two natives, one of whom was chopping wood. Then he saw about fifty or sixty Japanese who were widely dispersed. They opened fire instantly, while Peter Barr dropped to the ground shooting one of the native guides who was aiming a rifle at him. In an instant Corporal Rice's section went to ground and began returning fire while Lieutenant Mosley signalled up to two rear sections to spread out right and left and to engage. The professional and speedy nature of 16 Platoon's deployment and engagement with the enemy said a lot for Lieutenant Mosley and also for [Sergeant] John Audsley and the three section leaders, all of whom had trained for such moves back in the Dimra [Syria] days.

Captain Clowes, who had watched all this activity, turned around to Lieutenant Ted Logan, leading 17 Platoon, and ordered him to move out to the right to try and outflank the Japs, while 18 Platoon was to try and get above the Japs on the left. Leaving the track immediately, 17 Platoon had a hard time beating into almost impenetrable bamboo. The men cut and hacked at the bamboo with bayonets, no one at that stage having any machetes. Slowly getting forward, the platoon, after about ten minutes, broke out into fairly open scrub, but was forced to ground by Jap rifle, grenades, and L.M.G. fire. 18 Platoon was busy examining the hopeless-looking cliffs for a way up, the main burden of the task falling on Corporal Wally Mackay's 9 Section.

Meanwhile, up at 16 Platoon two men had already been hit. Private Frank O'Brien received a gunshot wound in the leg as he was engaging Japs, and Private 'Bluey' Beggs was hit in the body by a grenade fragment, as he was trying to get into a better fire position. Lieutenant Mosley ordered Corporal [Harold] Bradford to move around to the right and try to flank the Japs. 'Brick' Bradford ran back, grabbed the section L.M.G., called on his section to follow him, and, running and crouching, bravely tried to move further to the right. He was soon lost sight of by his men, but his L.M.G. was heard in action. Lieutenant Mosley now got up to move further around to the left of the large rock that most were using as cover when he fell to the ground, mortally wounded. He was seen to stumble, fall and roll over, clutching his body. At the same time three more men were hit with bullets – Privates Billy Kenny, [Alan] Steel and Corporal Billy Rice, their section leader.

At about this time Corporal 'Brick' Bradford was seen to be hit but nobody was sure how badly. Sergeant Audsley, who had watched this developing situation from some twenty or thirty yards in the rear, came forward to the rock to hear that Lieutenant Mosley had been hit, as had four or five of his men. Seeing that the OC, Captain Clowes, was now also behind the rock, John Audsley began to remove his gear intending to go to

CHAPTER 27

the aid of his platoon commander who could be seen moving. However, Captain Clowes yelled out, 'What do you think you're doing? This is your platoon now. Get them together and get them back into the scrub.' This Sergeant Audsley proceeded to do under fire, and with difficulty [he] succeeded in withdrawing the platoon some fifty yards to the rear.

To the right of 16 Platoon, 17 Platoon was now having a difficult time in the low belt of thick bamboo. Lieutenant Logan had swung wide down into a re-entrant and had reached the flank area of 16 Platoon's recent battlefield, only to be greeted with a blast of fire and grenades. The first volleys of fire sent a bullet grazing Sergeant [Francis] McTaggart's neck although not seriously wounding him. Private [Keith] McLeod, a Thompson machine gunner, was hit and hurled off his feet, backwards, down into a gully. There he lay still, with blood pumping from him. No others were hit in the storm of fire that was cutting the tree branches and swaying the bamboo. Lieutenant Logan's reaction was that they had walked into an ambush, and although none could see any Japs, he ordered the platoon to withdraw. Sergeant McTaggart passed Private McLeod by, convinced he was dead as he looked lifeless and covered in blood from a gaping wound in the neck. At about this time Captain Clowes ordered a general withdrawal of the company, ordering Lieutenant Cox to send somebody after 17 Platoon. This task Lieutenant Cox gave to Sergeant Crooks, who, borrowing a Thompson machine gun from an unhappy gunner, who was unwilling to give it up, and ordering Private Oliver Hawkins to cover his rear, blazed off into the scrub looking for 17 Platoon. He was very much relieved when not far into the bamboo he saw 17 Platoon coming out. With great difficulty D Company turned about and retired back some 200 yards, where, as it was getting dark, the company was deployed in an all-round perimeter position. The OC discussed the situation with Corporal Stan Ward, the attached intelligence section member, then sent him off to BHQ to acquaint the CO with what happened, and to give him the strength of the Japs as the OC was unable to communicate by wireless.[15]

Several men had been hit, and three were missing: the gravely wounded 28-year-old Lieutenant John Mosley from Sydney; 21-year-old Private McLeod from Medowie in New South Wales; and 28-year-old Corporal Harold Bradford from Sydney.[16]

* * *

Meanwhile, Captain 'Doc' Vernon was still located at his aid post at Dump No. 44, just behind the ridge. He reported events over the last few days in his report to ANGAU: 'At Camp 44. A hospital was run up and many cases of fever and dysentery were received and evacuated to WO Maxwell at Uberi where necessary. One N.M.Ord. was posted with our advance carriers about ¼ miles below Ioribaiwa village. I visited him on the 14th and 15th. This camp was close to the firing line and potentially in more danger, but we heard of no casualties among the carriers. The ground round camp 44 was soon fouled by the number of troops passing through and I cooperated with army officers in digging additional latrines .... Flies were numerous, and dysentery well established. It was at camp 44 that I received a copy of special orders calling for greater consideration for carriers and appointing two ANGAU officers on different parts of the line to ensure they were carried out. The A.I.F. troops were now wearing green uniforms'.[17]

Again, he was more expansive in his personal diary: 'Ioribaiwa was as rapidly evacuated by all but combatant troops as the other parts. I found Lieut. [sic] Kienzle established 1200 feet [400 metres] below in the ravine at Camp 44. Here the same old routine began again, building shelters, admitting fever and dysentery cases, with a daily trip up the hill to see if my forward orderly with WO Davies detachment was all right'.[18] Sometime that day, Captain Bert Kienzle himself wrote

CHAPTER 27

to his wife Meryl back in Sydney: 'Within firing range of the Japs and the constant thud of bombs, mortars and firing is a little disconcerting and at times uncomfortable'.[19]

* * *

That night, Brigadier Eather – who was at his headquarters, which was represented by a piece of 'canvas stretched over a boulder in a sea of mud'[20] – informed Major General Arthur Allen by telephone: 'I consider I have just arrived in time. I think it is going to take me all my time to stabilise the position for the present. Porter agrees'.[21] He also said that he had sent out 180 carriers with stretcher-cases and had none left for forward support. He urgently needed another 200 to be sent out and judged that supply by air drops was not possible, as they would be impossible to recover in the rough terrain and thick jungle with the Japanese so close.[22] Indeed, this would likely result in the Japanese being supplied by mistake, as rations would certainly fall into their lines.[23]

The end of the day found the centre of Eather's right flank penetrated with the Japanese hurriedly entrenching their newly won position. Eather was soon on the line and ordered Lieutenant Colonel Alfred Buttrose to move two companies of the 2/33rd Battalion west along the top of the ridge and drive the Japanese from the heights the next morning. At this point, Buttrose was still awaiting word from Captain Clowes and his men of 'D' Company located somewhere below the forward slopes of Ioribaiwa Ridge, having encountered the strong Japanese force just on dusk.[24]

Brigadier Eather recorded in his diary that night: 'Up 0530. Shooting and mortars. Dunbar still held up trying to outflank Japs. Towards 1400 hours Jap attacks right. Send Withy with two companies out from centre and one of Buttrose companies from right. Withy makes contact. Buttrose doesn't. Gunfight goes on all afternoon until dusk then quietens. May pinch him [Japanese] out during the night. Nothing much else doing'.[25]

* * *

Meanwhile, further north, 1st Lieutenant Hirano, leading *No. 3 Platoon, No. 1 Company, I/144th Regiment*, was heading for Isurava to bring back supplies for the men of his regiment and recorded in his diary: 'Crossed the peak of the Stanleys at 1000. Reached KUMUSI River [almost certainly Eora Creek] at 1230. Camped by an enemy provision warehouse'.[26]

* * *

Still lost north of the Australian frontlines was Lieutenant Colonel Geoffrey Cooper and his men of the 2/27th Battalion, who that morning made for Nauro. On finding it occupied by the Japanese, he and his men were forced to retreat further into the jungle to continue their journey south. Cooper had considered launching an attack against the village, with the aim of gaining food, but he quickly decided against it, considering the poor physical condition of the men, and the likelihood that the attack would merely result in being burdened with more stretcher-cases with little to show for the effort.[27] The battalion diarist recorded: 'Bn ready to move at first light but rested and waited reports of recce and observation parties. NAURO clearly visible from high ground near bivouac site. Capt. Skipper, watching village through field glasses for some time, reported that NAURO appeared to be practically deserted. Bn moved from high ground down steep slope and across river and approached to within approx. 800 yds of NAURO. Bn halted whilst recce patrols penetrated to outskirts of village. Patrols reported village had been reinforced and that there were well over one hundred Japs in occupation. As track did not skirt village but led directly to it, CO had no alternative but to return to site of previous night's bivouac and find another way around NAURO'.[28]

Corporal Clive Edwards recalled with frustration: 'It [was] certainly a tremendous job for men weakened as we [were] but the lads stuck at

CHAPTER 27

it valiantly and took the stretchers down a long hill and across a large river which we recognised as the one which flows past Nauro. It was a most difficult crossing and our hearts fell to zero when, just as the last stretcher had crossed, orders came from the front to retrace our steps across the river and up the hill as quickly as possible because the Japs were sighted in the front. It was an awful grind up the hill with the stretchers'.[29] Captain Harry Katekar also recalled:

> We hurried down the spur to the river – It took us some time to ford this deep swift flowing stream with the stretchers …. The men were absolutely exhausted and ravenous. Nauro was our hope …. Imagine our intense anguish and disappointment when the report came back that the Japs were occupying the village.

> We considered the wisdom of attacking the village, in the hope of capturing food; but our better judgement prevailed, as we still had to consider our wounded and we didn't want to be encumbered with more. It was late afternoon when the decision was made to retrace our footsteps and attempt to find a suitable track around the village further to the east. By this time the men were desperately exhausted, and it was a cruel blow to them to be told to about-turn. A couple of providential incidents occurred that day; it seemed as if God was taking a hand in the matter. While we were waiting outside Nauro it suddenly occurred to me that a lot of the food which had been dropped by our planes would be scattered far and wide in the undergrowth around the village ….

> I wandered off into the jungle in search of food. Something must have led me to the spot, for after a while I noticed that a branch of a tree had been broken off, as if by a fallen object, and so looking down I saw a bag-covered bundle. Almost tenderly I went down on my knees to tear away some of the bag, and lo

and behold inside was a perfectly good tin of Arnott's Biscuits. Some of the starving men were watching me, and it was as much as I could do to stop them from struggling to get a share. I managed to salvage some of the biscuits which were distributed amongst our wounded. In the meantime, a flank guard patrol of B Coy had stumbled over [a] 25 lb [11 kilograms] tin of Crowe & Newcombe's dried apricots, and these provided a handful for each man in the company, standing them in good stead in the days to come.[30]

The battalion war diary also records that at around 7 pm: 'A high-pitched voice from northern valley could be heard. "Are you there?" "Won't you answer me?" repeated several times in English. No answer was made'.[31]

28

'SPIRITED JAP ATTACK AGAIN CONCENTRATED ON D COY'

At first light on 16 September, the Japanese continued their attacks against the Australian positions. On the extreme right, and as ordered by Brigadier Kenneth Eather the previous night, Lieutenant Colonel Alfred Buttrose commanding the 2/33rd Battalion sent in a two-company attack against the Japanese left flank. Leading this assault was Captain Thomas Archer, a 26-year-old wool classer from Brighton in Victoria, and his men of 'B' Company. Behind them in close support were the men of 'C' Company, led by Captain Lindsay Miller, a 24-year-old librarian from Launceston in Tasmania.

By 8 am, these men were passing through the position of 'C' company, 3rd Battalion to drive the Japanese off the ridge. Earlier, having finally heard from Captain Trevor Clowes and his men of 'D' Company, Buttrose ordered them to take up an ambush position to target the Japanese as they withdrew. His men of 'A' Company were to remain in position to secure the east flank, but a platoon from this company was also ordered to patrol west below the ridgeline and to take up an ambush position in support of 'D' Company.[1]

Archer and his men of 'B' Company had some initial success launching their attack at around 9.30 am, having passed through the militiamen's position. Archer and his men were aware that somewhere below their position where the men of 'D' Company, and they were unwilling to throw grenades in their attack from fear of them exploding among these men.[2] As they charged the Japanese position, they managed to disperse the enemy with small arms fire, suffering just two casualties in the process; Japanese casualties remained unknown.

SAVING PORT MORESBY

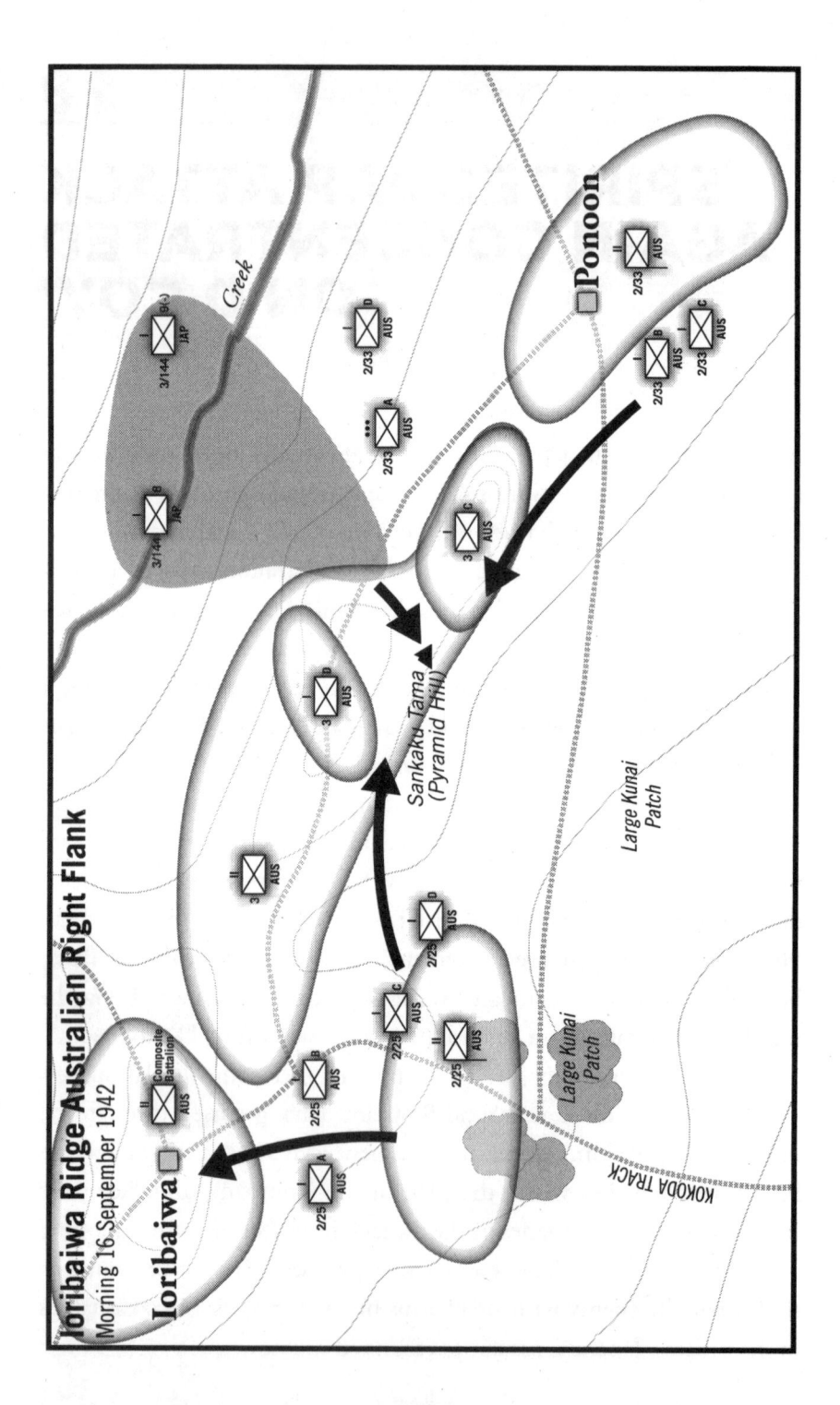

CHAPTER 28

However, soon after, the Japanese of *No. 8 Company, III/144th Regiment* counterattacked, forcing Archer and his men to fall back, enabling the Japanese to occupy a high point just east of the main Kokoda Track known to them as 'Sankaku Yama' (Pyramid Hill). This was a critical position, as it overlooked the spine of the ridge, allowing concentrated enfilade against the Australian positions; the centre right of the Australian position had been fully breached and was in danger of being turned. It was now that the men of 'C' and 'D' companies, 2/25th Battalion, under the temporarily command of Buttrose, launched another attack against the Japanese from the right, but little headway could be made against the entrenched Japanese who now held the immediate high ground. This was a turning point in the defence of Ioribaiwa Ridge, and Brigadier Eather knew it; his extreme right flank, represented by the 2/33rd Battalion, was in serious danger of being cut off from the rest of the brigade, and he had no idea of the strength of the Japanese he was facing.[3]

* * *

As this fighting raged, Lieutenant Bright, leading 9 Platoon, 'A' Company, 2/33rd Battalion, had reached an ambush position below the ridgeline at a watercourse near the Japanese left flank. Close by was Captain Clowes and his men of 'D' Company, who were also in an ambush position. Just before setting up their ambush, Clowes and his men were astonished but extremely happy to come across the wounded Private McLeod, who had been given up for dead during the previous days fighting. He was found by a scout of 17 Platoon on the track and although wounded in the neck, McLeod was still clutching his Tommy gun. He informed Captain Clowes that he had found and talked to the gravely wounded Lieutenant Mosley throughout most of the night before the officer fell unconscious.[4] Years later, Sergeant Crooks recalled McLeod's story of survival:

SAVING PORT MORESBY

Captain Clowes came up and began to question Private McLeod who, although pale and bloodied, was overjoyed at being amongst friends after his fearsome and lonely night. He told the company HQs group how he had regained consciousness in the dark and wet night and began to crawl about to try and find the pad but gave this up as hopeless. Later he heard some movement and moaning and called out: 'Who is it?' The badly wounded Lieutenant Mosley replied. Throughout the night they had talked to one another. Both wondered if the company had been ambushed and all killed or captured. As the night wore on to dawn, Lieutenant Mosley lapsed into silence, and at first light Private McLeod had begun to move to find help when Private Les Cook, 17 Platoon's leading scout appeared on the scene.

Quickly, Captain Clowes sent off Private McLeod with an escort, and ordered Lieutenant Logan to spread out and try to find Lieutenant Mosley and Corporal Bradford. At the same time 18 Platoon was ordered to move along the pad until the kunai of the previous day's engagement was reached. There it was to take up all-round perimeter defence positions and wait until B Company activated the Japs. Soon the Japs apparently heard movement and opened fire, but from a position above and to the left of D Company. This situation forced Captain Clowes to halt all further forward movement, and concentrate on forming a company perimeter, but at the same time sending off small groups to continue the search for the two missing men. However, this was a hopeless task. None was able to break out of the bamboo into the area of 16 Platoon's attack, as any such attempt brought instant Jap enfilading fire. So none saw either Lieutenant Mosley or Corporal Bradford. Captain Clowes then ordered all movement to stop and concentrated all in watching for any Jap movement.[5]

CHAPTER 28

It was not long, however, before these men of 'D' Company, and those of nearby 9 Platoon, 'A' Company received word from Lieutenant Colonel Buttrose that they were to withdraw from the failed attack to dislodge the now entrenched Japanese.

In the centre of the Australian position, the men of the composite battalion were still holding their forward positions and witnessed the fighting to their right. The diarist of the 2/14th Battalion recorded Japanese mortar fire was targeting their positions as the fighting broke out, while in the sector held by 'A' and 'B' companies, 2/25th Battalion the Japanese launched a frontal assault, but the attack was repulsed. Indeed, the diarist of the 2/16th Battalion recorded that they were still holding their forward positions, when to their right they observed the two-company attack by the men 'C' and 'D' companies, 2/25th Battalion, supported with additional attacks by the 2/33rd Battalion, and 'D' Company, 3rd Battalion who were all trying to 'squeeze out enemy's movements'.[6]

The men of the 21st Brigade were on their last legs, having been involved in continued bitter fighting over the last three weeks, under the most appalling conditions. The 2/14th Battalion recorded how these men were in desperate need of rest and reorganisation – even these 2nd AIF veterans of the Middle East had their limits. The men of the 2/14th and 2/16th composite battalion should have been taken out of the line days before with the arrival of the 25th Brigade group.

The 2/14th Battalion records for 16 September: 'The strain was beginning to tell on all members of the unit, and some of the lads in forward positions who had stood up to it well and had done a wonderful job right through, began to crack up. Lieut. [Lindsay] MASON, CSM [John] BARTLETT and Sgt [Dudley] WARHURST were evacuated for a rest. Enemy Mortar, MG and Field Piece continued to do deadly work on our forward positions all morning, and our casualties mounted. These forward coys were forced to withdraw further. The position on our flanks was still unsatisfactory and 2/31 and 2/33 bns and 3 Bn CMF were meeting with little success'.[7]

The 2/16th Battalion war diary also records: 'Japs attacked strongly on our positions forward of village. D Coy forced to withdraw and consolidate further up ridge'.[8] By 10.30 am, these men were relieved by 'A' and 'B' companies, 2/25th Battalion. Meanwhile, the men of 'A' Company, 2/25th Battalion were ordered to relieve the balance of the 2/14th Battalion on the forward slope of Ioribaiwa Ridge, while 'B' Company, 2/25th Battalion was ordered to send a patrol down the gully to the left of the Battalion Headquarters' position. These men would return hours later having encountered no enemy troops.[9]

Earlier that morning Corporal William Cousens and his mate, 23-year-old Private Victor Miles, from Subiaco in Western Australia, both with 'B' Company, 2/16th Battalion had returned to the company position having spent the night beyond the perimeter. The NCO recalled:

> We were glad when daylight peeped through the next morning to give us a spell until about mid-morning with plenty of activity from the mountain gun and mortars about us. The Jap seemed to be concentrating on this area from where all the Bren gun havoc came yesterday, wreaking so much devastation on his troops, and we were keeping out heads down low while all this was going on. Me and my two mates Joe Paul and Charlie Hubon were just below the brow of the hill with our biscuit tin of water next to us, while directly below us were two 2/14th chaps dug in about fifteen feet down, while to our right was our Command Post with Captain Bert Madigan in charge with a signaller.
>
> All of a sudden it happened, I heard a whoosh, whoosh, whoosh descending upon us quickly, the unmistakable sound of a mortar and with a mighty yell to the boys to cover up, there came a terrific explosion about us and bewilderment all round. We were knocked rotten by the blast and couldn't believe we were still alive after surveying the aftermath of it all. The two 2/14th

CHAPTER 28

chaps below us were killed, Captain Madigan and the signaller were badly wounded, our biscuit tin of water was riddled with shrapnel and we three were only concussed from the blast ....

It was soon after this episode when I noticed the presence of some smartly dressed fresh troops around us, who surely must be our long-awaited relief. These men proved to be the 2/25th Battalion of the 25th Brigade, here to push back the Jap from the spent conditions in which he now found himself, but not before dealing out a baptism of mountain gun fire on these fresh new soldiers .... I saw one of these new troops killed by a shell burst and the consternation of his mate, as we were pulling out and thought. 'It won't be long before you become as callous as us mate and take it in your stride'. That was the only way to deal with the situation in this type of warfare'.[10]

* * *

Meanwhile, to the west, the men of the 2/31st Battalion were under attack by a determined Japanese force from the *II/144th Regiment*. At around 8 am, Captain Charles Hyndman, a 29-year-old professional soldier from Ivanhoe in Victoria, with 'D' Company was able to repulse a determined Japanese attack resulting in around a dozen Japanese casualties.[11] The battalion diarist recorded: 'Spirited Jap attack again concentrated on D Coy. Attack lasted half an hour. D Coy suffered two casualties, Jap casualties 14. D Coy 2" Mortars inflicted unknown damage on Jap patrol, trying to bring up heavy MG on right flank'.[12]

Attacking the Australians here were the men of the *II/144th Regiment*. Sergeant Ito and Corporal Yokoda, leading their 16-man platoon, *No. 6 Company*, were to spearhead a charge against the Australian lines. Supporting them would be Lieutenant Horibe and another platoon, *No. 6 Company, II/144th Regiment*. These men had been ordered the previous night to renew their attack at dawn and drive the Australians off the ridge. In the fighting here, it is recorded that Sergeant Ito and his entire platoon was whiped out to a man, and the leader of *No. 6 Company* wounded. Command now passed to Lieutenant Kamimura, the leader of *No. 5 Company*, who was also soon after killed in the fighting.[13] It was now that an Australian mortar-bomb exploded among the party of Japanese who were trying to position a medium machine gun to enfilade the Australian right flank. Horibe recorded in his diary later that day: 'IORIBAIWA – 0700 – made a charge, met strong enemy resistance and failed. Coy comd was wounded – reported withdrawal of enemy'.[14] He also recorded of meeting 'stiff resistance', and in the fighting, Corporal Hosokawa was killed.[15] By the end of the day, Horibe and his men would be on the ridgeline and in communication with *II/144th Regiment* Headquarters. Horibe was now given acting command of *No. 6 Company*.[16]

* * *

By mid-morning, the usually calm Brigadier Eather was observed to be increasingly worried about being able to hold the ridge. At 8.35 am, he signalled Major General Allen: 'Enemy feeling whole front and flanks. Do not consider [that we] can hold him here. Request permission to withdraw to Imita Ridge if necessary. Porter concurs'.[17]

In a telephone conversation with Allen about an hour later, Eather stated he had no indication of the strength of the Japanese force he was facing, but it was greater than he had expected; nor could he give any details regarding Australian or Japanese casualty figures. He informed Allen that the enemy was moving around his flanks, and he did not think he could hold them, but he would do so if possible. He again asked Allen for permission to fall back to Imita Ridge if he assessed he could no longer hold Ioribaiwa. Allen replied that he must keep on the offensive and hold the enemy if possible, emphasising the importance of retaining Ioribaiwa, but he agreed that the ultimate decision rested with Eather.[18]

At 10.30 am at a commander's conference, Eather ordered a general withdrawal to Imita Ridge to commence at 11.30 am. Eather

CHAPTER 28

assessed that if he continued to hold the Ioribaiwa position, he would have to commit all his force to defensive tasks, losing any freedom of movement and restricting his ability to conduct offensive operations. Indeed, given that he was to cover Port Moresby, he needed to keep his force intact. Exacerbating the situation was renewed problems of supply, as he was totally dependent on Papuan carriers who had quickly disappeared, due to the threat to Eather's lines of communication. He judged that a withdrawal to Imita Ridge would give him time to establish forward patrols with a view to advancing again as soon as he had established a firm base along the ridgeline.[19]

* * *

Captain Burt Kienzle had already received orders to be fall back from Dump No. 44, as recorded in his report: 'The Japs had pushed on and taken IORIBAIWA by 16th Sept and by 1050 on that day an order was received to evacuate the camp in 2 hours. We moved back towards UBERI and established a camp south of IMITA Ridge to be known as [Dump] "66". It was a very wearying day, as 2 trips had to be made and the track was crowded with troops going fwd and carriers coming back with stores and wounded to safety. Rain set in heavily towards evening and some carriers could not get back into camp'.[20]

Nearby was Kienzle's good friend Captain 'Doc' Vernon, who recalled moving up to Ioribaiwa Ridge from his aid post at Dump No. 44 early that morning: 'I went up the hill to our advanced post early in the morning. A general evacuation of Ioribaiwa was in progress and we waited for some time in order to assist any wounded carriers. Large quantities of stores were again burnt. The last of our N.M.Ords and I returned to camp 44 at 1230, but the camp had been evacuated, a native constable being left to help us out. There was some delay in collecting stretcher parties to get some white men with dysentery out, but [by] about 1600, in heavy rain, I moved back in company with the last of ANGAU personnel. There followed a most exhausting walk to camp 66, the last hour of the journey being in complete darkness'.[21] In his personal diary, he provides a more vivid account of the withdraw from the ridge:

> A determined stand was made at Ioribaiwa between 11th and 16th Sept. We were heartened by accounts of the entire destruction of small Jap assault parties advancing from the ravine, but their successes were not lasting, and the occupation of the post was proving costly....
>
> We were again outmanoeuvred at Ioribaiwa, as the Japs were found one day to have worked round on the heights on our right flank, and the post already a costly position to hold, became untenable. But reinforcements were coming through every day; climbing up the hill now with these lads in green uniforms, created new conceptions of the Army, and you felt an air of comradeship about you as you toiled up in their company. I went up the hill on the morning of the 16th and met many carriers coming down as a withdrawal was imminent. Our forward dressing station in a dip at the top was not actually under fire. But I felt some concern about some parties of carriers still working in and around Ioribaiwa village, though Davies assured me they would be all right, and in fact we never heard of any casualties among them. When I got up quantities of food and clothing were being smashed up and burnt, and I heard afterwards that one cause of the Jap sickness rate was due to their eating the putrefying remains of what we had left, certainly by this time they had little enough of their own to eat. AIF units were coming out of the combat area, and I spoke to one officer, an able and intelligent chap, who told me that our men were not really fighting, not at least to the degree he had known fighting in the Middle East, adding that too many were picked off from long range from lack of skill in concealing themselves in the village. The Jap mortars were firing with deadly precision, and they had more powerful weapons with longer range than

CHAPTER 28

we had. This was merely the opinion of one man who went through the fighting at Ioribaiwa and was comparing it with another campaign where conditions were not the same. I asked an AAMC officer why we were evacuating Ioribaiwa, and he said it was because we were losing too many men. No doubt the position was too exposed to enemy fire, but it was the Jap outflanking movement probably that had made withdrawal necessary.

When there was nothing further to wait for, I picked up my orderly and got out with all my gear. There was a dangerously exposed grassy patch on the downward track from which we were ordered off onto a newly cut trail that for roughness surpassed anything on the whole track. How the stretcher-cases got through passes my imagination, as the walking files had to swing from branches and cling to the side of the precipice like flies. Camp 44 had folded up and gone on, but the OC had left a native policeman to help me out. There was a long delay in finding stretcher parties to take out a few dysentery cases, but [by] about 1600 hrs I finally moved on with WO MacCree ….

The heavens opened and amidst lighting and thunder sheets of water fell, wetting us through in a few moments. The big hill was so steep that in places you had to scramble up on hands and knees, and so slippery after the rain that you often slid back yards, while the steps once a great help in climbing, were now a hindrance, broken into fragments with potholes each containing half a foot of mud. But the most curious feature of the trek was the hordes of men all pressing on in the same direction, and as soon as we caught up the main body, we had to fall into a queue that extended for miles. Progress was by the slowest of crawls, step for step with the man in front. When night fell it was so dark that I could only keep on the track by placing my hands on the shoulders of the NCO in front of me, in fact most of the line was linked in this way to avoid a fall down a precipice of unknown depth. Finally, someone found a lantern and someone else a dry match, and eventually we reached our new camp and crawled into such shelter as it provides.'[22]

Earlier that morning, Major Moore had left his Red Shield post near Ua-Ule Creek, passing through Dump. No. 44 on his way forward to the top of Ioribaiwa Ridge to assess the situation: 'About 40 minutes along the track, I met a lad from the unit, Ian Clift, who gave me instructions from the unit to pack up and get out. Ian said the instruction was urgent. Lieut. Austin and a padre were with me. Reluctantly I packed up and loaded our gear on the natives who were glad to get moving smartly because the natives who were carrying out the wounded had been telling them of the rapid advance of the opposing forces. We kept out a couple of dixies and had them filled with coffee and with a tin or two of biscuits we were able to keep a cuppa going until 3.30 pm, when passing officers intimated that I must go quickly'.[23]

Indeed, as the men fell back from Ioribaiwa Ridge, the 25th Brigade war diary records: 'Conditions en route very bad. Hy rains during afternoon and track very muddy and slippery. Tps able to obtain cup of tea or coffee at SALVATION ARMY DEPOT at foot of fwd slope IMITA RIDGE'.[24] Major Albert Moore and Corporal Jock Inglis were soon after forced to evacuate their Red Shield Post, as recalled by the Salvation Army officer: 'We set off up [Imita Ridge] as briskly as we could. As we climbed it seemed that all heaven wept and poured its tears upon us as we were drenched through and through. The water coursed down the steps as we struggled to keep a foothold. Already I had decided as we climbed … that we would retreat to Owers' Corner, which was the head of the road, or had been before it got cut up with excessive traffic. We reached Uberi and decided to put the night in there. Next morning, I nearly wept when I discovered that in their rush up [Imita Ridge] the natives had lost two of my precious dixies'.[25]

29

'KEN. YOU'LL DIE THERE IF NECESSARY. DO YOU UNDERSTAND'

The withdraw from Ioribaiwa Ridge commenced at around 11 am. Lieutenant Colonel Alfred Buttrose and his men of the 2/33rd Battalion were still holding on the right, while Lieutenant Colonel Charlies Withy and his men of the 2/25th Battalion began to relieve the composite battalion in the centre forward position. By 11 am, these men from the 2/14th and 2/16th battalions were withdrawing, with orders to take up a reserve position behind Imita Ridge.[1] The 2/14th Battalion war diary records: 'At 1100 hrs the position became critical, and the relief of our unit by 2/25 Bn commenced. By 1215 hrs the changeover was completed and BHQ handed over to Lt Col DUNBAR and followed out after the rest of the unit. All were urged to push on as quickly as possible with the object of making UBERI by nightfall. The task proved very difficult on account of poor condition of troops and the bad track, and eventually the forward elements struck camp about half an hour off the objective at 1930 hrs. The 2/33 Bn acted as rearguard while 2/31 and 2/25 bns withdrew also to a position on the ridge between EURO–BIRA [sic] and UBERI where 2/1 Pioneer Bn had already taken up a position'.[2]

The 3rd Battalion moved out just behind the men of the composite battalion. These men were sandwiched in between the 2/33rd Battalion on the extreme right and the men of the 2/25th Battalion who had relieved the men of the 2/14th and 2/16th battalions. The war diary of the battalion reports that orders came through from the 25th Brigade

Headquarters that the battalion was to withdraw to Imita Ridge. Just before 12.30 pm, the Battalion Headquarters Company moved out, followed by 'B', 'A', the Battalion Headquarters' staff, and then 'D' Company just after 1 pm.

The men of 'C' Company, now isolated from the rest of the battalion, were to move out with the men of the 2/33rd Battalion.[3] Sergeant Kennedy with 'C' Company recalled moving out, 'of the saddle very late under the control of 2/33rd and spent the night uncomfortably on the main track overcrowded with retreating soldiers and carriers. No more depressing sight in the whole campaign was seen – a defeated force scaling the heights, held up by nightfall, lying down to sleep wherever they might, feeling in their haversacks for a cold tin of bully beef and moisture-laden biscuits. Ioribaiwa need not have been lost'.[4]

Two hours after the militiamen had withdrawn from the ridge, the men of the 2/31st Battalion on the left were also falling back, the battalion war diary records: 'Heavy rain, slippery track, darkness, fatigue – extra ammunition made the trudge back to IMITA RIDGE heavy. There was no hurry or panic. En route HURREL's and RYLAND's patrols rejoined the Bn. Main body of Bn gained the forward slope of IMITA and passed the night there'.[5] Twenty-six-year-old Sergeant Bruce Robertson, from Mortdale in New South Wales, with the 2/31st Battalion recalled the withdrawal: 'Our troops complied [with the order] with obvious discontent, their mood was sullen. The men were confident indeed that they could hold that ridge … quite a few considered the commanders must be stupid to vacate a position they considered able to stop a brigade of enemy'.[6]

By 2 pm the 2/33rd Battalion began to full back from their position as the rearguard, and a runner had been sent out to recall the patrol from 'A' Company led by Lieutenant Bright. The battalion war diary records that the 2/33rd and 3rd battalions moved to the 21st Brigade area near Ioribaiwa village, with 'C' Company, 2/33rd Battalion taking up an ambush position to the rear of the former 21st Brigade Headquarters' area. In front, the men of 'A' and 'B' companies, 2/25th

CHAPTER 29

Battalion were assigned as the rearguard to hold the centre of the ridge while the rest of Maroubra Force fell back towards Imita Ridge. Among these men in the forward position were Bren gunners, 34-year-old Private Victor Cruickshank from Rockhampton and 25-year-old Private Joseph O'Brien from Maryborough, both in Queensland. Each had been wounded earlier that morning, but they stayed in their forward exposed positions to help cover the withdraw.[7]

By 4.30 pm, the last of the 2/25th Battalion holding the centre of the position finally left Ioribaiwa Ridge and passed through the men of the 2/33rd Battalion, who were on the track, and took over as the rearguard for Maroubra Force. The 2/25th Battalion war diary concludes for 16 September: '2/25 Bn passed through 2/33 Bn, who took over rearguard. Darkness and heavy rain forced Bn to bivouac on track for night'.[8]

Private Waters with 'A' Company, 2/25th Battalion recalled the withdrawal: 'Late in the afternoon came the word to evacuate the position. We were to pull back to another high feature called Imita Ridge. This position was to be held to the last man .... Then the trek began .... On the apex of Ioribaiwa Ridge was a clearing which evidently had been a native village. As there appeared to be only one track from one side of the ridge to the other, all traffic had to pass over this clearing which was a bald spot among the jungle .... This was undoubtedly the place where many casualties occurred'.[9]

Meanwhile, as the Japanese approached the seaward side of the ridge, an ambush was sprung by the men of 'C' Company, who then quickly withdrew through 'A' Company, who had established a fallback position further down the track. Soon after, these men leapfrogged south through 'C' Company, leaving these men forward of the Australian main lines. Captain Lindsay Miller and his men of 'C' Company were in another ambush position just behind Ioribaiwa Ridge, at the 'Kunai Patch' about one kilometre north of Dump No. 44 along the Kokoda Track.[10] With darkness, 'C' Company had formed the rearguard while the rest of the brigade struggled wearily over a slippery, muddy track through the night, which was made darker by heavy rain.[11] Miller placed 15 Platoon on the right, 13 Platoon on the left and 14 Platoon to the rear in support. About 300 metres to the rear, a two-inch mortar team was in position to help cover the company from any Japanese attack. A further 100 metres south from the mortars were their cobbers from 'D' Company.[12]

That night, the men of the 2/33rd Battalion bivouacked along the main track north of Imita Ridge, while the Battalion Headquarters' personnel and Headquarters Company moved on to Imita Ridge to establish their fallback positions.[13]

* * *

On hearing of the withdrawal, Major General Arthur Allen told Brigadier Kenneth Eather later in the afternoon that he must fight out the battle on the Imita Ridge. Indeed, Lieutenant General Sydney Rowell was concerned on hearing the news and sent a message to Allen: 'Confirm your orders to EATHER. Stress the fact that however many troops the enemy has they must all have walked from Buna. We are now so far back that any further withdrawal is out of [the] question and EATHER must fight it out at all costs. I am playing for time until 16 Inf Bde arrives. In [the] meanwhile and to guard against worst case following are being moved forward, HQ 14 Inf Bde, 39 Bn and 36 Bn .... It is absolutely essential that we retain control of KOITAKI-ILOLO area covering direct route into MORESBY'.[14] Allen himself reported the situation throughout the day:

> 0930 hrs – telephone conversation: I told him [Eather] that he must hold enemy as long as possible and try to keep on the offensive. I left the final decision to him, as to whether to withdraw or not, but impressed on him the importance of holding IORIBAIWA if he could. He expressed his realisation of the importance of holding IORIBAIWA. I told him that it was important for him to keep his force intact, but that, if

CHAPTER 29

he withdrew, he should leave ambushing troops to harass the enemy ....

Approx. 1300 hrs: EATHER stated that he had acted quickly and was in the process of extricating himself .... EATHER stated that PORTER concurred in his views .... PORTER was then on his way back with a recce party ....

1435 hrs: I rang UBERI at 1455 hrs and spoke to Col SPRY, GI 7 Aust Div. I acquainted SPRY with situation and instructed him to go fwd and contact EATHER and get the complete picture. I instructed SPRY to tell EATHER that he must get into posn quickly and strike at the enemy. He (EATHER) must contact enemy, and prevent him, by offensive action, from encircling his posn. He should [protect] the gap at UBERI, preferably by 21 Aust Inf Bde, and fight it out on the IMITA posn. He should keep a reserve as a striking force, preferably a Bn. I told SPRY to impress on EATHER the seriousness of the situation, and the extreme danger of coming further back. SPRY was asked to ring me after he had met EATHER.

1635 hrs: Gen ROWELL rang me. He instructed me to fight the battle out on the IMITA RIDGE. I rptd the instns I had given Col SPRY to convey to EATHER. He expressed approval .... He also stated that if things went wrong fwd, I must hold KOITAKI and ILOLO at all costs. He would, if necessary, send 39 Bn, 36 Bn and HQ Inf Bde up to me.[15]

Allen reportedly informed Eather on hearing of his intent to withdraw: 'There won't be any withdrawal from the Imita position, Ken. You'll die there if necessary. Do you understand'[16]

* * *

On hearing of the withdraw, Rowell held a briefing with his senior officers and was keen to get Brigadier Arnold Potts' advice on the current situation; he was greatly concerned about the abandonment of Ioribaiwa Ridge. Rowell asked Potts about the withdrawal, even though Potts was unaware of the situation facing Brigadier Eather, or the reasons why he was forced to fall back onto Imita Ridge: '"What's your view? Can you suggest any reason for it?" Potts replied: "No tactical reason. Eather wouldn't have been driven back." Rowell asked: "Would you say that the Jap could by-pass him, and attack down the Goldie River?" Potts thought for a few seconds before responding: "That would be unwise, because the Jap would expose a vulnerable line of communication. He was in a better position to do that to me above Menari. If any elements do come through, they'll be confined to weak patrols, and you can deal with those from this end". Rowell asked: "Do you think the Jap has any chance of pushing Eather off the Imita position?" Potts did not hesitate in his response: "None. Unlike Ioribaiwa, there's no room for deployment. There's only one approach, along a narrow track, and the range is precipitous anyhow." "Could the enemy move around the right flank?" persisted Rowell. "He could, but physically, he's not capable of it"'.[17]

All agreed, including Brigade Major Hugh Challen, that the Japanese had made their last assault in the direction of Port Moresby. With this, Brigadier Porter led a reconnaissance party onto Imita Ridge to assist Eather in selecting his defensive positions along the ridge.[18]

* * *

Later that day, the unknown leader of the signals unit with *No. 2 Infantry Section, No. 1 Company, I/144th Regiment* had moved up to occupy the ridgeline. He merely recorded in his diary: 'Occupied IORIBAIWA Ridge'.[19] Meanwhile, 1st Lieutenant Hirano, leading *No. 3 Platoon, No. 1 Company, I/144th Regiment* had made his way across the highest part of the range and was approaching Isurava with the task of bringing back supplies to the forward troops. He recorded in his diary: 'First Battalion continued on. Reached IORA

CHAPTER 29

[*sic* Eora] at 1000 and Isurava at 1500. Drew provisions, but the amount received was surprisingly small. We are to return to the front almost empty-handed'.[20]

Meanwhile, still trying to make his way south from Myola was Private Teruoka Akira *Tanaka Unit, 55th Division Medical Unit*. He had originally been too sick to advance south, but over the last few days, he had been struggling south on the track, with others of his unit. He soon fell out beside the track, but a few hours later, two other privates came back for him and carried his equipment until they caught up with the others of his platoon. There the sick private was given a once-over by a medic and provided with some medicine and allowed to rest. It was not long, however, before he and the others were again on the track heading south to join the rest of their unit.[21]

* * *

The day before, a rescue party had been organised to try and bring in the lost men of the 21st Brigade. While these men lacked maps of the area to be searched, Captain Kienzle, who had established Dump No. 66 just forward of Uberi, provided several aerial photographs that morning to Sergeant Hylton Jarrett, a 31-year-old engineer from Ramsey in Victoria, also with ANGAU, who was to lead the party in search of Lieutenant Colonel Geoffrey Cooper and his lost men of the 2/27th Battalion.[22]

East of the Kokoda Track, Cooper and his large party were still trying to reach the Australian lines. At first light, they moved out from the eastern outskirts of Nauro. The battalion war diary records that the battalion moved up to the high ground that was occupied the previous night and halted while the intelligence section tried to find a track that led around Nauro. One platoon of 'B' Company moved with this section to be used as runners to the CO with information regarding the direction and condition of track. The Battalion Headquarters with one platoon of 'D' Company, and 'B' Company followed the track taken by the intelligence section.[23]

Meanwhile, 24 men from Battalion Headquarters and two platoons of 'D' Company remained on the high ground approximately two kilometres east of Nauro with the stretcher-cases. The CO left the stretchers there in the care of Captain Thomas Gill, a 28-year-old chartered accountant from Largs Bay in South Australia, with a strong party to help defend them. Cooper would push on to bring back help.[24]

Cooper and the remaining men took the track leading down from the high ground to river flats through undergrowth overgrown with creepers, but the going was not too difficult as they followed the Naoro River in a southerly direction. The river crossed the track about five kilometres from Nauro, and the men headed along the course, climbing to a high grassy plateau southeast of the village. Cooper now sent six men forward to get a message through to Port Moresby about their whereabouts. That night, Cooper found more yams and taro in a deserted hut and sent Corporal Edgar Penney, a 24-year-old labourer from McLaren Flat, and Private Clyde Ramsey, a 24-year-old farmer from Cowell, both from South Australia, back to Gill's group the next morning to tell them that yams and taro were being sent to them. They ran into a Japanese patrol, however, and Private Clyde Ramsey was killed. Corporal Penney was then forced to return to Cooper's party not being able to contact Gill's men who were defending the wounded.[25]

* * *

Further north and east of the Kokoda Track was Lieutenant Maurice Treacy and privates Frederick Rockliffe and William Avent, 2/14th Battalion, who were still on the narrow native track heading southeast. The young lieutenant recorded in his report: 'At the end of two days after travelling across a series of dry lakes, it was obvious that the party could not reach the main track by moving S.W., owing to the extremely rough country and the absence of tracks leading in the desired direction. Weakness through lack of food was also a big factor

CHAPTER 29

which influenced the decision to remain on the track which was already being used and it was felt it would eventually lead to a native village. By this time, the hope of getting assistance to Capt. Buckler's party had almost vanished. Once that party moved it would be impossible to trace it, the heavy rains obliterating tracks'.[26] Indeed, Treacy and his men were now heading south-east towards Dorobisolo, about 50 kilometres east of Uberi, as the crow flies, taking them further way from the Australian lines; the only consolation was that it was also distancing them from the Japanese.

Meanwhile to the west, Lieutenant Robert McIlroy, leading his lost patrol of the 2/14th Battalion, had sent four of his men forward to reach the Australian lines. This party was led by Lance Corporal Stewart Gedye, a 25-year-old farmer from Blackburn in Victoria, with privates Douglas Brown, a 29-year-old labourer from East Devonport in Tasmania, and Wilbur Cahill, a 26-year-old linotype operator from Toowoomba in Queensland, both from his platoon. With them was a recent addition to his party, Private Alexander Matschoss, a 34-year-old labourer from Waikerie in South Australia, with the 2/27th Battalion. They were to reach the Australian positions and send back help.[27]

Lieutenant Robeert McIlroy later reported: 'On 16 Sept as the injured man's knee had not responded to treatment, I decided that one man and myself would remain with the two unfit men and the other four men went on with the compass to try and get through to our lines if possible, sending bearers back. At this stage due to the weakening effect of the native food and the rough nature of the country, the party was not strong enough to carry a stretcher'.[28] McIlroy, and his men were helped by some local Papuans who supplied them with yams and taro and drew them water; however, Corporal Gedye and his small party would not be so lucky when encountering the next group of Papuans.[29]

30

'EVERYONE KNOW HIS JOB...'

During the early daylight hours of 17 September, Captain Lindsay Miller and his men of 'C' Company, 2/33rd Battalion were still in their ambush position 40 minutes north of Imita Ridge (900 metres above sea level), representing the rearguard of Maroubra Force's retirement.[1] One of Miller's men, Private Norman Stokes, a 24-year-old clerk from Sydney, recalled: 'No sleeping during the night was the order. We cursed those who couldn't sleep without snoring. An occasional firefly floated and darted weirdly against the blackness. Perhaps they should have been long anxious hours, but we were tired and fatalistic'.[2]

At around 6.30 am, a patrol from Miller's mates of 'D' Company passed through their position heading north, but within minutes had returned with news that the 'enemy were moving down the track'.[3] These men continued to the rear to regroup with the rest of 'D' Company. It was not long before Miller and his men observed advancing elements of Horii's force coming down the track towards their position. The battalion history records that at 6.45 am., around 'fifty Japs, led by a sword-carrying officer, and all chattering and laughing, moved into the kunai. Some carried tins of Aussie food. None of them was on guard for possible surprise. They were all bunched up together in two files'.[4] These were the Japanese of *No. 9 Company, III/144th Regiment*. Captain Miller and his men held their fire until the Japanese were less than 50 metres from their position. Miller felled the officer with a single shot from his rifle at 30 metres, and the ambush was sprung. In his report of the battalion, the young officer recalled:

CHAPTER 30

At 0645 hrs several men appeared on the track and a burst of LMG fire hit 14 Platoon area [rear of company]. Another burst was fired sometime later which cut the trees low over the heads of the Coy. The enemy hesitated and did not advance further. In the trees he [Japanese] 'chattered' considerably, and remarks in English, such as 'They have all gone' were heard by the fwd sections. Later the enemy advanced again, the leading man, who appeared to be an officer or NCO, took the 'detour' notice from the stick, read it, beckoned to his companions, and came on. The Coy then opened fire ....

As the Coy opened fire ... the enemy was greatly disconcerted, and retired in haste to the trees, leaving approx. a dozen casualties visible. An enemy mortar which had been firing to the right of 15 Pl was beginning to find the Coy posn. The Coy comd ordered 15 Pl comd 'out'. At this time 15 and 13 pls noticed enemy flanking movements in the gullies to either side of their posns. The Coy threw a few grenades, but the enemy was allowed to continue here. When 15 Pl withdrew [on the right], the enemy again attempted to advance into the clearing and was again thrown back by 13 Pl [on then left]. 13 Pl was then ordered to withdraw, and as the enemy in the gullies was now level with the rear of the Coy posn, he was a good target for grenades thrown from both sides. This was done, and quite a number of grenades was thrown amongst them. From the mvt of the undergrowth, the strength of the enemy in either gully would be about 20. Coy HQ moved after 13 Pl; 14 Pl [was] to move when comd considered it necessary (the flanking movements had been stopped by grenades). 14 Pl held the enemy until Coy HQ was clear, then withdrew, Pl comd giving the 'open fire' signal to the mortars as he moved out.[5]

Miller's company, by platoons, leapfrogged back to their fallback position at Dump No. 44, with the Japanese not far behind. However, it was now that the Australian mortar team to the rear of the company began to lob their two-inch mortar-bombs into the advancing Japanese, which halted the Japanese advance. As 14 Platoon withdrew, Miller recalled that the Japanese were caught in the open, targeted with exploding mortar-bombs. Miller and his men were able to break contact with the enemy, and the remainder of 'C' Company moved in good order through those of 'D' Company. Sergeant William Crooks, with 'D' Company, recalled they waited 'anxiously peering up the track watching for Jap movement. Visibility was almost zero from D Company's position on a reverse slope in [the] rear of the Kunai Patch. This part of the track traversed a wild, overgrown razorback ridge, so there was little room for deployment'.[6] The men of the 2/33rd Battalion soon fell back, taking up another ambush position halfway between Dump No. 44 and Major Albert Moore's recently evacuated Salvation Army post about one kilometre further south.[7]

The diary of the Australian 7th Division Headquarters records the situation at around 11 am. All battalions, less the men of the 2/33rd Battalion, were in positions along Imita Range. astride the main Kokoda Track. The 2/33 Battalion was approximately halfway between Ioribaiwa and Imita ridges, having inflicted serious casualties against the forward elements of the advancing Japanese. These men would soon take up a position to the right rear on Imita Ridge. The 2/33rd Battalion war diary records on the retirement after their ambush, stating that few Japanese were observed, and no firing was heard.[8]

* * *

Brigadier Kenneth Eather had been seeking a firm footing along Imita Ridge; his men lacked shovels and were forced to dig-in using their bayonets and helmets, at all times screened by offensive patrols, which were about 50 strong. The men of these patrols carried five days rations, at least one Bren gun and seven submachine guns. They were to harass the Japanese and keep them off-balance, particularly along the lines of communication. Eather was now finally getting the breathing space

CHAPTER 30

he needed as the Japanese had not followed his withdrawal to Imita Ridge in any strength.[9] Journalist Geoffrey Hutton had been based in Port Moresby but had moved up the line and was now on Imita Ridge; he wrote of these patrols:

> There is no such thing as static, positional warfare in the jungle, and near the front you are continually meeting fighting patrols, business-like parties of men walking in single file and carrying Tommy guns, rifles and grenades. The frontline was a scattered series of positions running along a mountain stream below Ioribaiwa. The village was unhealthy because it was being persistently shelled by a Japanese mountain gun. Our mortars were steadily replying, setting the valley echoing as the bombs volleyed into the air. Machine guns rattled, and you could occasionally hear the report of a sniping rifle. As the column of troops approached their positions, word was passed down the line to keep quiet. Everyone knew his job, and the troops deployed through the trees and began to dig-in quickly and efficiently, each man scooping out a little hole with as clear a line of vision as possible.[10]

Earlier that morning, Lieutenant Colonel Allan Cameron issued orders that his militiamen were to dig-in along the established defensive line along the ridge, covering the right flank. His men of 'C' Company were located forward, with 'B' Company on the right, Headquarters Company in the centre, and 'A' Company of the left (its left flank resting near the main track), while 'D' Company was in support at the right rear. Soon after, Cameron sent orders to the remaining rear elements instructing them to send forward one section of his medium machinegun company as well as additional mortar teams.[11]

By 11 am, the main deployment on Imita Ridge had been established. Cameron and his men of the 3rd Battalion were covering the right flank, the 2/25th Battalion was in the centre astride the main

Kokoda Track, while the 2/31st Battalion was covering the left on a small knoll. The 2/1st Pioneer Battalion were on the left rear, while the 2/33rd Battalion would fill the right rear position when their withdrawal was complete.[12]

Early that morning, the Lieutenant Colonel Charles Withy commanding the 2/25th Battalion made a reconnaissance of his position along the saddle and track in the centre of the brigade's position on Imita Ridge. He ordered 'B' Company forward with its 12 Platoon in an ambush position at 'Hayden's Lookout', about two kilometres forward from the main ridgeline overlooking the main track. The men of 'A' Company were astride the main saddle of the ridgeline, with 'D' Company on the left and 'C' Company on the right. The Headquarters Company and staff were located to the rear, just in front of Brigade Headquarters. To date, the battalion had suffered ten men killed, with two officers and 34 other ranks wounded and one man missing.[13]

The men of the 2/31st Battalion to the left of the 2/25th Battalion were digging weapons pits. At around 9 am, Lieutenant Alexander Black, a 30-year-old schoolteacher from Rockhampton in Queensland, had returned from a patrol reporting no sign of the enemy. Another patrol was sent south to reconnoitrer the track back to the RAP and Brigade Headquarters to ensure no Japanese infiltration parties were present to their rear. By 1 pm, the men of the 2/1st Pioneer Battalion moved through the battalion's area to take up a position on their left flank. It was now that the battalion's front was extended with 'A' Company moved to 'B' Company's position on the left. The battalion diarist concluded: 'Local patrols reported no enemy activity but noise from own area – talking and digging. By nightfall Bn was well dug-in [in] coordinated positions. Patrols reported impossibility of approach to within 50 yds of our positions (visibility very limited in country encountered to date – 5 yds being therefore much more than average field of vision). A quiet night followed'.[14]

CHAPTER 30

To the south, the men of the 21st Brigade consisted of just 272 men, with most being from the 2/16th Battalion, as the 2/14th Battalion could only supply ten officers and 77 men because most of their men remained missing when swept into Eora Creek Valley with the fighting at Isurava Rest House over two weeks before. This composite battalion was now moving south to Uberi.[15] The 2/14th Battalion war diary records:

> Reveille at 0600 hrs, and at 0630 hrs unit pushed on to objective where hot meal and tea were enjoyed by all. The morning was spent in washing and cleaning. Capt. RHODEN reported in at 1000 hrs and Major HEARMAN with 4 ORs were evacuated at 1400 hrs. Information received that 2/33 Bn had accounted for 50 Japs without casualties to own troops in the rearguard action; and that other Bns were in position. CO was informed that the Bn would remain in present position for the night 17/18 September and ordered that all equipment, weapons, clothes, and personnel be checked and returns handed in. In the afternoon, opportunity was taken to thoroughly clean up and rest. Instructions at 1600 hrs, that the following day, our unit was to occupy high feature North of UBERI … it was decided to commence moving unit following morning at approx. 0900 hrs. At 1800 hrs Lieut. GREENWOOD with 15 ORs reported into area after having been 'bush' since 9 September. They had been [part of a] stretcher party, but when relieved by the natives, had split up to make their way back.[16]

Their mates of the 2/16th Battalion were also on the move, as recorded by their official diarist: '0945 Reached Biv area 500 yds NORTH of UBERI. 1530 Coy Comds Conference. Bn to take up position on feature NW of UBERI tomorrow morning'.[17] Indeed, commanding the men of the composite battalion, Lieutenant Colonel Albert Caro, CO of the 2/16th Battalion, was all too aware of the appalling physical and mental state of his men; later that day, he penned a note to the 21st Brigade Headquarters:

2/14-2/16 Comp Force.
17 Sep 1942

HQ 21 Aust Inf Bde.

1. With respect the following points are submitted for your consideration in relation to the fitness both physical and mental of the troops under my Comd for active ops. In submitting this case it is realised that tactical considerations are largely paramount but yet it is respectfully submitted that the points listed hereunder may outweigh the tactical considerations.

2. The 2/14 and 2/16 Bns having been actively engaged against the enemy since 27 Aug to 16 Sep have been whittled down until the combined strength amounts to 272 all ranks.

3. This prolonged period of action has also greatly dissipated the general health of the remaining men and has deprived most of them of that very necessary reserve of nervous energy.

4. Any reinforcements received have been men who were evacuated early in the campaign and owing to a long difficult L of C have had little or no rest and have returned to their Units in an unfit condition to fight. In this regard it is anticipated that should this Force be committed in the near future many NCOs and men will be left out. Also, many of these reinforcements are men who have been missing in the bush for a period of days and of a consequence are of little operational use.

5. On the 11 Sep 2/14 and 2/16 Bns were combined to form a Composite Force under my Comd. The combined

CHAPTER 30

strength was 307 all ranks. Although it is understood this is purely a temporary arrangement due to the exigencies of the Service, yet it is feared that if our losses from battle, sickness and Allied causes continue at the same rate not only may this temporary arrangement cease but our entire identity may be swamped by other existing Units.

6. In conclusion the combined effect of all these causes makes it my plain duty to the troops under my Comd to submit for your consideration a plea for the early relief of the Combined Unit from active ops to an area where the separate units may recuperate, refit, and reorganise.

7. In support of this appended is the report submitted to me by the RNOs of the 2/14 and 2/16 Bns.

Signed Albert CARO Lt Col
OC, 2/14-2/16 Com Force.[18]

The supporting report below was penned by Captain Donald 'Don' Duffy, a 27-year-old physician from Hawthorn in Victoria and RMO with the 2/14th Battalion, and Captain Henry Steward, RMO with the 2/16th Battalion – both men were veterans of the fighting from the very start:

REPORT FROM RMOs 2/14-2/16 Bns

17 Sep 1942.

CO 2/14-2/16 Aust Inf Bns.

In our opinion the mental and physical condition of the a/n Bns is such that at present they are unfit for active operations.

This is occasioned by the fact that for the past three weeks there has been a constant mental strain associated with the most severe conditions of exposure and appalling living conditions which have considerably undermined the health of the remaining tps.

We consider that they are in urgent need of rest and relief from active field ops.

Signed H D Steward   Capt. 2/16 BN (RMO)

Signed D G Duffy   Capt. 2/14 Bn (RMO).[19]

Porter agreed to this sensible and frank advice by the unit's CO and medical officers, but he would not be able to relieve them for another week. Lieutenant Colonel Albert Caro's welfare for his men was such that by early 1943, he was relieved of command when he and several other battalion commanders complained of the way their men were being slaughtered during the battles for the Buna–Gona beachheads – he would not be the only battalion CO to be dismissed while defending his men from the incompetency of some more senior Australian commanders.[20]

* * *

That day, while on their way back to Port Moresby, Salvationist Major Albert Moore and Corporal Jock Inglis searched out their mates of the 2/14th Battalion, who they knew to be located somewhere close by:

> Just as we were leaving Uberi I contacted some of the 2/14th Battalion, also some of the 2/16th. I spent some time with them and again the old 2/14th came to my rescue. From Ray Ross, of Broadford and Syria days, I received two dixies to make up for my loss. The journey back to the road head was a nightmare. The record from my diary runs thus: 'Thursday, September 17, 1942. Departed for Owers' Corner quite late and ploughed through indescribable conditions of mud .... Set off through the centre to try and save time but had to give up as I went down well over my knees in mud and just could not pull my legs out so had to try and climb along the edges and wherever I could get a foothold'. We spent the night in a signals tent. I took off my socks, wrung the mud out of them and hung them

CHAPTER 30

on the tent. In the morning I pulled them on again, wet, cold and muddy. I had no socks at that stage for the men, so there was none for me.[21]

* * *

During the early morning hours, Captain 'Doc' Vernon, in charge of the health of the Papuan carriers along the Kokoda Track, had finally reached Dump No. 66, just south of the Golden Stairs. The 60-year-old veteran had been up and down the track several times already having suffered dysentery and malaria, but he was up early to look after the welfare of his men. He recorded in his report to ANGAU for 17 September: 'Spent at Camp 66. Our retreat ended here. Our native staff had behaved very well throughout the operations and 90 per cent of our stores had been saved. On the 17th I returned within a mile or so of Camp 44 carrying dressings to cover large carrier parties sent forward to rescue stores'.[22] Again, he was more expansive in his personal diary, including a discussion he had with Brigadier Eather:

> I returned along the same track to cover large parties of carriers who had been sent forward to retrieve stores from Camp 44. When these men got close to the dump, Jap scouts were already exchanging shots with our rearguard and the carriers were hustled back out of the way. On the way home I called to the Brigadier's shelter (Gen. Eather), merely a strip of canvas stretched over a boulder in a sea of mud and asked. 'How do you think we ought to reward the carriers after the war?' The question caught me unawares; the end of the war on that particular day seemed so remote that all I could think of saying in reply was, 'Give them all the food and smokes they would wish for now and leave the final reward [for later]'. However, the question got me thinking, and while there are much larger considerations at stake, I think our carriers should share in the distribution of a New Guinea war medal if one is to be given to the white troops.

A further point was the consideration shown by the Brigadier in the welfare of the boys at a time when he must have been deeply involved in military matters. As a matter of fact, our carriers had shown up remarkably well during the whole retreat and the command was pleased with their work.[23]

* * *

Finally, also about to be relieved were the medical personnel of the militia 14th Australian Field Ambulance and the 2/6th Australian Field Ambulance, who had manned several aid posts along the track. That day, they received word that the medical personnel of the 2/4th Australian Field Ambulance, 7th Division, commanded by Lieutenant Colonel Arthur Hobson, a 35-year-old physician from Sydney, would soon take over the forward duties of all aid posts along the track. These men of the 2nd AIF had served in the Middle East between 1940 and 1942. Now the exhausted and often sick medics of the 14th and 2/6th field ambulance who had been serving for months along the track could finally think about retiring to Port Moresby. However, the men of the 2/4th Field Ambulance would not begin their advance into the Owen Stanleys until the first days of October.[24]

* * *

Still heading north leading his party of 47 men was Captain Sydney Buckler. Along the way, he managed to recruit three Papuan carriers to help carry the stretcher-cases. As they travelled, his men were able to shoot a wild pig, and it was quickly dismembered and stewed. The captain recorded in his report: 'Nothing was wasted ... the three carrier divided up the entrails'.[25] Even the hide was carried for two days, scorched and eaten. Soon after, a nearby village provided sweet potato, sugar cane and bananas, and the Australians troops regained some of their strength. But these Papuans refused to help with the stretchers.[26]

31

'THE SEA! LOOK! IT'S THE SEA OF PORT MORESBY!'

The 25th Brigade war diary records the situation as Brigadier Kenneth Eather assumed full command of Maroubra Force on 17 September. His force now consisted of the 2/25th, 2/31st, 2/33rd, and the 3rd battalions, along with the 2/1st Pioneer Battalion, elements of the 2/14th and 2/16th battalions, and 'C' Patrol, 2/6th Independent Company. The bulk of this force was now distributed along Imita Ridge with the 2/25th Battalion forward and astride the main track on top of the ridge with its right flank covered by the men of the 3rd Battalion. To the left of the track was the 2/31st Battalion, who were supported by the men of the 2/1st Pioneer Battalion positioned just behind them. The 2/33rd Battalion was still located just forward of Ioribaiwa Ridge as a blocking force. The surviving elements of the 2/14th and 2/16th battalions were at Uberi as a reserve force. In addition, two 25-pounder field guns, from 'B' troop, 53rd Battery, 14th Field Regiment, were being manually dragged to Owers' Corner, to provide much welcomed support to the defence of the ridge. With the artillery forward observer was Private Samuel Kenny, a 30-year-old stockman from Mackay in Queensland, with 'B' Company, 2/25th Battalion, who recalled the selection of the site for these two guns: 'I was with the party that accompanied the forward observation officer [who] was searching for a flat area which was clear of foliage and from where the shell could clear Imita Ridge. When he finally selected a suitable plot of ground, the FOO used my bayonet to mark the exact spot for the gun'.[1]

SAVING PORT MORESBY

Two three-inch mortars were in the sector held by the 2/25th Battalion with an additional mortar team allocated to the 3rd Battalion. Two medium machine guns were also dug-in to support the centre of the line held by the 2/25th Battalion, and one additional medium machine gun was in the sector manned by the 2/31st Battalion and another supporting the 3rd Battalion. Five-days emergency rations were distributed among the men with another five-days of reserve rations held in each company sector. Orders were issued that these men were to hold Imita Ridge and prevent any infiltration by the Japanese – this was the end of the line – there would be no going back from here.[2]

* * *

The Japanese troops of the *144th Regiment* were anxious to gain the top of Ioribaiwa Ridge; each man struggled up its slopes and through its thick scrub and jungle – to gain the heights not only meant victory, but more importantly, food. The Australians, however, made sure there was precious little if anything left to aid the enemy in this regard. Food that was left behind was made useless and often purposely contaminated, and those that ate the tainted food were put out of action by disabling gastric ailments such as amoebic dysentery and severe diarrhea. Now entrenching on Ioribaiwa Ridge was a disappointed and hungry Lieutenant Hayashi with Horii's staff, who recorded that they had suffered substantial casualties in taking the ridge: 'Tremendous litter of the dead ... the stench ... is almost unendurable'.[3]

First Lieutenant Horibe, acting commander of *No. 6 Company, II/144th Regiment*, was also on the ridge, likely just east of Spotter's Hut. He recorded succinctly in his diary: 'Constructed positions – todays strength 50 men'.[4] His battalion commander, Major Horie Tadashi, told him to start digging air raid trenches and erect tents, as well as evacuate his wounded. The company now had a strength of just 50 men, about a third of its nominated strength. The young lieutenant

CHAPTER 31

was forced to reconfigure his company into just two platoons each to be led by two NCOs: Sorimoto and Fukumoto.[5]

Meanwhile, Japanese war correspondent Okada Seizo was finally nearing the end of the Kokoda Track. That morning, he made his way onto Ioribaiwa Ridge and recorded in his unpublished manuscript:

> We covered about four-fifths of the distance across the range in twenty-three days, and at daybreak on September 17 reached the top of Mt. Efogybia [*sic* Ioribaiwa]. It was the last of the series of high mountains that blocked our way towards the southern coast. When we got to the rocky summit, which had been taken by the combined attack of the whole detachment, the endless waves of mountains upon mountains that had wearied our eyes had suddenly vanished, and we saw through the trees a wide expanse of green wood gradually sloping away before us, and beyond that a sheet of misty light.
>
> 'The Sea! Look! It's the sea of Port Moresby!'
>
> Wild with joy, the soldiers who were stained all over with mud and blood, threw themselves into each other's arms and wept …. But when we reached Mt. Efogybia [*sic* Ioribaiwa], we had lost nearly 89 per cent of our men killed or wounded in action or disabled by illness. Besides, we had almost run out of food and ammunition. Our supply line by way of Buna had been particularly cut off by enemy submarines and airplanes. We were indeed in a hopeless position. The only thing that kept up the morale was the thought of Port Moresby.[6]

Also on the ridge, and somewhere close by, was Lieutenant Nakahashi Kokichi, with the *55th Mountain Artillery*. He recalled the closeness of their objective: 'Gazing out from the summit in a southerly direction, there was not even one mountain to obstruct our range of vision. A dense, overgrown sea of forest, rising and falling like ripples on the water and, far off one was able to see – as the sun's rays came through a break in the clouds, a glitter and a sparkle – without doubt it was the sea! Over there was Moresby the object of our invasion, which had become an obsession. Officers and men alike embraced one another, overcome by emotion. During the battles so far, the South Seas Force had, to date, suffered 1000 dead and wounded. The line of captured positions more than atones for their blood, even if they will never be able to gaze upon this splendid spectacle .... At night we are able to see the lights of Port Moresby, and the beam of the searchlight of Seven Mile Airfield on the outskirts'.[7] Indeed, the veteran of the campaign, Staff Officer Tanaka Kengoro, records that night in his history of Japanese operations in Papua: 'From these highlands, they could get a distant view of the lights on the Gulf of Papua and in Port Moresby'.[8]

Close by would have been 2nd Lieutenant Onogawa and his men of the *No. 2 Platoon, No. 3 Company, I/144th Regiment* who recorded in his diary that the general offensive was to begin on 20 September. He was determined that he and his men would be part of the advance, having spent much of their time at Myola, and missing out on most of the action since Isurava.[9] Heading south from Isurava with their meagre supplies for the frontline troops was 1st Lieutenant Hirano, *No. 3 Platoon, No. 1 Company, I/144th Regiment*, who recorded in his diary that day: 'With renewed spirits, reached the foot of Stanleys from ISURAVA in one push. Camped there at 1400'.[10]

Somewhere on the Kokoda Track was Private Teruoka Akira, with the *Tanaka Unit, 55th Division Medical Unit*. He and the men in his unit were becoming increasingly sick and weak as they struggled south. He recorded early that morning that five men from each sub-unit went out in search of food, but all 'came home disgusted'.[11]

* * *

That afternoon, Captain Sims, with 'A' Company, 2/27th Battalion and his lost party came across Lieutenant Peter Gorrie, a 26-year-old monitor from Swanbourne in Western Australia, who was

CHAPTER 31

the intelligence officer of the 2/16th Battalion. They had finally reached the Australian lines. Sims later recalled their journey: 'In the late afternoon we went up a steep spur and bivouacked opposite a waterfall where we had a shower. I had pneumonia and tropical ulcers. By the time I got back I had dengue too. Every man in the company lost 2 to 3 stone'.[12] Gorrie sent a message to divisional headquarters advising them that he had found some of the men from the 2/27th Battalion and had been provided information regarding Cooper and his men. Within days, the other parties of Captain Sims' group, which had divided into smaller parties on 10 September, also arrived in the base area.

Meanwhile, Geoffrey Cooper and a small advance party were heading along a track, and strung out well behind them were the rest of his men. With the battalion CO was Captain Harry Katekar, who recalled: 'We set out early and when our hopes seemed lowest, at about 1200 hrs a ray of hope beamed, for there at a point where 2 tracks met was a stick on which a piece of paper was fastened saying "92 this way" – that meant that Brigade HQ was along that track, somewhere. An hour later, we got the thrill of our lives when we suddenly met a patrol of the 2/14th Bn at Nigabaifa. We were saved; we could get food at a camp 3 hours away'.[13]

Within hours, they had made their way to Jawerere, where Major Alfred Robinson, commanding Jawforce, was based. Soon the remainder of the battalion were coming in. Robinson and his men provided Cooper and his men with available food and made them as comfortable as possible.[14] Among these men was Corporal Clive Edwards who recalled years later: 'I can't possibly describe fully the hopes and fears, achievements and disappointments, the sheer determination and will to survive which was all that kept us going during some of the harder stages …. Each night I used to think of the mob at home and pray for them, and myself'.[15]

* * *

SAVING PORT MORESBY

In Port Moresby, Australian journalist George Johnston reported: 'The Japs are down to Ioribaiwa Ridge, only 32 miles [50 kilometres] from Port Moresby. If they can cross two more ridges, they are clear of the southern foothills of the Owen Stanleys and on the motor road that leads to the garrison. When the wind is in the right direction you can faintly hear the rumble of battle now from Moresby itself'.[16]

* * *

Back in Australia, Blamey presented his review of the operations in Papua New Guinea before the Advisory War Council, stating that the strength of the Allied forces in New Guinea was approximately 30,000. He estimated that the Japanese had up to 10,000 men and that most of these troops were in the forward area, consisting of two regiments with a combined strength of around 6000, outnumbering Eather's force. However, the Japanese facing the Australians probably numbered fewer than 2000, considering ongoing casualties and sickness. Blamey outlined that in addition to the 25th Brigade, which had recently arrived in Port Moresby, the 16th Brigade was at sea, one squadron of light tanks would soon arrive with another squadron scheduled to follow, and that three field gun regiments, a mountain battery and one pack horse transport unit were also being sent to Lieutenant General Sydney Rowell. In addition to this, a US regiment, numbering around 3500, would soon arrive in Port Moresby in 'accordance with the desire of the Commander-in-Chief S.W.P.A. that American troops should obtain experience in operations and in the development of supply arrangements in this area'.[17]

Blamey concluded by saying 'Lieut. General Rowell, Major General Allen and the troops are confident that the Japanese will not be able to take Port Moresby from the land' and that he shared their confidence.[18] In a broadcast to the Australian public, he also stated he had full confidence in Rowell, his officers and troops: 'Every man is confident that the Japanese will never take Port Moresby. There is the very highest degree of optimism among all ranks …. I found the

CHAPTER 31

gloomy view of the Japanese advance over the Owen Stanleys, which is prevalent on the mainland, peters out as one moves forward'.[19]

As such, it must have been a shock to Prime Minister John Curtin when later that day General MacArthur spoke to him by telephone from Brisbane, informing Curtain that he was worried about the situation in New Guinea. MacArthur said that he considered the reason for the unsatisfactory situation was due to the lack of efficiency of the Australian troops; he argued that the Australians were in superior numbers to the Japanese in the Owen Stanleys, but despite that, they were still withdrawing. He believed the Japanese must be having similar difficulties to the Australians, but they were not withdrawing, they were advancing. MacArthur felt that, if the Japanese advance continued, the Allies in New Guinea would be forced into a defensive posture with no offensive operations being launched against the Japanese. The invaders, he said, had not pushed across the Owen Stanleys with a significant force, and this was causing him serious concerns.[20]

The American warlord was at pains to point out that he did not share the views of the Australian High Command, who were confident of their ability to now push the Japanese back across the mountains. MacArthur argued he was proposing to send American troops to the area by air or sea to stem the Japanese onslaught in Papua, and within a week, he expected to have 40,000 men in New Guinea. If these men were prepared to seriously take on the Japanese all would be well; however, if not, having 100,000 men would make no difference to the outcome – a Japanese victory.

MacArthur also informed Curtin that he had been assured of receiving substantial air reinforcements soon, but the commitments of the British and American fleets had left him without adequate naval support. He stated his hands were tied due to the American position in the Solomons – the fighting on Guadalcanal – which he thought was not favourable. Overall, he 'considered, his problem was reduced to one of fending off the Japanese for some months, and the fight to this end must be made in New Guinea'.[21] MacArthur was wrong on all counts. Brigadier Potts' fighting withdrawal across the Owen Stanleys had fended off any serious threat to Port Moresby by the Japanese from the Kokoda Track, while Major General Cyril Clowes and his men of Milne Force had comprehensively defeated the Japanese invasion of Milne Bay; while the US Marines on Guadalcanal would soon dominate the island.[22]

MacArthur's immediate concern was that Blamey should go back at once to Port Moresby to personally take command of the forces there to 'energise the situation'.[23] MacArthur said he would speak to Blamey, although he had no authority to direct him; he asked Curtin to speak to Blamey on his behalf. Curtin agreed and said he would tell Blamey that he considered that he should go to New Guinea and take command there. Indeed, MacArthur and Curtin had earlier made a packed in which MacArthur had stated: 'You take care of the rear, and I will handle the front'.[24]

Curtin affectively signed off enabling MacArthur to do whatever he wanted when it came to handling the Australian High Command and Australian troops. While Blamey's earlier address to the Advisory War Council made no hint of any lack of confidence in Rowell, he now readily agreed with Curtin and MacArthur and prepared to go to New Guinea to take charge of the situation. Ever the opportunist, rather than supporting Rowell and his commanders in the field, Blamey now prepared to take command of NGF when the Japanese advance had essentially come to a halt, and Australian offensive operations were about to commence.[25]

* * *

Also back in Australia was Damien Parer – his ground-breaking file – *Kokoda Front Line* –was released that day. In the opening to this film, the war photographer provides a sobering introduction to the fighting along the Kokoda Track, and a necessary wake up call to the Australian public of the realities of war:

CHAPTER 31

Eight days ago, I was with our advance troops in the jungle facing the Japs at Kokoda. It's an uncanny sort of war, you never see a Jap even if he's only 20 yards away; they're complete masters of camouflage and deception. I should say about 40 per cent of our boys wounded in these engagements haven't seen a Japanese soldier, a live one anyway. Don't underestimate the Jap, he's [a] highly trained soldier, well-disciplined and brave, and although he's had some success up to the present, he's now got up against him some of the finest and toughest troops in the world – troops with a spirit among them that makes you intensely proud to be an Australian. I saw militiamen over there fighting under extremely difficult conditions alongside the AIF and they acquitted themselves magnificently. When I returned to Moresby, I was full of beans. It was the spirit of the troops and the knowledge that General Rowell was on the job and now we had a really fine command.

But when I came back to the mainland, what a difference. I heard girls talking about dances, men complaining about the tobacco they didn't get. Up the front they were smoking tea some of the time. There seems to be an air of unreality, as though the war were a million miles away. It isn't. It's just outside our door now. I've seen the war and I know what your husbands, sweethearts and brothers are going through. If only everybody in Australia could realise this country is in peril, that the Japanese are a well-equipped and dangerous enemy, we might forget about the trivial things and go ahead with the job of licking him.[26]

Kokoda Front Line would go on to win Australia's first Academy Award in 1943. Meanwhile, the apathy of the Australian Government saw Damien Parer snapped up by the Americans. He would be killed two years to the day of the release of his film – 17 September 1944 – while filming the US Marines fighting on the small Pacific Island of Peleliu.[27]

32

'RETURNED EMPTY-HANDED'

On 18 September, still forward of Imita Ridge, the 2/33rd Battalion sent out a patrol from 'A' Company to Dump No. 44. The rest of the Battalion had been expected to fall back to Imita Ridge, but they got word that they were to hold their position until they were attacked or flanked by the Japanese. With this, 'B' Company was positioned to the font overlooking Dump No. 44, while 'A', 'D', and 'C' companies were located further back at the former Salvation Army Red Shield post. Now another patrol, led by Lieutenant William Innes, a 36-year-old pharmacist from Narromine in New South Wales, with 'C' Company, was sent out to contact the Japanese.[1]

By noon, the battalion 2i/c arrived from Imita Ridge with a resupply party. By 5 pm, Innes returned having seen no enemy troops, but he reported coming across a patrol early that day led by Captain Robert Dodd, a 29-year-old theatre manager from Toowoomba in Queensland, with the 2/25th Battalion. That night, nine 'emaciated, hungry and exhausted' men led by Lance Corporal Mannion from Captain Charles Sims' 'A' Company, 2/27th Battalion, stumbled into the perimeter of the 2/33rd Battalion having spent nine days lost in the jungle.[2] Sergeant William Crooks recalled: 'All were taken to the CO, who had a bully beef stew cooked for them, and sent them back with a guide, very much heartier in spirit'.[3]

By 7 pm, revised orders arrived from Lieutenant Colonel Alfred Buttrose stating that the battalion less 'C' Company was to now fall back at first light next morning onto Imita Ridge. They would occupy a position on the right rear supporting the militiamen of

CHAPTER 32

the 3rd Battalion. Captain Miller and his men of 'C' Company were to remain overlooking the former Salvation Army station to cover the approach to the ridge. They would remain in this position until 23 September when they were finally relieved by the men of 'A' Company. The move of the battalion would be covered by a patrol of 14 Platoon led by Lieutenant Innes, who was now returning to the battalion having been relieved by Captain Dodd and his patrol from the 2/25th Battalion.[4]

Indeed, earlier that day, the first of the strong, long-ranging patrols was led by Captain Dodd, commanding 50 men from the Headquarters Company, along with men from 'C' and 'D' companies. They had moved through the forward position held by the men of the 2/33rd Battalion and were tasked with harassing any Japanese they encountered between Imita and Ioribaiwa ridges. Dodd recalled leaving Imita Ridge at 10.45 am and quickly pushing forward reaching the former rear Brigade Headquarters' area at 'Grassy Plateau' by 5 pm, located about 800 metres north of the Kunai Patch (and two kilometres north of Dump No. 44). No Japanese were seen by the forward scouts, but there was 'plenty of sounds, mainly chopping and voices, indicating their presence in our immediate front'.[5] Dodd's men took up an ambush position forward of the Grassy Plateau and rested until midnight.

* * *

On Imita Ridge, it was now that a Forward Observation Officer (FOO) with the Australian artillery from Owers' Corner arrived at the Battalion Headquarters to help coordinate their bombardment against the enemy entrenching along Ioribaiwa Ridge.[6]

To the left of the 2/25th Battalion, the men of the 2/31st had been ordered to stand-to just before 5 am with the men continuing to deepen, strengthen and camouflage their weapons pits. A five-day 50-man patrol was sent out, led by Lieutenant Robin Upcher, a 35-year-old orchardist from Drover in Tasmania. These men were armed with Thompson submachine guns, Lee-Enfield rifles, grenades, and dehydrated rations and were tasked with harassing enemy communications and to attack any opportunity targets in the Ioribaiwa sector, west of the main track. Meanwhile, other small patrols were sent out and returned later that day with nothing to report.[7] By now, a kitchen was operating on the reverse slope of the ridge, providing welcomed hot meals to the men; the battalion recorded a 'quiet night'.[8]

Covering the left flank on the ridge, the CO of the 3rd Battalion, Lieutenant Colonel Allan Cameron, reported that by 12.30 pm his mortar and medium machinegun teams from the rear echelons were now established along Imita Ridge in support of the battalion. At this point, the battalion strength was 22 officers and 352 other ranks. At 6 pm, Cameron ordered 16 Platoon, 'D' Company to support the men of 'C' Company who were in the left forward position of the battalion.[9]

Just south of the ridge, the 21st Brigade Headquarters had by 10 am moved south to Uberi. The men of the composite battalion moved out from the village to take up a supporting position for the men of the 25th Brigade. Among the first to move out was the battalion's intelligence officer, Lieutenant Stan Bisset, and his section. The diarist of the 2/14th Battalion recorded that rations for two days, plus two emergency rations were issued per man, and at 8.15 am, Bisset and his men, acting as an advance party, arrived at a native hut on top of a forward ride at 10.15 am. The Battalion Headquarters was established, and a reconnaissance patrol was made of the positions to be occupied by the battalion. 'A' and 'B' companies of the 2/16th Battalion and 'D' Company of the 2/14th Battalion were placed in position and later – after Brigadier Porter and Lieutenant Colonel Caro had moved through to a forward observation post, about three-quarters of an hour north – the remaining men were placed in position. It was decided by the battalion CO that only picquets were necessary and that the men could take the opportunity to rest. In the meantime, 'information

CHAPTER 32

was received that forward units were adopting an extensive patrolling programme, but so far little contact had been made with enemy'.[10]

At this point, the 2/14th Battalion's strength was reduced to just ten officers and 77 other ranks, barely representing a single company; indeed, the combined strength of all four rifle companies of the battalion was just two officers and 30 men.[11] The men of the 2/16th were marginally better-off; the war diary reports that by 9 am they were on the move to cover the high ground just north-west of Uberi: 'Coys move to new position, very steep climb. 1200 coys report being in position. 1400 BHQ moved to new area. 1600 Brig Porter visits area'.[12]

The 25th Brigade war diary records from midday:

> Bde Comd's Policy – An aggressive deep patrolling policy to be adopted under arrangements of Bde HQ. Strength each patrol – 50 men, lightly eqptd and carrying 5 days rations, one Bren gun and minimum of 7 TSMGs, grenades. Will move under verbal orders with object of locating and harassing the enemy and his comms. Patrols to remain out till object achieved or for 5 days. Orders issued by Bde HQ. Patrol (DODDS Patrol) moved out astride main track 1000 hrs .... One patrol from 2/31 Bn to move out left of main track at 1400 hrs. 2/33 and 1/3 Bns each to be prepared to move one patrol out on right flank. C Patrol 6 Independent Coy moved out 0900 hrs 18 Sep NE along IMITA RIDGE en route IORIBAIWA. Situation quiet – no contact. Units strengthening their posns IMITA RIDGE. Patrolling activities from all units. Spasmodic LMG and mortar fire from direction IORIBAIWA during morning. DODD's Patrol reported struck enemy opposition IORIBAIWA and attacked JAPS. No firing during night.[13]

* * *

Still located between Uberi and Imita Ridge at Dump No. 66 was Captain Burt Kienzle and his carriers from ANGAU; he recorded that day: 'We established a camp, shelters, latrines, hospital in 24 hours and in addition continued with [the] normal routine of carrying supplies to forward troops. I also established a forward camp rear top of Imita Ridge with WO Davies in charge of 60 carriers. They were to carry fwd to units and bring back any wounded'. He also reported the next day that five 'exhausted' carriers from Warranted Officer Ronald Preece's lost party arrived at Dump No. 66.[14]

* * *

Further south, having spent the night in a signallers tent, somewhere close to Owers' Corner. Salvationist Major Albert Moore was up early and put on his wet, cold and muddy socks. He recalled: 'I decided to leave Jock to return to the foot of [Imita] ridge and set up there, and Austin and I pushed [south] to secure socks and other things for the men as they came out from the front. We could not secure transport so had to walk to Ilolo, about 9 miles [15 kilometres]. On arrival at Major Groves area, I found Woodland and Geddes there and they, with Austin, set off into Moresby to meet my needs. I again fell into the creek to wash off the mud with which I had been caked for days. The following day, Geddes returned with socks and other comforts'.[15]

* * *

On the Japanese right flank of Ioribaiwa Ridge was Lieutenant Horibe, now acting commander of *No. 6 Company, II/144th Regiment*. He had assembled his two platoon commanders and inspected their position along the ridge. His orders from his battalion CO were to send out scouts to seek information on Australian dispositions as well as to forage for food. Those remaining in camp were to strengthen the defensive positions, erect cooking facilities, and deepen the air raid trenches. They were told not to cut down any trees near their positions, as this would merely help the enemy in identifying their positions along the ridge.[16]

CHAPTER 32

Close by was Lieutenant Sakamoto Atsushi, commanding *No. 2 Machinegun Company, II/144th Regiment*, who also recorded in his diary the need to find food and obtain rations: 'How will we live in our present position without food. Inspected tents and noticed scarcity of food'. He went to the company headquarters to report the situation, but they had no answers and the officer returned, scribbling: 'Returned empty-handed'. Soon after he recorded: 'Entire company turned out to find food'. He also judged who was ultimately responsible for the impossible turn of events, writing: 'Wonder what General HQ are doing. Patients will die and we will soon starve. How can we fight against this?'[17]

Still heading south was 1st Lieutenant Hirano, leading *No. 3 Platoon, No. 1 Company I/144th Regiment* and his small carrying party. The Japanese officer was in a hurry to rejoin his battalion in their advance to Port Moresby. He recorded in his diary: 'Crossed Stanleys and reached KAGI at 0700'.[18] Essentially, his mission to bring back supplies had come to nothing, as they were virtually consuming the small amount in their attempt to reach their comrades at the frontline.

Still on the Kokoda Track was Private Teruoka Akira, with *Tanaka Unit, 55th Division Medical Unit*, along with several other sick and exhausted men of his unit. They had finally made their way to their medical aid post at Menari. They woke early that morning and were told that they were to help assist with the evacuation of the wounded. Even so, some men went out in search of food, and this time they were luckier than the previous day, as several men returned with a large quantity of small, sweet potatoes. Given the increasing presence of enemy aircraft, however, they had to be careful about fires, and strict orders were issued – no fires at night.[19]

* * *

That day, further south, Lieutenant Colonel Geoffrey Cooper, the CO of the 2/27th Battalion, sent a message to Captain Thomas Gill, who was with the stretcher-cases, that he was back in the Australian lines south and east of Nauro. Gill was ordered to move the stretchers to a nearby garden to be picked up later, while he and the rest of his men were to rejoin Cooper's party. Gills would not receive the order until the next day.

* * *

Meanwhile, Lance Corporal Stewart Gedye, along with privates Douglas Brown, Wilbur Cahill and Alexander Matschoss – who had been sent out in advance of McIlroy's lost patrol of the 2/14thn Battalion – were attacked by local Papuans carrying spears. Cahill and Matschoss were killed, while Brown was wounded with a spear in the back. Brown and Gedye were tracked by these Papuans for two days before the locals finally gave up the pursuit. It was not until 3 October that Lance Corporal Stewart Gedye and the wounded Private Douglas Brown stumbled into the Australian lines.[20]

33

'GOOD MORNING, HONNER, YOU'VE JUST ARRIVED FROM AUSTRALIA, HAVE YOU?'

Just after midnight, on the 19 September, Captain Dodd and his 50-man patrol from the 2/25th Battalion advanced from their ambush position towards Ioribaiwa village in the moonlight. Dodd recalled as they approached their objective:

> Progress was restricted to this path, which became steeper as the objective was approached. Approx. 100 yds from A Coy's former rear HQ posn an enemy post was located on a posn overlooking the steps. As it commanded the narrow walk, an attempt to stalk the sentries was made, but the fwd troops could only get within 4 yds. 10 men were located in this posn by voices, coughs, and shuffling. There were probably many more. Voices were also heard in the gully on the immediate left about 20 yds away, and further movement 100 yds ahead could be heard from time to time. This, it was presumed, was a HQ – possibly a Coy HQ – in the same posn. As A Coy formerly had their HQ coughing in the same vicinity indicated that troops were in posn on the high ground in the rear.
>
> Since it was pitch dark by this time, it was decided to bring up one section to the fwd scouts; wait until first light, rush this post, then lead the remainder of the patrol to the posn in the rear. Unfortunately, as the men of the section were getting into

posn., a steel hat was dropped. When the sentries started to give the alarm, the post was rushed with grenades, TSMGs and bayonets. A number of casualties were inflicted.

Orders were heard being issued, and it being pitch dark, it was decided to withdraw to an ambushing posn., forward of GRASSY PLATFORM. At 0430 hrs the men were in posn.

At 0700 hrs movement by our troops gave our posn away to his fwd scouts, and the ambush did not succeed. His fire was engaged, but a flanking movement continuing, it was decided to withdraw to a previous recce posn where a further ambush could be affected. We withdrew and took up the new posn at about 0800 hrs., allowing the Jap, to take GRASSY PLATFORM. Our scouts reported that he did not come further down the track. We waited until 1030 hrs., and again pressed fwd, hoping to surprise him in his new posn. Partial surprise was obtained by the sections up the path, and some casualties inflicted.

A section sent on a flank did not progress far, owing to difficulties of ground; the other section was stopped by superior fire power. 2 LMGs on the path prevented any further advance. His strength was about 1 Pl. which, at this stage, was equal to our own, owing to 15 of HQ Coy being sent back because of fatigue and diarrhoea.

After pressing the attack for 45 minutes, the patrol was taken back to a posn 200 yds back, where once again they adopted an ambushing posn. By this time, all the patrol was fatigued, and it was decided to let all rest until that night or the following morning.[1]

* * *

Japanese Lieutenant Horibe, now the acting commander of *No. 6 Company, II/144th Regiment*, had been warned of Dodd's patrol and

CHAPTER 33

as a result strengthened his guard that night and sent out a five-man scouting party. His men returned hours later stating they could not locate any enemy troops.[2]

The unknown leader of the signals unit with *No. 2 Infantry Section, No. 1 Company, I/144th Regiment* was withdrawing and merely recorded in his diary that day: 'Changed location of communication post'.[3] Meanwhile, 1st Lieutenant Hirano, leading *No. 3 Platoon, No. 1 Company, I/144th Regiment*, was still heading south and was shattered to hear that the advance had halted: 'Met 1st Lt. NOSE and learned that the general attack has been postponed. We were all disappointed'.[4]

Private Teruoka Akira, *Tanaka Unit, 55th Division Medical Unit*, had been tasked that night with guard duties around the aid post. Soon after, he was also assigned at dawn to assist with the evacuation of the wounded. Later, he went down to a nearby creek to wash his clothes and on returning he was assigned to assist the aircraft spotter. While he was utterly exhausted, he felt he was now starting to regain some strength, writing in his diary: 'Perform satisfactory duty day and night after my long illness'.[5]

* * *

For the remainder of Maroubra Force, it was a quiet day. At 3.15 am, the 3rd Battalion reported hearing 11 explosions and small arms fire from the vicinity of Ioribaiwa – obviously Dodd's attack. At 6.15 pm, additional explosions and small arms fire was heard from the same direction. During these early morning hours, movement was also heard to the front of 'D' Company's position covering the right flank of the battalion; patrols were sent out just after 6 am to investigate. At 5.30 pm, Lieutenant Colonel Allan Cameron ordered the remainder of 'D' Company to advance to strengthen the forward position of the battalion. One platoon from 'B' Company moved towards the old 'D' Company position with a standing patrol of one NCO and five other ranks, situated 500 metres east from 'C' Company. Just after 10 pm, the battalion war diary records enemy planes passing overhead, probably heading south for Port Moresby.[6]

Meanwhile, the 2/33rd Battalion had made its way to Imita Ridge from the forward position near Dump No. 44. They were to take up a position on the rear right flank, behind the militiamen of the 3rd Battalion. As ordered the night before, the men of 'C' Company would remain overlooking the former Salvation Army aid post to act as the rearguard. Before noon, the rest of the battalion was met by Brigade Major Thomas Cotton, a 35-year-old insurance agent from Suffolk in England, who directed them to their positions behind the militiamen. It was now for the first time in well over a week that the men were served a hot meal of bully beef stew. It is unlikely, however, that 31-year-old Captain Cyrus Peach from Sydney with the Headquarters Company or his men were able to get much of a feed, as by 2 pm, the captain was leading a 50-man fighting patrol out to the east to harass any Japanese troops in the Jawerere–Nauro track area.[7]

Covering the left flank, the 2/31st Battalion recorded the welcomed arrival of an additional mortar detachments: '1230 hrs: Patrols from C and D Coy reported nil enemy – nil smoke but noticeable noise from IMITA RIDGE. Two 3" Mortar positions established at natural opening in the almost continuous line of trees, vines and other jungle growth. Lack of shovels hindering progress of weapon pits. These were dug with bayonets and steel helmets'.[8]

Brigadier Kenneth Eather and his men of the 25th Brigade, with supporting elements, were now concentrated with a strength of 132 officers and 2492 other ranks. This excluded the headquarters of the 21st Brigade and the composite battalion, which were near Uberi in support. Eather's frontline force now consisted of:

	Officers	**Other Ranks**
HQ 25th Brigade	8	70
2/25th Battalion	27	520
2/31st Battalion	26	502

CHAPTER 33

	Officers	Other Ranks
2/33rd Battalion	28	490
2/1st Pioneer Battalion	20	540
3rd Battalion	22	347
2/6 Independent Company 'C' Patrol	1	23

* * *

The diarist of the 2/14thn Battalion recorded little fighting took place in the sectors to their front, which was held by the 25th Brigade group: 'Except for one small patrol engagement with enemy in which there were a few casualties on both sides, no contact was made at the front with the enemy'.[9] The 21st Brigade war diarist recorded: 1200 Brig PORTER after recce of the 3000 ft [1000 metre] feature decided to move 2/14 [and] 2/16 bns on to tracks leading East and West from UBERI'.[10]

Later that day, the 21st Brigade got the gratifying news that the composite battalion would be broken up and the 2/14th and 2/16th battalions reconstituted. They would begin their movement south for much-needed rest, reorganisation and replacements. That day, the 2/14th Battalion could muster just 101 men of all ranks, while the 2/16th Battalion was marginally better with 142 officers and men.[11] The 2/14th Battalion war diary records: 'A ring from Lt Col CARO advised Bn to send out 2 patrols to check up on the topography of the feature on which we were stationed. Both patrols carried out thorough recces of the feature; one moving in S.W. direction, the other more to N.W. No. 1 patrol reported in at 1200 hrs, No. 2 patrol at 1400 hrs. Both reports were sent to Bde IO .... Brig PORTER advised us that the following day our unit would once more be split into the two identities and move down to respective positions; 2/14 Bn South-west of UBERI and 2/16 Bn North-east of the village, leaving on the feature one Pl of the 2/16 Bn as a standing patrol .... Mail was received'.[12] The diarist with the 2/16th Battalion recorded with greater brevity: 'Received instructions from Bde to leave 1 officer & 15 ORs on feature as standing patrol. The rest of Bn to move back to UBERI in morning'.[13]

The historian of the 2/14th Battalion perhaps best summed up the situation for not only these men but for all the men in the 21st Brigade, along with the militiamen of the 39th Battalion:

> By external appearances, the battalion presented a bitter contrast with the bronzed columns which entered the ranges a few weeks earlier. Gone were the spick and span khaki shorts and shirts, gone the heavy bronze from faces and limbs, gone the well-padded muscles bulging their shirts. Instead, suits of brindled khaki-green caked with mud, hung from bony shoulders, and faces shone with the unnatural waxy pallor of the sunless jungle. Clinging to all was the peculiar stench of mud and death. But there was an unconscious pride in their bearing-engineered of the self-respect of the man, who feels, rightly or wrongly, that he has faced the worst and triumphed .... The full story of the great struggle will never be known. Many of the greatest deeds were done by men who fought and fell unseen by their comrades behind a wall of green. Many who could have told of the deeds of others were themselves killed a few days later, their stories untold.[14]

This, however, would not stop Blamey from using these officers and men as scapegoats to hide his own significant failings. The general, who remained in Australia during the first two months of the campaign, would make an appearance in the Port Moresby area a few weeks later to review these men who had borne the brunt of the fighting along the Kokoda Track where his reputation as an ignorant bully would be confirmed. Lieutenant Colonel Ralph Honner, commanding the 39th Battalion – who for a time also commanded Maroubra Force, as he and his men of the 39th Battalion stood alone against the advancing

CHAPTER 33

Japanese – was told to report to Blamey when he eventually arrived in Port Morseby. Honner 'spruced himself up; and presented himself to Blamey, who on being introduced to his field commander said: "Good morning, Honner, you've just arrived from Australia, have you?" "No, sir," replied an astonished Honner, "I've been in Papua for some time." He recalled justifiably with some bitterness years later: "I'd been his commander of the Australian forces opposing the Japs. He didn't know who I was. He didn't know, he didn't care"'.[15]

Blamey was not finished, as on 9 November 1942, he would order the men of the 21st Brigade to parade at the Koitaki Cricket Ground. There, one of the most damning episodes in Australian military history would occur when Blamey gave his infamous 'Run Rabbit Run' speech, accusing the men of Maroubra Force of cowardice in the face of the enemy.

* * *

Meanwhile, Lieutenant Colonel Honner and his men of Honner Force, who were attempting to cut the Japanese lines of communications between Menari and Nauro, had reached the vicinity of Omaramara. Most of the men were physically spent from the trekking through the rough terrain covered in thick jungle. The promised horse and aerial resupply all predictably were a no-show. Honner was told that he was not to go beyond the limits of rations carried; however, by now he had approached that limit without having encountered Japanese. Clearly, the force was in no condition to conduct offensive operations and were forced to retire back to Port Moresby to resupply.[16]

* * *

At Ilolo, Major Albert Moore was keen to get his supplies forward to re-equip the newly established Red Shield post now relocated just south of Imita Ridge: 'We set off for "Ned" Kelly's and found a truck to transport my load. I was promised a jeep by 8 pm, but it did not eventuate and at 10 pm we decided to camp for the night. Geddes was with me, but we had no blankets, and it meant a night in the open. We found a heap of bales of wet sandbags, and we sank down on these and pulled some strips of wet hessian over us. It was a cold long and cheerless night with nothing to eat or drink'.[17]

* * *

Sometime that day, Captain Thomas Gill received Lieutenant Colonel Geoffrey Cooper's message that he was to move the stretchers to a nearby garden, while he and the rest of his men were to rejoin Cooper's party. Gill provided what food he could for the sick and wounded, settled them in a cultivated area several kilometres east of Nauro, and left them in the care of Corporal John Burns and Private Alfred Zanker, 27-year-old labourer from Wild Horse Plains in South Australia. Both men volunteered to stay and look after these men.[18] Burns later wrote in his report for the battalion war diary:

> 0810 hrs on Sat 19 Sep 42 found Pte Zanker and myself left in charge of the wounded, seven stretchers and nine walking cases, our sole supplies being ten shell dressings, bottle of morphia and syringe and a garden of yams. The garden was several miles east of NAURO and right on the northern bank of the NAURO [sic Naoro] CREEK. When Don Coy moved off, Zanker and I set to and built shelters to help protect the lads from the terrific heat and rain which was to come later in the day. At approx. 1000 hrs an American bomber appeared overhead and circled several times. I tried hard to attract his attention, but he made off and dropped rations about five miles [eight kilometres] north-east of us. It was a great disappointment to us all, but our prayers and thoughts went to better luck on the morrow. We experienced very heavy rains all afternoon and by the time it ceased at 1700 hrs we were all looking like half drowned rates. At 1800 hrs I lit a small fire and boiled enough yams for us all. We then baked enough on the coals to last us the following day.[19]

CHAPTER 33

* * *

By now, Lieutenant Maurice Treacy and his two men, who were originally with Captain Sydney Buckler's party, were still on the track that would eventually take them to the village of Dorobisolo and safety; Treacy recorded: 'On Saturday 19th October [*sic* September] after moving for a week in a general S.S.E direction the party came to a native garden. During this week the party had been without food other than that carried. No native gardens or village had been seen nor had any game been encountered. During two stages of two and three days respectively, no water could be found, and the party was suffering from lack of food. Several meals of corn were had that evening. About one half pound of rice was left. It was 16 days since leaving Capt. Buckler's party'.[20]

Meanwhile, further north, the lost Corporal Ernest Lang, a 22-year-old printer from Williamstown, and Private Lawrence Scott, a 25-year-old from sorter from Yarraville, both in Victoria, having separated from Lieutenant Colonel Key's party after a Japanese ambush a fortnight before, were by now utterly exhausted and close to starvation. They had, however, struggled on and sometime that day came across a lone Japanese soldier walking down the track carrying two packets of Australian biscuits. Scott tried to bail him up with his pistol, but the enemy soldier, seeing the weak gaunt figure, attacked him with a machete. Scott warded off the blow with his arm while Lang hurled a piece of wood at the soldier, giving Scott time to fire his pistol, which killed the soldier. They took the biscuits and staggered on. It would be another 18 days before a party of Australian engineers would come across them. Lang said that Lieutenant Colonel Arthur Key and the five men with him were suffering from hunger and exposure when he last saw them.[21]

34

'... TO SEE AN OLD WARRIOR OF THE SPEAR AGE PLACE DOWN HIS WEAPONS AND SHOULDERED A BREN GUN'

By 20 September, most Australians along Imita Ridge noticed that the enemy remained relatively quiet. Indeed, the next three days would be marked by Japanese inactivity due to the heavy bombardment from the guns of the Australian 14th Field Regiment, sited at Owers' Corner along with the ongoing patrols conducted by Australian troops.[1]

That day, General Horii was located at Nauro and called together his senior commanders, telling them how things stood. He praised his officers for the way they and their men had succeeded in crossing, 'the so-called impregnable Stanley Range', and explained that the reason for the halt was to regain their strength, so they could 'strike a crushing blow at the enemy's positions at Port Morsby'.[2]

Given the poor condition of his men, casualties suffered and lack of supplies, how they could accomplish this feat must have been considered by some as questionable, even with the victory fever that had entrenched most Japanese officers at this time.[3] Horii had received orders from the *17th Army* that Major Koboyahi Asao, commanding the *III/41st Regiment* had been transferred to duties in Japan; he left that day for Buna, having ordered the major to: 'Clearly communicate to *17th Army* command the actual overall conditions

CHAPTER 34

of the battle'. He would not arrive in Rabaul until 6 October and reported that day: 'The supply situation for the South Seas Force has already reached a crisis. The number of troops who are collapsing continues to rise. Allied pressure mounts daily with no improvement in sight. I would like to see an immediate transport of supplies by destroyer to ease the situation'.[4]

Unknown to Horii at the time, General Imamura Hitoshi – who was based in Rabaul and who had just taken commanded of the newly created *Eighth Area Army*, which was responsible for the Solomons and New Guinea Campaigns – on hearing that Ioribaiwa Ridge had been occupied by the men of the *South Seas Force*, ordered Horii to evacuate the position and to fall back north of Menari.[5] On the earlier order dated 28 August, the Japanese *17th Army* commanders had not defined what line Horii was supposed to take. On finally getting this order weeks later, Horii was approaching Ioribaiwa Ridge and justifiably defined Ioribaiwa Ridge as his defensive position. Horii would not receive his revised orders from General Imamura to withdraw north of Menari until 24 Sewptember.

Still on Ioribaiwa Ridge was Japanese war correspondent Okada Seizo, who in his unpublished manuscript of his experiences during the Kokoda Campaign, recorded:

> According to the plan laid down before we left Rabaul, another detachment (a regiment) was to be landed near Port Moresby [Milne Bay] about the time we reached Mt. Efogybia [*sic* Ioribaiwa] and attack the town in cooperation with us. So, the Horii detachment encamped on the mountain to prepare themselves for the last drive. It was about this time that news of the landing of American marines on Guadalcanal came from behind, as if borne by the wind. But none of us could then realise the vital significance of the news. Far away in the mountains of New Guinea, we did not bother much about it. The soldiers of the Horii detachment were thinking of nothing but Port

Moresby, which kept up their fighting ardour. Soon, however, the blackness of the situation at Guadalcanal began to dawn upon us. So that was why no Japanese planes nor supplies came to New Guinea! An atmosphere of uneasiness stole over the position on the mountain like the fog that gathered noiselessly every morning.[6]

* * *

Captain Robert Dodd with the 2/25th Battalion reported from his forward patrol that since dawn he had observed a distinct reduction in Japanese activity to his front. He noticed throughout the day that the Japanese showed little signs of advancing. He headed back to report to the battalion, as recorded in the battalion war diary:

> At 0800 hrs. 20th Sept., the patrol pressed forward once more to GRASSY PLATFORM. The enemy had vacated their posts that they had dug in, and all the posts had been camouflaged. We pushed further on.
>
> Fwd scouts reported sounds of digging about 250 yds from the PLATFORM. Alarm traps had been placed across the track to prevent a repetition of the earlier nights surprise. Men detailed to watch the spur to the immediate left, from which direction sounds of chopping and digging could be heard, reported considerable movement. As many as 100 men passing one point in five minutes. No further noises had been heard on the right spur since the first night, indicating that the troops there had been brought over to reinforce those on the track.
>
> After waiting for two hours for the Jap to come and meet us on our own ground, it was decided to return with what information we had collected.
>
> The patrol left vicinity of GRASSY PLATFORM at 1430 hrs., and reported to 2/25 Bn HQ at 1000 hrs 21st Sept.

CHAPTER 34

A number of the troops developed coughs as the patrol progressed and had to be kept well in the rear. At night it was difficult to conceal movement and positions, as the coughing was always accentuated by the night air. The Japanese soldiers seemed also to be suffering from the same trouble, as his positions were picked up by the coughing.[7]

Meanwhile a patrol from 'B' Company, 3rd Battalion led by Captain William Atkinson, a 32-year-old clerk from Canberra, and Lieutenant Edward Dullards, a 26-year-old public servant from Canberra, had been sent out. Atkinson and his patrol, totalling around 60 men drawn from all companies of the battalion, left at around 11 am with the objective of locating and harassing the Japanese lines of communications. They were to remain out until this objective had been accomplished or they were close to exhausting their five day's rations. About two hours later, another much smaller patrol lead by Lieutenant Noel Rumble, a 24-year-old schoolteacher from Carlingford in New South Wales, and five men left to reconnoitre and report on the existence of a line of advance from the battalion's forward defensive lines to the main Kokoda Track around 400 metres north, and if encountering any Japanese, they were to attack them and push them back from the slopes. As darkness set in, Rumble and his patrol returned, and while encountering no enemy troops, they did find a suitable route from this position that led west to the main track.[8]

The men of the 2/31st Battalion covering the left flank awoke to an early morning mist, which lessened visibility. It was not long before three four-man patrols were sent out to reconnoitrer their immediate frontage, and all returned later in the day with nothing to report. That day, a church service was conducted on top of the ridge by Donald Redding, a 44-year-old priest from Adelaide.[9] Meanwhile, it was a relatively quiet day for the men of the 2/33rd Battalion, as their diary merely records for 20 September: 'COs recc. of Coy posns and liaison with other Bns'.[10]

SAVING PORT MORESBY

* * *

Heading south, Japanese 1st Lieutenant Hirano, leading *No. 3 Platoon, No. 1 Company, I/144th Regiment*, finally rejoined his unit; however, in terms of supplies, he had little to show for his efforts. He recorded in his diary for 20 September: 'At 1520, finally reached Battalion HQ Detachment. Commander was surprised at the small amount of food we brought back and ordered 1st Lt. YOKODA to carry back the provisions'.[11]

Meanwhile, Lieutenant Sakamoto Atsushi, commanding *No. 2 Machinegun Company, II/144th Regiment*, also recorded in his diary the ongoing significant supply crisis: 'Never till now, did I realise the true meaning of the saying "A full belly counsels well". Not a single grain of rice left. Taros sufficient for only another day. From tomorrow we will have to chew grass or bark'. Even so, he remained confident of ultimate victory, concluding his diary entry that day: 'The battle we are fighting now is an important one. The eyes of the whole world are upon us'.[12]

* * *

The composite battalion of the 21st Brigade was no more, and the men of the 2/14th and 2/16th battalions began their respective withdrawals south of Uberi. The diarist of the 2/14th Battalion recalled that at 7.15 am, Captain Phillip Rhoden and his intelligence section moved out to contacted Brigade that was about 100 metres forward of Uberi. Lieutenant Norman Cairns, who was among the first to observe the Japanese at Brigade Hill during the early morning hours of 8 September, now guided this party to an area 300 metres south-east of the village. The balance of the battalion arrived during the next hour and were placed in position.

Meanwhile, lieutenants Ken Evans, a 24-year-old law clerk from Moonee Ponds, and 30-year-old John Crameri, from Collingwood, both in Victoria, along with 38 men were sent forward from the

CHAPTER 34

rear echelon; they had arrived the previous day and made their way forward to the rest of the battalion. At the same time, Rhoden and part of his intelligence section made a reconnaissance of the area and tracks.

At noon, a patrol was ordered to go out for two days, but just as they were preparing to leave, they were ordered to standby pending a conference that was to be held at Brigade Headquarters. Captain Rhoden reported to brigade at around 5 pm and received the information that the brigade was going to be relieved by the 14th Militia Brigade, consisting of the 3rd, 36th, and 55th battalions. It was stated that the relief would not take place for a few days until the 36th and 55th battalions were in the area and had been given the benefit of the veteran's 'knowledge of Jap and country'.[13] The brigade was to move back at 8.00 am the next day. It was also recorded that Brigadier Porter spoke 'very highly of our unit and asked Capt. RHODEN to pass on to the offrs and NCOs his congratulations and commendation of effort throughout the campaign'.[14]

The official diarist with the 2/16th Battalion recorded that at 7.30 am, the 2/14th and 2/16th composite battalion split with each regaining their original identities. The 2/16th Battalion took up position north of Uberi to patrol the Goldie River area to the east. The 2/14th Battalion was to conduct similar operations to the west of Uberi.[15] The brigade war diary concludes that at 4.30 pm, Brigadier Porter received orders to return to Port Moresby for a debrief the next day.[16]

* * *

Still in the Ilolo area, having spent the night sleeping in wet strips of hessian, Major Albert Moore and his fellow salvationists, Charles Geddes, enjoyed the first rays of sunshine as they got ready their supplies for transport towards Owers' Corner. From there they would advance into the jungle further north along the Kokoda Track where Corporal Jock Inglis had established the Red Shield post days before, at the base of the southern slopes of Imita Ridge. Moore recalled: 'By 7 am we had a jeep to transport out goods to Owers' Corner and there we secured natives to carry the load to the foot of the ridge, where we found Jock Inglis with a fully functioning Red Shield centre. There were a number of sick and wounded men on the track, and they were happy to find us. The 2/14th had been relieved and so I went forward to locate their position, so that I could take some socks and other comforts to them on the morrow. It was after dark when I returned and discovered that the day was Sunday. How oft under such circumstances did we lose count of the days?'[17]

* * *

Captain Sydney Buckler and his lost party of men from Maroubra Force had by now made their way to Sengai – below the northern slopes of the Owen Stanleys. They were welcomed by the local Papuans, who fed and sheltered them. They advised the Australians about the country which lay ahead. Buckler recorded that some of the Papuans carried four-metre-long pig spears. Two local headmen, Faria and Ewoki, promised Buckler that he and his villagers would care for the wounded, as the country he was intending to travel through would be difficult even without having to carry and care for these men. Before leaving, Captain Buckler dutifully recorded in his diary the names, rank, next of kin and addresses in his diary of those to be left behind: Corporal John Metson and privates Ambrose Mayne; 21-year-old Stanley Hunter from Korumburra; 22-year-old James MacDonald from Seddon (all in Victoria with the 2/14th Battalion); Tom Snelgar, a 27-year-old miner from Subiaco in Western Australia (also with the 2/14th Battalion); 30-year-old Private Charles Yeo from Mt Lawley in Western Australia (2/16th Battalion); and Raymond Knights, a 22-year-old storeman from North Williamstown in Victoria (39th Battalion).

CHAPTER 34

Private Thomas Fletcher, a 41-year-old farmhand from Wareek in Victoria, volunteered to stay behind to look after these men. His remaining medical supplies amounted to two shell dressings, 14 field dressings, towels and soap – there was no morphine.[18] The Australian official historian of the campaign wrote: 'In the fighting at Isurava, Private Fletcher, a medical orderly, had shown himself to be a brave and devoted soldier. During the days and nights since the 30 August he had nursed the wounded and sick with great care although he was weak himself and, like the rest, always hungry. Now he volunteered to stay at Sengai and look after the seven men whom it was decided to leave there for the present'.[19] Unknown to all, however, was the fact that they were close to the Japanese garrison at Gorari, just seven kilometres north-east of the village. Indeed, Sengai was also located on a track that led to Gorari, which was likely to be patrolled by the Japanese.

Buckler and the others, who were to continue their way back across the Owen Stanleys, paraded before Fletcher and his little group and in 'tribute to the fortitude which they had shown over the past three weeks and to the courage with which they faced what was to come, the rest of the party formed up on parade to "Present Arms" to Pte Fletcher and his sick and wounded'.[20] The Australian official historian recorded that it was at 'once a salute and, had they but known it, a final farewell, for later the Japanese came and killed these helpless men: the chivalrous Fletcher; young Metson, who for nearly three weeks had crawled through the bush; Mayne, who had crawled away to die rather than burden his comrades; and the others who had endured so much'.[21]

Captain Buckler and Lieutenant Charles Butler, a 25-year-old quantity surveyor from Melbourne, and their 39 men moved out and dropped down towards Wairopi, then headed upstream along the swiftly flowing Kumusi. Leading them were several Papuans. Buckler wrote: 'It is a queer sight to see an old warrior of the spear age place down his weapons and shoulder a Bren gun over the waters'.[22]

SAVING PORT MORESBY

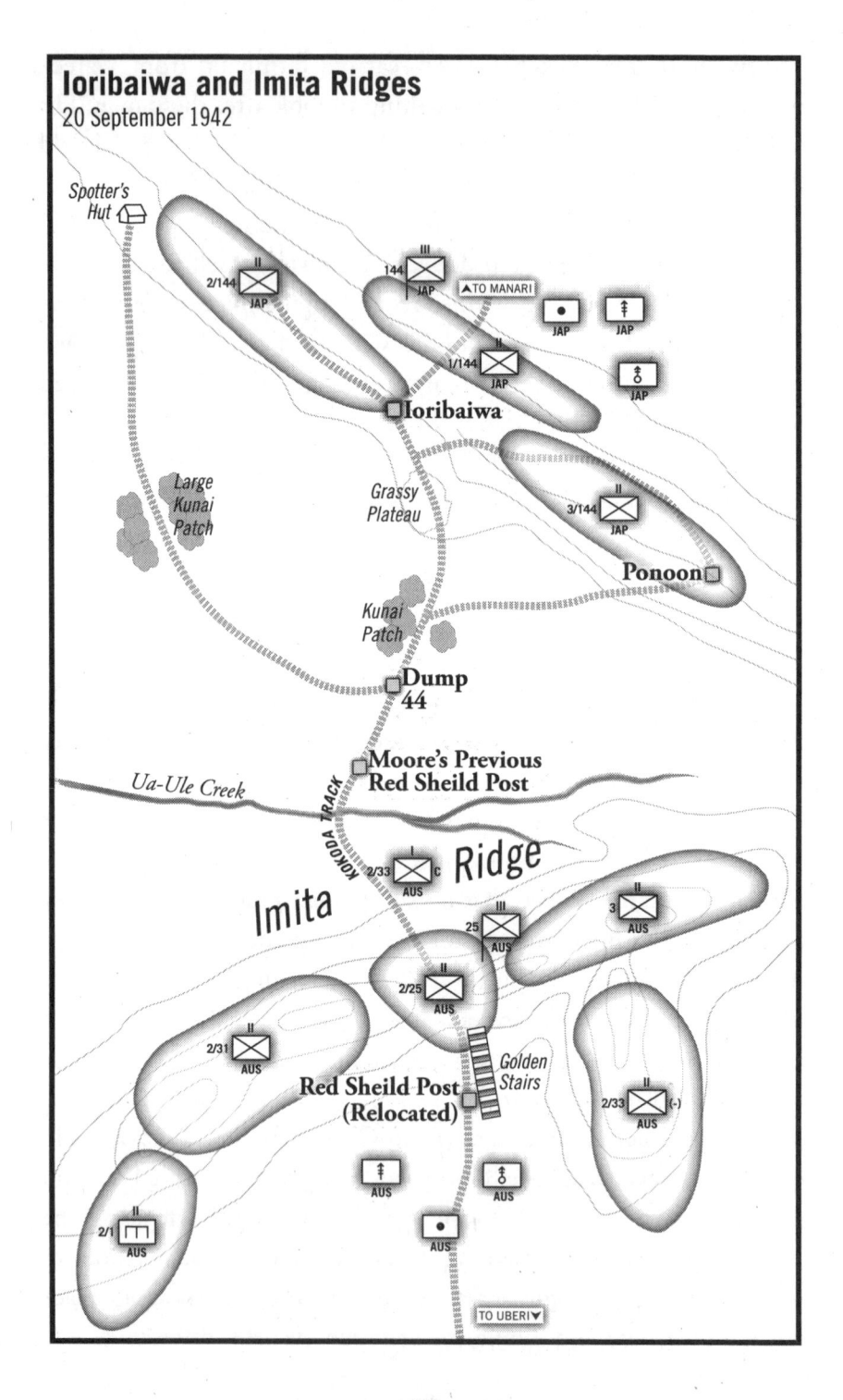

Ioribaiwa and Imita Ridges
20 September 1942

CHAPTER 34

* * *

Further south, Buckler's young officer, Lieutenant Maurice Treacy, and his two men were on a track heading east. While they were unaware they were now just two days journey from the large village of Dorobisolo, they knew they were finally approaching the Australian lines. The young officer wrote soon after in his report: 'After resting, the party set off on Sunday 20th and arrived at a native village. The native inhabitants gave the impression that the party was only two days from ITIKI where the Bn's 'B' echelon had been established. A guide was obtained, and the party set off with the intention of travelling S.W. The guide deserted after a few hours travel disappearing into the jungle'.[23]

Meanwhile, still located on the northern banks of Naoro Creek, likely less than ten kilometres west of Treacy's party, were the stretcher-cases and walking wounded who had been with Lieutenant Colonel Goeffrey Cooper and his large party of men. Corporal John Burns and Private Alfred Zanker were still located next to the yam garden and looking after these men; Burns recorded:

Sunday arrived after a night which seemed more like an eternity. The lads had a very bad night, Zanker and I spent no more than 2 hrs [asleep]. The planes were over again at dawn – three of them – I again tried to attract their attention, but all efforts were in vain. We gave the wounded a sponge over during the morning, it was the first they had had since they were wounded some thirteen days previously and it brightened them up no end. We tried hard to erect a more substantial shelter for them but when the rains came in the afternoon a party of drowned rats would have been a more apt title than human beings. A party of bush natives visited us during the afternoon, and I tried to get them to bring some food but I'm afraid my 'pidgin' was bad .... They departed very hurriedly when Nippon fired a rifle on the other side of the creek'.[24]

SAVING PORT MORESBY

* * *

It was only now that Blamey informed Rowell of his imminent arrival and basing in Port Moresby, writing to Rowell: 'The powers that be have determined that I shall myself go to New Guinea for a while and operate from there. I do not, however, propose to transfer many of Adv HQ Staff and will arrive by aeroplane Wednesday evening, I hope with [Ronald] Hopkins. At present I propose to bring with me only my PA, Major Carlyon, two extra cipher officers and Lieutenant [Raymond] Lawson. I hope you will be able to house us in your camp and messes. I hope you will not be upset at this decision and will not think that it implies any lack of confidence in yourself. I think it arises out of the fact that we have very inexperienced politicians who are inclined to panic on every possible occasion, and I think the relationship between us personally is such that we can make the arrangement work without any difficulty'.[25] Rather than making an 'arrangement', the situation would result in serious infighting within the Australian command system. Given that Blamey was bringing few staff with him, indicated that he would be relying on Rowell's staff. This, however, did not bode well for the future relationship between them, resulting all too soon in the pro-Blamey and pro-Rowell forces remaining extremely hostile and bitter even well after the war.

35

'OUR 25-POUNDERS OPENED FIRE ON JAP POSITIONS'

On the 21 September, the diarist of the 2/31st Battalion recorded: 'Rain began during early hours of morning making conditions unpleasant. Shelters were made from groundsheets and branches. Mist lessened visibility. By this time some of the posts had improvised warning devices consisting of empty tins suspended on vines. Diarrhoea is becoming prevalent. To establish a food reserve priority has been given to canned beef, biscuits, tea, sugar, and salt'.[1]

Meanwhile, a patrol reported to brigade that a large enemy force of around 600 men was observed on a ridge west of Ioribaiwa.[2] The battalion war diary for the 3rd Battalion also records: '0630 hrs: Situation normal. Quiet night. 1410 hrs two dull explosions heard from South. A third OP has been established on Northern flank of C Coy providing a good view of our left flank. 1800 hrs a large volume of smoke observed beyond IORIBAIWA bearing 42 degrees'.[3]

That afternoon, to the great satisfaction of the men on Imita Ridge, all heard the first rounds of Australian artillery explode against Ioribaiwa Ridge. The forward observation artillery officer was in the vicinity of the 3rd Battalion to help direct the firing of these guns (based at Owers' Corner) against Ioribaiwa Ridge, ten kilometres in the distant. It took around 25 seconds for each shell to reach its targets, passing over the Australian positions on Imita Ridge.

Over the next few days, more than 700 shells would target the Japanese positions along the ridge. At the same time, another

25-pounder from the Australian 53rd Battery, 14th Field Regiment was being transported intact down to the Goldie River. The gunners were assisted by men of the 2/1 Pioneer Battalion, who with ropes dragged the gun three kilometres into the range, as well as bringing forward its supply of ammunition. It would be a week before it was established close to Uberi.[4] By then, however, the Japanese would be well out of range.

Additional welcomed news arrived that Australian and American reinforcements were arriving in Port Moresby, as recalled by Sergeant William Crooks: 'Two guns of the Militia 14th Field Regiment that were man-handled in pieces up to Owers' Corner opened fire on Ioribaiwa. That night we were also given news that the Americans were arriving in New Guinea, together with the 16th Brigade of the Australian 6th Division. All this was morale-raising news'.[5]

Operational orders were issued at 2.30 pm to the 2/25th Battalion that they would attack and capture the Ioribaiwa Ridge the next day. The orders stated that Japanese patrols had been observed as far forward at the 'Large Grass Patch' on the main Kokoda Track, which was the former position of the 25th Brigade Headquarters on Ioribaiwa Ridge, about two kilometres north of Dump No. 44. This information had clearly come from Captain Dodd's patrol, who had reported back to the Australian lines earlier that day – Dodd had been wounded as part of this patrol but remained on duty to command the patrol.

The men of the 2/25th Battalion were to move forward along the main Kokoda Track as a harassing party and secure the dump. From there they were to patrol forward to secure the 'Kunai Patch' and the ground leading up to the Ioribaiwa Range. The orders provided details of the intended advance, which was to commence at 7 am with 'D' Company leading the advance with 25 metres between platoons. They would leave from Major Moore's former Red Shield post; following closely behind them would be 'A' Company, Battalion Headquarters' staff, the Headquarters Company, 'C' Company, a brigade signallers party, and 'B' Company bringing up the rear.[6]

CHAPTER 35

Meanwhile, the men of the 2/31st and 2/33rd battalions were preparing to push forward in support. Indeed, the men of 'C' Company, 2/33rd Battalion were still positioned forward of the brigade near the former Salvation Army aid post.[7] The brigade war diary records: 'If 2/25 Bn successful, 2/33 Bn to move fwd and pass through them. 2/31 Bn to [also] move. General patrol activity throughout the day. Slight JAP activity. JAP working party using timber, building operations vicinity old HQ 25 Bde. Arty commenced ranging with smoke shells, from vicinity OWERS' CORNER approx. 1500 hrs and successfully registered on ridge west of IORIBAIWA'.[8]

* * *

Still on Ioribaiwa Ridge was 1st Lieutenant Horibe, acting commander of *No. 6 Company, II/144th Regiment*, who had sent out another five-man patrol. These men later returned with a Bren gun and reported hearing trees being felled on the other side of Ua-Ule Creek, but they could not see the Australians. A meagre ration was supplied to each man, consisting of a third of a litre of rice, soy sauce and wheat paste, which was to last for two days. It was not enough, but the Japanese officer recorded: 'Life is tenacious'.[9] He received a few replacements that morning, raising the strength of his company to sixty men. He concluded in his diary that the *Takasago Formosan Naval Volunteers* (workmen) arrived with more ammunition and were given a day to rest before heading north back down the track. Thirty-six of the 238-man party had malaria.[10]

Now back with the battalion, 1st Lieutenant Hirano, leading *No. 3 Platoon, No. 1 Company, I/144th Regiment*, recorded in his diary: 'The Battalion was attached directly to the HQ. Reconnoitred the position. Saw the Detachment Commander and Staff Officers. The regimental commander [Colonel Kusunose] seemed to be improving from his sickness'.[11] With him was Private Watanabe Toshi with *1 Section, No. 3 Platoon, No. 1 Company, I/144th Regiment*, who had been fighting the Australians as part of the original force that invaded Gona on 21 July.[12]

He was among the half-starved and sick somewhere on the Ioribaiwa Ridge. Sometime that day he wrote what was to be the last entry in his diary: 'Since landing the battle has continued for two months in the mountains. In the combat between KOKODA and ISURAVA our No 1 Coy of 180 men was reduced to only 80 men. Our Coy had the heaviest casualties in the 144th Regt. Coy Comd OGAWA was killed in action at KOKODA, and his successor, HATANKAKA, was killed at ISURAVA. Due to heavy casualties, every coy is reduced to about half strength'.[13] It was likely that this diary was retrieved by advancing Australian troops from is dead body somewhere on Ioribaiwa Ridge or the track just beyond.

The unknown leader of the signals unit with *No. 2 Infantry Section, No. 1 Company, I/144th Regiment* was trying to maintain communications, recording in his diary for 21 September: 'Repaired the line which connects Bde HQ with Regt HQ'.[14] Still at Menari was Private Teruoka Akira with the *Tanaka Unit, 55th Division Medical Unit* who recorded in his diary that he was gaining more strength and was finally recovering from his illness. He also recorded that for the first time in ages, he and his comrades had a decent meal, which was likely provided by the recent discovery of some mis-dropped supplies by the Australian and American biscuit bombers. The meal that day consisted of dry bread, beans, mash, oil and meat. He concluded in his diary: 'I was happy because supper was very tasty'.[15]

* * *

Meanwhile, Brigadier Porter was making his way back to Port Moresby to report while the men of the 2/14th and 2/16th battalions were finally about to be taken out of the line. These men, however, still had a few days of patrolling the rugged terrain and thick jungle of the Owen Stanleys ahead of them, searching for Japanese infiltration parties either side of the main Kokoda Track.

The 2/16th Battalion was conducting patrols east of the track, and soon came across more stranded survivors of the 2/27th Battalion.

CHAPTER 35

The unit diarist recorded that at 8.15 am, 'A' Company moved along the northern bank of the Goldie River to establish a standing patrol three hours out from Uberi, while 15 minutes later, 'C' Company, who were patrolling, crossed the river to conduct a reconnaissance of a knoll south of the main bivouac area. At noon, reports came in from 'A' Company that they had 'ran into Jap patrol' two hours from Uberi. About an hour later, the 'Jap patrol' reported by 'A' Company turned out to be four men from the 2/27th Battalion who had been cut off for ten days. Their 'emaciated and dirty unshaven condition accounted for the error'.[16]

To the west, the men of the 2/14th Battalion were soon joined by the much-admired Major Albert Moore of the Salvation Army who had been affiliated with the battalion since their days in Syria. A month earlier, he had been forced to fall back from Myola to Port Moresby to ensure that their Red Shield post was erected further down the track to ensure the men of Maroubra Force were supplied with a hot cup of tea, and cake, as they withdrew back towards Port Moresby. The battalion diary records: 'An issue of comforts by Adjt MOORE of the Salvation Army was welcomed. The issue included socks, tobacco, cigarettes and chewing gum'.[17] Indeed, Major Moore also recalled: 'We found remnants of the 2/14th and we were able to issue a pair of socks and other comforts to each man. That day the Australian 25-pounders opened up, but only with smoke shells to get the range of their shooting. I was told that an artillery officer had gone ahead and had a position in a high tree overlooking the invader positions on Ioribaiwa Ridge'.[18]

Also returning to the battalion that day was Sergeant William Irwin, who had last been seen as part of Lieutenant George Pearce and 16 Platoon's action on 29 August when they had gone out to silence a Japanese machinegun position just forward of the Australian northern perimeter at Isurava. Irwin had been lost in the jungle for three weeks and had collected along the way the seven men from Lieutenant McIlroy's 'C' Company. It is likely these were the men mentioned in the diary of the 2/31st Battalion war diary who struggled into their lines that day: twenty-two men of the: '2/14 Aust Inf Bn walked into A Coy lines'. These men had been separated from their unit 31 days before and had survived by eating any fruit or crops they could find. The diarist also recorded they had: 'One meal of fish from a pool ([using] grenade last one) and the leavings of Jap depredations on our abandoned food dumps. Japs had forced them off the tracks but fortunately they had a compass'.[19] The 2/14th Battalion war diary records the condition of Irwin and his men as they returned to their battalion: 'All members of the party were in poor physical condition owing to the trying conditions experienced'.[20]

The diarist of the 2/14th Battalion also recorded a patrol was sent out that morning led by Lieutenant Ken Evans and 13 men, including Private Raymond Wild, a 23-year-old schoolteacher from Maryborough in Victoria, of the intelligence section. They were to patrol the track north-west of Uberi. At 12.30 pm, the battalion received a hot midday meal. During the afternoon, it was again noted with satisfaction that the Australian artillery was heard carrying out registration shoots against Ioribaiwa Ridge.[21]

At Dump No. 66, Captain Bert Kienzle, with ANGAU, also recorded the welcomed news of the Australian guns: 'I went down with an attack of dysentery and fevers. Our 25-pounders opened fire on Jap positions at IORIBAIWA ridge from OWERS' CORNER'.[22] With Kienzle at Dump No. 66 was Captain Geoffrey 'Doc' Vernon, who recorded in his report to ANGAU: 'At camp 66 we reaped the aftermath of a hurried retreat. Dysentery became very prevalent owing to the almost universal fouling of the track in the darkness of the withdrawal. Many white troops had left Ioribaiwa with dysenteric symptoms on them and had deposited infected matter along the road. I sent a letter into the base suggesting that the remainder of the LOC be well supplied with latrines in case a further retreat became necessary, and later a party of pioneers came out and cleaned up the whole area. Besides dysentery, we had many cases of malaria and pneumonia, the hospital was always full'.[23]

CHAPTER 35

* * *

Further east, having rested for a few days at Jawarere, Lieutenant Colonel Geoffrey Cooper and his large party of the 2/27th Battalion were now on the last stage of their journey, striking the main road beyond Owens' Corner, where they were met by trucks and quickly moved to a convalescent camp on the outskirts of Port Moresby. However, the ordeal of the stretcher parties was not yet over.[24]

Still located along the northern banks of the Naoro River, Corporal John Burns and Private Alfred Zanker were still attending the seven stretcher-cases and nine walking sick and wounded. Their sole supplies consisted of ten shell dressings, a bottle of morphia and a syringe – and a garden of yams. They made crude shelters and sweltered in the heat and were soaked by the afternoon rains. Aircraft flying above their positions raised their hopes, but they could not attract their attention.[25] There was firing nearby, and Burns wrote for 21 September:

> Monday arrived after a terrifically cold night. The boys were very restless all night, especially Cpl [Leonard] Williams and I sat with him for the best part. The heat and flies were so bad that they almost drove us to the first stages of insanity. The heat was terrific and the flies I think we had all that were in New Guinea. The lads all received a wash and shave this morning, my shaving gear, face washer and toothbrush serving everyone. Cpl Williams and Pte [Thomas] Burke felt the going very hard from this stage on. We had to be with them day and night. At 1425 hrs we hear heavy mortar fire or artillery fire SW of us and were really thrilled as it was the first time, we had received any idea as to where the front was. Cpl Arthur brought his New Testament to light during the afternoon, and it used to help fill in many of our long, hungry hours. At 1510 hrs a mortar [bomb] landed very close to our place of abode and it gave the lads a terrible fright. We did our best to cheer them up but a couple of them appeared to lose all nerve. For about an hour it continued on

various tracks round the area. At dusk we lit the fire again and cooked the long-awaited boiled yams.[26]

* * *

Meanwhile, south-east of Burns and Zanker were Lieutenant Maurice Treacy and privates Fredrick Rockliffe and William Avent, 2/14th Battalion, who were determined to make their way to Itiki. Treacy later reported: 'The following day a native garden was reached, some potatoes were being gathered, when a patrol of the 6th Independent Coy arrived. Badly needed food was eagerly accepted. The members of the party by now were fairly weak being able to move for only ten minutes and resting for ten minutes'.[27]

The three commandos who found them were part of a patrol led by Lieutenant William Nichols, a 24-year-old manager from Elwood in Victoria. He was commanding an eight-man unit from 'D' Patrol, 2/6th Independent Company, which was tasked with covering the tracks around Jaure in search of Japanese infiltration parties. That day, 22-year-old Corporal Clive Westbrook, from Melbourne, with two of these men were conducting a patrol when they came upon Treacy and his men, as recalled by 24-year-old Lance Corporal Alfred Miller, from Richmond in Victoria: 'Late in the afternoon we came across footprints in the mud and decided to follow them. Not knowing whether friend or foe. Night-time overcame us so we decided to rest up for the night. Come dawn, and we took up the tracking of the footprints until we came upon a party scrouging in a native garden. We surrounded the garden from a distance, having them well covered with our firepower, when one person jumped up and in true Aussie language rejoiced at finding something to eat. So, we made our presence known. Turned out to be three members of the 2/14 who were completely lost'.[28] Treacy and his men were led back to Jaure where they were looked after as best as they could with the limited supplies available.

* * *

CHAPTER 35

Back at the Japanese beachhead at Sanananda, Father James Benson was approached by the Japanese interpreter Susuki who informed him that he was to join a patrol to Gona, what was its intention he had no idea, but he described it was a great diversion from sitting around the Military Police Tent which was his current place of abode. He wrote after the war:

> Nothing of particular interest happened during this trip, except that I had to take a high hand with one of the infantrymen; a nasty bumptious little fellow who wanted me to carry him pick-a-back across the Garara lagoon.
>
> It was low tide when we began the crossing, and soon I was in the water up to my waist; suddenly this nasty little man grabbed hold of me and made signs for me to bend over while he climbed on to my back. I looked at him coldly.
>
> 'No, you don't, my lad!' I said with all the dignity I could muster, 'I am not a coolie.'
>
> Susuki, the interpreter, called back: 'What are you then?'
>
> I don't know what prompted me to reply with some pomposity: 'I am a priest and a gentleman!'
>
> But it turned out to be an inspiration. Susuki translated the remark to Sergeant Furokoa, the N.C.O. in charge of the patrol, and he at once told the soldier in no uncertain terms to leave me alone. He then turned the word 'gentleman' over and over for quite a while, repeating incessantly, 'Oo Benson oo, Gentleman oo?' He seemed most impressed; so from that day I took to playing the 'High *Seigneur*' as well as I knew how, and there were several occasions when it got me out of tight corners, face slappings and other indignities – though I must place on record that, apart from my first meeting on the Kokoda road, when I was punched and kicked, no

Japanese laid a hand on me or hit me in the three years I was a prisoner.

One other incident on the patrol that I remember was a demonstration of the very beautiful Japanese courtesy which I had so often read about. It was alongside the ruins of our beautiful All Souls School at Gona that Furokoa saw a young officer whom he knew, and he rushed up to him with a great demonstration of joy. They stood apart and bowed to each other from the hips; then, laughing, they embraced, and held each other at arm's length, each looking the other up and down in admiring approval; they talked nineteen to the dozen for about ten minutes; then bowed twice with solemn gravity, saluted each other and parted. Susuki explained to me that they were old school-friends who had not met for many years. It was a gracious and dignified expression of friendship; and I could not help thinking at the time how much beauty and dignity has been lost by our Western abandonment of some of the courtesies still practised by the Japanese.[29]

36
'I NEVER TOLD THEM ABOUT THE WEEVILS'

On 22 September, with the Japanese seemingly cowering under the Australian artillery barrage, the men of the 2/25th Battalion were by 7 am carefully advancing from Imita Ridge down the Kokoda Track. Leading the way were the men of 'D' Company, followed by 'A' Company, the Battalion Headquarters and Headquarters Company, with 'C' and 'D' companies bringing up the rear. They arrived at Dump No. 44 meeting no opposition and established the dump as a base for operations. At 2 pm, patrols from each of the forward companies were sent out to reconnoitre the way forward. Meanwhile, the men of the 2/1st Pioneer Battalion took up their former position on Imita Ridge.[1]

Just after 7 am, Lieutenant Upcher and his strong 50-man patrol of the 2/31st Battalion returned after almost four days of patrolling with nothing to report, having made no contact with the Japanese. An hour later, the men of this battalion observed aircraft strafing and bombing Japanese positions north-west and north from their positions.[2] The men of the 3rd Battalion also observed Allied aircraft at around 8 am bombing Nauro Village, with large columns of smoke reported, which was followed an hour later by additional Allied sorties against the village area.[3]

It was likely now that Sergeant Bede Tongs, a 22-year-old carpenter from Queanbeyan in New South Wales, with 10 Platoon, 'B' Company, 3rd Battalion saw his first Japanese: 'We went back on patrol to Ioribaiwa on Imita Ridge and the 3rd Battalion they asked for 50 volunteers and we volunteered myself, and my good friend and I were in charge of this No 2 section and … four Japanese came from our right .... they never knew we were there, and they were just sort of walking along side by side because the ridge at that particular spot was wide enough for people to move like that. And we dealt with the four of them'.[4]

In front of the 3rd Battalion to the east of Ioribaiwa, Captain Atkinson was still with his men of 'B' Company, 2/33rd Battalion conducting a long-distance patrol. Lieutenant Edward Dullard and three other men were killed in a Japanese ambush that afternoon. Atkinson's report provides a fascinating account of the close-quarter fighting while on a long-range patrol:

> At about 2 pm the forward scout came upon a well-established track which apparently connected Spotter's Hut to Ioribaiwa village. There was a yellow signal wire running along the track. The scout saw three Japanese. One ran left and one ran to the right. Lieut. Dullard moved left along the track with the leading Section. Enemy LMG's opened fire from the crest of the ridge beyond the track. I ordered the 3rd and 4th Sections of the patrol around the left flank to deal with any enemy approach from that direction. I then moved around to the track on the right flank, could not contact Dullard, cut the signal wire and ordered the second Section into an ambush position covering the track on the right flank. I heard a voice giving orders in Japanese about 50–100 yards further up the ridge across the track, and then shortly after another voice shouting out in Japanese about 150–200 yards down the track to the right. I sent the Bren Section to that flank to fire in the direction of the voices. This they did and the voice was heard again.
>
> In the meantime, the second Section observed four enemy approaching up the track from the right and fired at about 20 yards range. The Sec. Comd. thinks his Section got the lot. The two Sections on the left found their progress barred by an almost sheer drop and then moved forward through a bamboo patch

CHAPTER 36

to the track. The movement of the bamboo drew enemy fire, but no one was hit. Sec Comd. of No. 3 Sec. jumped into an unoccupied circular weapon pit on near side of track and threw a grenade at an enemy post on the opposite side of the track. Enemy chatter ceased abruptly and later LMG fire appeared to come further back across the patch. A man from No 1 Sec. reported a party of enemy moving across left flank. I then blew a whistle blast, which was the prearranged signal to withdraw ….

No 1 Sec. Comd. reported one man badly wounded near track and Lieut. Dullard possibly hit rolled down gully to left. I sent the Sec. Comd. to see if Dullard was ok: he called down the gully and I heard the reply 'I'm all right' [he was never seen again]. I sent two men with the medical orderly to get the wounded man out …. They reported that the man was dead. Pte Gash the Sec. Comd. No 1 Sec. then stated he was not sure if all his men were out or not. The Sec Comd. called out but got no reply and after waiting some minutes I withdrew the last few men. I waited about half an hour at the rearranged rendezvous in creek bed and then continued down-stream, where about half an hour later we met a patrol from 25 Bn. moving up the creek, and later the main body of 25 Bn. Information gained was communicated both to the outgoing patrol, and later the CO of 25 Bn.[5]

Australian journalist George Johnston was still on Imita Ridge and reported on the Australian advance – finally, the fighting withdrawal had come to an end. He also reported on the Australian patrols that were now harassing the Japanese and their forward positions on the southern slopes of Ioribaiwa Ridge:

The Australian retreat has ended. Today our forward troops began to advance from Imita Ridge back along the track to Kokoda. No more than a couple of miles have been gained, but the news is refreshing after the constant story of retreat and withdrawal .... Scouting patrols are now well forward of Ua-ule [*sic* Ua-Ule] Creek. One patrol came back today after having pushed several miles behind the Japanese lines in search of the 75-millimetre gun that had been sending over a few shells lately in answer to the salvoes from our 25-pounders. They sighted it perched on a high knoll, which they couldn't reach. They could see the crew lolling around waiting [for] orders from an officer standing with binoculars to his eyes. The Australians took careful aim and killed the officer and half the crew before they were forced to withdraw to escape a stronger enemy patrol sent out to get them. Another patrol ambushed a Jap patrol on the southern slopes of Ioribaiwa and killed six of them.

These patrols of ours are bringing in a heap of valuable information and we don't lose any time in exploiting that knowledge. Some of our fellows went out last night and found the Japs frantically digging new trenches and weapon pits on the southern slopes of the Ioribaiwa spur. Behind them coolies ... impressed 'chain gang' from Rabaul – were working on a network of new tracks through the flanks and widening the old main track.

The information was sent back to our 25-pounders and within a few minutes, shells were whistling through the driving rain across two mountain ridges and bursting nicely dead in the centre of the new enemy positions. All activity ceased and the Japs scuttled back across the protecting spur.

Behind Imita new Australian forces are assembling. There's a constant procession of hard-faced, green-uniformed Australians moving northward along the slimy mountain track. Hundreds of muscular natives are slithering through the rain with great loads of machinegun ammunition and mortar-bombs and

CHAPTER 36

grenades. The Japs certainly won't get any closer to Moresby. And if our advance goes as it's planned, they may be many miles farther away before this week is out.[6]

* * *

Meanwhile, the men of the 2/16th Battalion were still conducting patrols east of the main track while those of the 2/14th Battalion were doing likewise to the west.[7] That day, these men of the 21st Brigade finally got the welcomed news that Lieutenant Colonel Geoffrey Cooper and the bulk of the mates from the 2/27th Battalion had finally been located. The 21st Brigade diarist recorded that at 8.30 am, Brigadier Arnold Potts visited Brigade Headquarters. While there, word came in that Lieutenant Colonel Cooper, twelve officers and 151 other ranks from the 2/27th Battalion were now at Jawerere; the stretchers with 'D' Company were still four days out.[8]

The war diary of the 2/14th Battalion also records that the men were still anxiously waiting to be relieved: 'The weather was particularly good this day, assisting in keeping up the spirits of the men. News summaries giving world news were received by wireless and promulgated to all ranks'. At 10 am, a patrol of one NCO and seven men from the Headquarters Company had moved out along the west track following the Goldie River, on a five-hour patrol. Lieutenant Colonel Albert Caro passed on the information that Cooper and around 160 of his men had been located. Brigadier Walter Smith, a 47-year-old engineer from Sydney, accompanied by the CO of the 55th Battalion visited Cara, but he could provide no information regarding when the brigade would be relieving them. Captain Phillip Rhoden and Lieutenant Stan Bisset then made a reconnaissance of the river and track crossing the area south of Uberi. It was decided the unit would likely occupy an extricating position in that area, to enable the 2/16th Battalion and other forces to withdraw. At 4 pm, a patrol by men from the Headquarters Company returned with nothing to report. Thirty minutes later, a 'distribution of Canteen supplies was made. The cost of these was defrayed by the Ladies Welfare Fund. The comforts were appreciated by all ranks'.[9]

* * *

Still on Ioribaiwa Ridge was 1st Lieutenant Horibe and his men of *No. 6 Company, II/144th Regiment*. When he inspected his men's weapons, he supervised a cleaning session. As per usual, he sent out a patrol to reconnoitrer the area south of their position. At 2 pm, the *No. 4 Company* sentries fired, and his comrades of the *II/144th Regiment* stood-to in the defensive positions on 'the resistance line'.[10] From 3.30 pm, *No. 4 Company* was shelled, and Horibe ordered his men into their air raid trenches. Later, one day's dry bread ration was issued.[11]

* * *

Still behind Imita Ridge was Major Albert Moore and Corporal Jock Inglis manning their Red Shield post; Moore recalled:

> September 22, the 25-pounders opened up. I was told it was the first time in the campaign that the Australians had superior fire power. Now that our supply line was so much shorter and the supply line so much longer for the opposition it was not surprising that the tide was about to turn.
>
> I managed to secure a large biscuit tin and got the idea of trying to build it into the mud for use as an oven. Having secured a bag of flour I thought that I would try my hand at making some scones for the men. It has to be remembered that they had not seen anything else for many weeks but hard rations – bully beef and hard 'dog' biscuits. When our 'oven' was ready to function, I lit a fire under it. With a small chimney working the boys thought it reminded them of the Spirit of Progress. Perhaps it did, but the progress was a long way behind the time. I tipped a heap of flour into a mixing bowl and went away to secure some water, sugar and salt. When I returned, the flour had come to

CHAPTER 36

life! It was moving, and there was a hill in the centre on top of the flour. As I took a closer look, I discovered that the cause of the life was weevils. There were millions of them. I scooped them out by the handful and tossed them outside, then I waited for the next swarm to work themselves into a heap and out they also went. That went on for quite a while. I felt sorry for the weevils, but it had to happen!

In went the ingredients, they were all mixed up, and perhaps there was more body in those scones than in any other fare ever served to the fighting forces! The colour of the finished article was not so good! They were not 10-minute scones, but when you are hungry even sour grapes can taste sweet. As the fellows passed by and had their cuppa, I dished them up a scone with a little margarine and jam and, boy, the comments!

The result, however, was so encouraging that I decided to produce more for the remnants of my battalion. I also made a few jam tarts and took the cooking forward with a thousand or two sheets of paper and envelopes which the men had requested. As I arrived at their position the rain was falling a deluge and they were partaking of a cuppa. Phil Rhoden and Cliff Thompson were having their cuppa out of a jam tin. Huddled under a groundsheet and without a complaint in the world they were thrilled to partake of scones and jam tarts. I never told them about the weevils, but I am sure that they would have eaten the scones just the same.

I took a quantity of writing paper and envelopes into Uberi for the use of men in the area there and continued to make scones for the men on their way out. Our oven had come apart somewhat as it was only soldered, and the solder had melted. However, we managed to stick it together with mud and it continued to function.[12]

* * *

That day, the men of 'B' Patrol, 2/6th Independent Company, covering the extreme left flank of Maroubra Force, where ordered to mount offensive patrols towards Efogi as far as their supply situation would allow. These commandos were also to patrol around the Mount Asquith area about ten kilometres west of Efogi, as the crow flies. They were to supply as much intelligence as possible, including the capture of prisoners. They were also ordered to aggressively attack any Japanese force they encountered.[13]

* * *

Now back behind Australian lines, Lieutenant Maurice Treacy wrote in his report for 22 September: 'After receiving kind treatment from the commando patrol the party set off the following morning and arrived at Dorobisolo at 1330 hrs .... A wireless message was sent to 7th Division HQ giving the story regarding Capt. Buckler's party and asking for assistance'.[14] For three days, the three men rested in the village before making the final leg of their voyage back to Port Moresby.

* * *

Meanwhile, along the northern banks of the Naoro River, Corporal John Burns, Private Alfred Zanker and their wounded mates were anxiously waiting for rescue. However, the appearance of Allied aircraft and the bombing and strafing of their immediate area did not bode well for their imminent rescue: 'The 22nd finds our air force over us again at dawn and they bomb and strafe right opposite us on the northern side of the creek, we made it a habit to keep very low when these little acts were put on. During the morning I found four pigs whilst on the top end of the garden so if ever the real emergency had arrived, we could have killed one of them. The flies were around in millions again by lunch time or should I say midday and several of the lad's wounds became blown at this stage. It gave the lads a very rough

CHAPTER 36

time. Heavy rains during the afternoon spoilt all our wood for the evening fire and we could only boil a very limited quantity'.[15]

* * *

Still lost in the heart of the Owen Stanleys was Lieutenant Robert McIlroy and his small party – they had been lost in the jungle for three weeks now. He was not to know at this point that the four men he had sent forward to bring back help had already been attacked by Papuans, with two of these men being killed. He later reported: '16–22 Sept: During this period, I made contact with 2 natives from the mission village, who although unable to obtain bearers were prepared to wait until the injured man's leg recovered. Also, some other natives came into the garden and on 22 Sept they carried the injured man for ½ a day to another garden (ESAOLO)[xxiii] and then stated that they would have to get additional bearers and would pick us up again in two days' time, however, they didn't appear again'.[16]

---

xxiii The author could not locate this village, and it may be the village of Enivilogo located about two kilometres north-west of Brigade Hill.

37
'GENERAL TURNED ON A GOOD BREW'

On 23 September, the staff of the 7th Division reported that the troops had suffered unnecessary casualties by Japanese mortar, artillery and heavy machinegun fire from not adequately digging-in. It also recorded: 'The Jap, on the other hand, is an adept and soon after penetrating into our posns was very quickly well dug-in. Our tps must not be allowed to think that because they are concealed from view, that there is no necessity for them to dig-in. The necessity for digging-in is equally important after having seized an objective'.[1]

That morning, Lieutenant Colonel Alfred Buttrose, commanding the 2/33rd Battalion, ordered 'D' Company to send out 50 men on a long-distance patrol to reconnoitre a route east of Ioribaiwa village. That night, these men, led by Captain Trevor Clowes, would bivouac at Dump No. 44. It was also now that the men of 'C' Company, who had since 19 September been in the forward position overlooking Major Moore's former Salvation Army post, were finally relieved by 'A' Company.[2]

Meanwhile, on Imita Ridge, the men noted that their artillery continued to register and pour harassing fire onto Ioribaiwa Ridge. This fire was having excellent results, as another patrol of the 2/25th Battalion returned at around 2.30 pm and reported no sign of the enemy on the main track 300 metres south of Ioribaiwa.[3] The diarist with this battalion recorded: 'Active patrolling by Bn carried out towards ridge. No enemy contacted, and patrols report very little enemy activity fwd of IORIBAIWA RIDGE'.[4] The 7th Division war diary also records that the Australian artillery continued their

CHAPTER 37

harassing fire against Ioribaiwa Ridge during the night and that a Japanese listening post had been located on the forward slopes of Imita Ridge, which was attacked and driven back by the men of the 2/25th Battalion.[5]

At 12.30 pm, Captain Atkinson's standing patrol with the 3rd Battalion, which had moved out three days before with orders to find and harass the Japanese lines of communications, returned to the battalion. Atkinson reported that they had been ambushed by a Japanese patrol west of Ioribaiwa village, and while they had inflicted several casualties against the enemy, he too had suffered casualties with four men killed, among them Lieutenant Edward Dullard.[6]

* * *

On Ioribaiwa Ridge was 1st Lieutenant Horibe and his 60 men of *No. 6 Company, II/144th Regiment*. That day's patrol returned with the familiar report that trees were being felled across the creek, to the front of *No. 5 Company*. At 3.30 pm, his position was shelled by Australian artillery for an hour, but no damage or casualties were reported.[7]

Still trying to maintain communications between headquarter units was the unknown leader of the signals unit with *No. 2 Infantry Section, No. 1 Company, I/144th Regiment* who recorded in his diary that day: 'The line between Bde HQ and Regt HQ was again damaged but was repaired by the YOSHIDA sec'.[8] First Lieutenant Hirano, leading *No. 3 Platoon, No. 1 Company, I/144th Regiment* recorded in his diary: '1st Lt. CHIGAWA is suffering from kidney ailment'.[9]

* * *

At this time, more men from the 2/27th Battalion had arrived back behind the Australian lines, and the men of the 21st Brigade received word that within days they would finally be moving back to the Port Moresby area for much-needed rest and reorganisation. The 2/14th Battalion war diary records for 23 September that Lieutenant David Rainey and seven men took over from the 2/16th Battalion standing patrol, who were on the high ground north-west of Uberi. Captains Phillip Rhoden and Clifford Thompson, along with Lieutenant Stan Bisset contacted Lieutenant Colonel Albert Caro at Uberi at 9.30 am and received word that another 119 men of the 2/27th Battalion had arrived at Itiki. Caro was then joined by Lieutenant Colonel Oscar Isaachsen, a 31-year-old barrister from Malvern in South Australia, who had been a former 2i/C of the 2/27th Battalion, having served with the battalion in Syria. He had since been promoted to lieutenant colonel and now commanded the 36th Militia Battalion. Isaachsen informed Caro that he was to relieve the 2/14th and 2/16th battalions within the next few days.[10]

Isaachsen was described as a quiet, calm, and an efficient officer, and these qualities would come to the fore during his time leading the battalion; he quickly gained the respect and admiration of his men. During the fighting in Papua, it was noted that the Japanese targeted officers, and it became imperative that officers be indistinguishable from the rank and file. As described in the battalion's published history: 'At 36th Battalion Headquarters, the commanding officer and his batman, Private [Clyde] Jacobs looked like two ordinary soldiers and were called "Ike" and "Oigle" respectively. There innovations were welcomed by all and strengthened the spirit of the unit'.[11]

Most of the rough shelters used by the 2/14th Battalion were quickly seen to be unsatisfactory, as demonstrated after the first heavy downpour that afternoon. At 10 am, a patrol composed of 12 men led by Lieutenant John Crameri started out to relieve the patrol led by Lieutenant Ken Evans. The battalion war diary records with gratitude: 'A surprise visit to the Bn paid by Adjt MOORE of the Salvation Army. The scones he brought were thoroughly enjoyed by all. Active patrolling was still being carried out but little or no contact made. Our artillery continued to shell EURO-BIRA [sic], while aircraft resumed strafing and bombing of track'.[12]

* * *

CHAPTER 37

Further north, Corporal John Burns and Private Alfred Zanker of the 2/27th Battalion, along with their wounded mates of the 21st Brigade must have become increasingly concerned, as they were still well behind enemy lines with no sign of any rescue party:

> Wednesday 23rd was one of our hardest days. The sun was fiercer than ever, and it took a lot out of the lads. Cp Williams spent a terrible night and when Zanker and I had washed the lads we decided to put him on a new stretcher and put the first fresh dressings on his wounds. It was a terrific job, but we succeeded in the end. Both Zanker and I had a couple of blackouts during it. We had now used two of our last three dressings, so we felt none too happy. About 10 o'clock a bush-native came to us with a bunch of vegetables bananas, we managed to rake up 1/7d for him. In the evening we duly boiled them and enjoyed a tasty few mouthfuls. The lads had run out of smokes too and I collected a few likely looking leaves but I'm afraid they weren't quite the right type. Diarrhoea broke out during the day, and we were lifting the poor lads for the next twenty-four hours without respite.[13]

* * *

Lieutenant Robert McIlroy and his three men, along with the two Papuans from the Seventh Day Adventist Mission Station in the vicinity of what was likely Enivilogo village, were still waiting on the return of the Papuan bearers, who told them they would be back the next day. He later wrote in his report for the battalion war diary: 'Remain in the garden, the two mission natives assisting us to dig yams and taro and drawing water each day'.[14]

* * *

Meanwhile, General Thomas Blamey had arrived in Port Moresby during the late afternoon and took over command of NGF. By now,

Brigadier Kenneth Eather had already begun to edge forward. The Japanese numbers west of Ioribaiwa were estimated by his patrols to number just 600 men.[15]

Lieutenant General Sydney Rowell quickly made it clear to Blamey that he did not appreciate his presence, and it quickly became evident to all concerned that Blamey's 'arrangement' of mutual support would not work. Rowell had good reasons to be suspicious as Blamey's track record for using the success of others to boost his career while blaming others in order to cover his own shortcomings were legendary, even among the average digger. That morning, Rowell sought advice from a good friend, 37-year-old Lieutenant Colonel George Fenton with NGF, who oversaw censorship and was all too familiar with current political machinations. Like Rowell, Fenton did not trust Blamey – after his appalling behaviour during the campaigns in the Middle East, the Mediterranean, and North Africa, and as recorded in his diary:

> Rowell sent for me. He said, 'You have mixed in high politics a good deal in the past two years. I wonder if you can help me in my problem. I have here a letter from the C-in-C. He is coming up tonight to take charge – and I'm not going to tolerate it. I have no confidence in him; he has let me down twice already, and I've already told him I'll never serve under him again on the field. I'd like to tell him to go to hell, but I'm a professional soldier with no private income, and I have my family to consider. Without boasting I think I can say a bad show has been pulled together since I came up and now, we seem likely to go ahead. I have a good team here, and they trust me. More than that; the other services trust me; I don't think anyone else could have done as much as I have done to get cooperation with the Air Force. The Americans are playing with us 100 per cent; last night quite off his own bat Willoughby sent a signal to GHQ and Sutherland [MacArthur's Chief of Staff]

CHAPTER 37

is going to do something. I don't trust the C-in-C; I don't think he's a good soldier, and I don't think he's going to do the show any good here. So far, he's done nothing at all; LHQ [Land Headquarters] just hasn't played the game in sending stuff up here … the position is better now; we've pulled it round, and so he's coming up to take over. There's no need for him here, I shall tell him so. He says here that Curtin has sent him up. Do you know anything of the political background? Unfortunately, I haven't played politics; haven't friends in Parliament; I have only been a soldier doing my job.[16]

Indeed, Major General Charles Willoughby stuck his neck out in support of Rowell and cabled MacArthur at the time: 'Visited Corps and Divisional HQ and forward areas. Situation entirely in hand. Local offensive movement in progress objective Ioribaiwa. Have just heard from Rowell of projected move of Landops [Allied Land Forces Headquarters]. Cannot see that any benefit is likely to arise from interposition of another headquarters at the present juncture. Rowell enjoys the confidence of all command echelons'.[17]

Rowell later wrote in his memoirs: 'Blamey arrived in Port Moresby on the 23rd September. During that evening and each of the two following, we had a full and frank discussion of my position as general officer commanding in view of his arrival. At times the discussion was pretty acrimonious as I believed, and said, that the confidence he had expressed in me in his national broadcast a week earlier no longer existed. On several occasions he said that this was the first "bump" I had had in my service, and this clearly indicated to me that, for all practical purposes, I was being supplanted in command. The main theme of these discussions, apart from the question of loss of confidence, was an endeavour to find a working arrangement suited to the circumstances. Blamey had no staff, so inevitably sooner or later my own staff would be called on to serve two masters, and I would merely become a figurehead'.[18]

SAVING PORT MORESBY

Indeed, Rowell was still concerned about MacArthur's ongoing preoccupation with large-scale flanking manoeuvrers and informed Blamey that the American 'Wanigela concept' was 'all of a pattern with the American idea of outflanking the mountains', which he strongly opposed as a waste of men and resources.[19] He told Blamey any large-scale movement towards Wanigela – located on the north coast approximately 100 kilometres, as the crow flies, east of Buna – would merely result in the Japanese sending more troops to Buna. He also stressed that the ability to supply such a large-scale manoeuvrer was beyond their abilities, even if such a movement through the mountains east of the Kokoda Track were possible – which seemed most unlikely. Within days, however, and without conferring with Rowell, Blamey went about instigating this very plan – it would be the last straw for Rowell.[20]

By the time Blamey arrived on 23 September, Rowell's command was organised around a two-divisional structure with the arrival of the 6th Division headquarters in Port Moresby on 19 September, commanded by Major General George Vasey. The 6th Division had only recently reformed since leaving the Middle East, as two of its brigades had been held in Ceylon, Sri Lanka, while the other had been posted to the Northern Territory. While the division's headquarters had been re-established, the division itself remained fragmented, as its 16th Brigade was in Port Moresby, its 17th Brigade was bound for Milne Bay, and the 19th Brigade was still in the Darwin area.[21]

Within the 7th Division were the 25th Brigade Group (including in addition to its normal battalions, the 3rd Battalion, and the 2/1st Pioneer), the militia 14th Brigade (36th, 39th and 55th battalions), and the 16th and 21st brigades. The role of the 25th Brigade was to continue holding Imita Ridge, the 14th Brigade had the task of keeping the lines of communications between Ilolo and Uberi secure, the 16th Brigade was to cover the approaches to Port Moresby through the Eilogo–Hombrom Bluff area, while the 21st Brigade would be held in reserve to gain some much-needed rest and reorganisation.[22]

CHAPTER 37

The 6th Division, in Port Moresby, now consisted of the 16th Brigade (2/1st, 2/2nd, and 2/3rd battalions)[xxiv], the militia 30th Brigade (49th and 53rd battalions and Honner Force), the 2/6th Independent Company less detachments, the US 128th Infantry Regiment, 32nd Division, the NGF Carrier Group, the 13th Field Regiment, the 2/6th Field Regiment and a battery of the 14th Field Regiment with some anti-tank guns. Indeed, the men of Honner Force were soon assigned to reinforce the 50-man 'B' Patrol, 2/6th Independent Company, covering the extreme left flank of the Australian positions along the Kokoda Track. Finally, the exhausted and battle-weary men of the 21st Brigade and the militiamen of the 39th Battalion came under the command of the division; this also applied temporarily to the 7th Divisional Cavalry and 2/1st and 2/5th field regiments, which were moving to Port Moresby. Major General Vasey had been ordered to stop Japanese infiltration parties from penetrating into the Port Moresby base area through the country west of Hombrom Bluff. As such, his men were to patrol offensively to the Japanese western flank, as well as be prepared to defend Port Moresby against invasion.[23]

Private Frederick Ashford, 2/6th Independent Company, recalled the arrival of 'Bloody George', as Vassey was affectionately known throughout the AIF. Vassey arrived to inspect the company then at Laloki and to discuss operations west of the township. Later, he hitched a ride from there to inspect the men of Honner Force and 'B' Patrol, 2/6th Independent Company. Ashford was his assigned driver, trailing behind the jeep with a trailer of rations. At some point, the jeep lost a nut from its steering arm and was immobilised. Vassey walked back along the track and by some miracle found the nut. On returning to the jeep, Ashford offered to boil the billy for a cup of tea. Vassey replied, 'You unload your Jeep and I'll make the bloody tea'. Ashford recalled that the 'General turned on a good brew'.[24]

[xxiv] The 16th Brigade would be temporarily assigned to the Australian 7th Division for the duration of the fighting along the Kokoda Track and the assaults against the Japanese beachheads.

Meanwhile, moving north to inspect his men was Brigadier Arnold Potts. On the way up, he passed both Blamey and Rowell. After a brief discussion, Potts moved out along the track. Blamey now asked Rowell, 'What do you propose to do about Potts?' Rowell replied, 'I don't see any reason why he shouldn't go back to his brigade. This was his first command, and he was out there on his own. He's had a very trying time. I think that Potts will profit a lot by the experience he's had'.[25]

38

'IF WE DON'T GO FORWARD, WE WILL ONLY STARVE TO DEATH'

On 24 September, Major General Arthur Allen, commanding the Australian 7th Division made his way to Imita Ridge to confer with Brigadier Kenneth Eather. By 8 am, the Japanese were firing at least one mountain gun as reported in the brigade war diary: 'Flashes observed on ridge'.[1] On the way, he had called into Major Albert Moore's Red Shield post just behind Imita Ridge, no doubt appreciative of the cuppa and scone and jam provided. The salvationist recalled: 'September 24, General "Tubby" Allen called in and was warm in his expressions of appreciation for what we had done for his men. The general called again the following day and again had a cuppa and a chat'.[2]

That morning, Lieutenant Colonel Charles Withy, commanding the 2/25th Battalion, had ordered several patrols to be sent out. Among them was a patrol from 9 Platoon 'A' Company led by Lieutenant Stanley Steel and his 2i/c Lieutenant Bruce Walker, a 24-year-old clerk from Clayfield in Queensland. This patrol soon engaged entrenched Japanese near the former Brigade Headquarters around the Grassy Plateau along the main track. Steel attacked one flank by using grenades, forcing the Japanese to withdraw, and now he and his men dug-in occupying the position. During the fighting, 22-year-old Private James Eltherington, from Brisbane in Queensland, and 25-year-old Private Spencer 'Spinner' McDonald, from Kelvin Groove in Queensland, were wounded. It was not long before the Japanese were trying to flank Steel and his men, and Lieutenant Bruce Walker, a 24-year-old clerk from Brisbane in Queensland, leading his men of 17 Platoon went forward to support the men of 9 Platoon. Later that day, Captain Thorlough O'Bryen, a 26-year-old branch manager from Emerald in Queensland, leading 'B' Company, 2/25th Battalion was ordered to support Steel and Walker at the former Brigade Headquarters' position just behind the main ridgeline of Ioribaiwa.[3]

On the left flank, Lieutenant Colonel Colin Dunbar, commanding the 2/31st Battalion had also sent out several patrols. All returned with nothing to report. Japanese mountain guns, however, continued to shell Imita Ridge, close to his position, but inflicted no casualties. The battalion war diarist wrote that day of hearing spasmodic machinegun bursts and 'mortar and/or grenades bursting' in the distance east of Ioribaiwa during the morning hours. This was soon followed by Australian artillery fire from the 14th Field Regiment against the ridge, which continued for most of the day.[4] Even so, the battalion's aid post had enough wounded already to deal with, as recalled at the time by stretcher-bearer, Private Hamlyn-Harris:

> In a little clearing in the rain sodden jungle the RMO, Captain Forbes Mackenzie, had set up the RAP. Here stretchers are on the ground and sufferers upon them, men who have been in action for days, bearded and dishevelled bloody men with a blue deathly pallor, either in [fever] ... or shivering under a cold sweat; and all in pain. Boys like me are there too, to perform various duties, and Padre [Donald Redding] is giving comfort to everyone, but Doctor is the saver of life if there be but one chance in a hundred. His operating table is a collapsible affair on which the bearers can set a stretcher supporting the badly wounded boy who needs immediate attention. No matter what the time be, whether daytime, evening, midnight or in the still small hours of morning. Doctor is ready at that table if need be, and he never gives way to fatigue. Surrounded by

CHAPTER 38

dripping jungle and reeking mud, not far perhaps from places of pestilence and infection beyond description, where the enemy had not paused to bury its dead sufficiently or where the almost dead with dysentery or fever have been abandoned, Doctor has managed to make a clean space and perform most difficult and unusual operations, in wind, rain and appalling cold.[5]

Covering the right flank were the men of the 2/33rd Battalion. That day, Captain Cyrus Peach from the Headquarters Company and his 50-man patrol returned from their reconnaissance of the Jawarere–Ta-ai–Nauro Track to the east, having encountered no Japanese, nor had they discovered any suitable tracks in that area that led directly to Nauro. Meanwhile, Captain Trevor Clowes and his 50-man patrol from 'D' Company left Dump No. 44 that morning, as they advanced towards Ioribaiwa Ridge. Sergeant William Crooks recalled that by 10 am, these men were passing through elements of the 2/25th Battalion, which had been edging forward since 21 September from the Kunai Patch where the men of 'C' Company, 2/33rd Battalion had sprung their ambush during the afternoon of 19 September. Here, Lieutenant Colonel Withy provided information to Captain Clowes of the low barricades the Japanese had set up along the lower slopes of Ioribaiwa Ridge.[6] Clowes, who was using the Kunai Patch as a base of operations, now sent out several sections on two-hour patrols north, east and west of the track to ensure there were no ambushes up ahead. One of these patrols came across the men of 9 Platoon, 2/25th Battalion who had dug-in at the former Brigade Headquarters' position on the Grassy Plateau, having dislodged the Japanese just hours earlier. It was not long after that the rains began beating down and the mists rose from Ua-Ule Creek, making observation almost impossible. Most patrols had returned by 5 pm, having encountered no sign of the enemy.[7]

* * *

Captain Bert Kienzle at Dump No. 66, who had been suffering from fever, that day wrote to Meryl: 'Feeling ever so much better, a little weak, after that second attack of dysentery and fever, thanks to a rest and Dr Vernon's care. I know what I do need – a complete change & rest away, preferably in your care my dearest. I have been in this campaign longer than any other man in the forward area & in this sort of country as tough as I am, must admit that it has played up with me. Pray that I come through this all safe & sound as I am living only for the day when I can be with you and the children again. Do take care of yourself my dearest. My life with you is the happiest period in it & it has a long way to go yet God willing'.[8]

* * *

During the early morning hours, Horibe and his men of *No. 6 Company, II/144th Regiment*, on Ioribaiwa Ridge, where again targeted by the Australian artillery, with the first shells exploding at 3 am. These shells were exploding in his bivouac area, indicating that their position had been observed and noted by the enemy. Three of his men were wounded before the rest of his company dived into their air raid trenches. The bombardment continued into the later morning. Later, one of his men was evacuated due to sickness, so the company strength was now 56 men. Half of Koiwai's *II/41st Regiment* moved up on to the right flank of the Japanese position at Ioribaiwa, likely close to Spotter's Hut.[9]

Sometime that day, Japanese Major General Horii Tomitarô received orders from the area army commander, General Imamura Hitoshi, that he was to withdraw his entire force back over the mountains; offensive operations were on hold for now. The order states: 'The commander of the South Seas Force will assemble his main strength in the Isurava and Kokoda areas and secure these as a base for future offensives. In addition, the defenders in the Buna area will be strengthened'.[10] Reinforcements from Rabaul would be used to help bolster the beachhead defences and would be shipped to Papua from Rabaul.[11]

CHAPTER 38

On reaching the Ioribaiwa Ridge, Horii must have already known he was at the end of his resources; he and his men could no longer get supplies from the other side of the rugged cloud-covered mountain range. Added to this, the consistent sorties flown by Allied airmen did effective work at the Wairopi crossing where they destroyed the bridge as fast as the Japanese could rebuild it. Also under continued bombing and strafing was the Japanese airstrip at Buna.[12] Samuel Milner, the American official historian of the United States involvement in the New Guinea campaign, recorded:

> As fast as the Japanese naval construction troops at Buna filled in the runway, the Fifth Air Force would see to it that it was pitted again; and efforts of the *15th Independent Engineers* to keep the Wairopi Bridge in use were being continually set at naught by Fifth Air Force and attached RAAF units that would roar in at low levels to demolish it. Because of the relentless air attacks, Japanese supply trains were virtually forced off the trails. Food, as a result, though still available to the Japanese in the rear areas was not getting through to the frontlines. Whole battalions of the *South Seas Detachment* were foraging everywhere along the trail for food. Native gardens along the line of march were being stripped of sugar cane, taro, yams, pumpkins, melons, and everything else that was editable, but there was not enough food on that poor upland area to feed such a host for long. By September, the frontline ration was down to less than a cupful of rice per day. By 17 September, the day after the Japanese seizure of Ioribaiwa, with the beach at Port Moresby almost visible from the height on which the Japanese found themselves, there was not a grain of rice left on the ridge for issue to the troops.[13]

By now, the Japanese High Command felt that it should concentrate all its energies on holding Guadalcanal, especially given the disaster at

Milne Bay inflicted upon them by the Australians. Horii and his men were now ordered to fall back to the more easily defended Buna–Gona beachheads. After the reversals at Guadalcanal had been brought back to the Japanese favour the main Japanese effort would be redirected against New Guinea and, in concert with a fresh move to seize Milne Bay and another coastal approach to Port Moresby, Horii would once more cross the mountains.[14]

Horii ordered Colonel Kusunose Masao and the men of his *144th Regiment*, who had originally spearheaded the invasion of Papua on 21 July, to act as the rearguard. The men of the *I* and *III* battalions would hold Ioribaiwa Ridge, while two companies of the *I/41st Regiment* would be in support at Nauro, and another in reserve at Kokoda. Horii ordered the men of the *II/144th Regiment* to hold Ioribaiwa Ridge for as long as possible before retiring northward; these men would then be relieved by troops positioned at Isurava and Kokoda. The remainder of Horii's force, defined by the main body of the *41st Regiment*, would fall back to the coast to bolster the beachhead defences while the regard of the *144th Regiment* would fight a delaying action against the advancing Australians.[15]

Colonel Kusunose was furious on hearing the order to retire and argued with Horii's staff officer, Lieutenant Colonel Tanaka Toyonari, exclaiming: '"How many men have we lost to come this far? We shall not go back a single step. Tanaka, you people are only drawing lines on paper with a ruler." Tanaka replied they had been given orders, and they must be obeyed. Kusunose responded, "We have come this far after all. Let's continue to strike. There is food over there. If we go back along the track, there is no food. If we don't go forward, we will only starve to death." Tanaka responded. "We still have some food which allows us to manage to go back. If we charge into the enemy, we might as well commit suicide." The more junior officer could hear the conversation and now their commanding officer, Horii intervened, "Wait. A telegram with the same content as the one from the area army has come from the Imperial Headquarters. This is the order from

CHAPTER 38

the Emperor." On hearing this, Kusunose clicked his heels and stood still – the issue was resolved – they would fall back as ordered by the high command in Tokyo'.[16]

Close by was Japanese war correspondent Okada Seizo, who provided a vivid picture of Horii on hearing the news and discussing it with his senior commanders:

> On the 24th a wireless message from General Imamura, commander of the Area Army, who had his headquarters at Rabaul then, came to Major General Horii, saying 'Stop attacking Port Moresby and wait for further instructions at present position'. It was followed by a second message that night: 'Withdraw from present position to some point in the Owen Stanley Range which you may consider best for strategic purposes'. The detachment headquarters was upset.
>
> In one of the small, thatched huts which we had hastily built on the mountain side, Sato [a fellow journalist] and I had just finished our usual scanty evening meal of sweet potatoes, when we scented some important change in the situation. We hurried out to see Major General Horii in his tent that stood on a little uncovered elevation. The night was wearing on in the profound stillness of the mountains. On a thin straw mat in the tent, the elderly commander was sitting solemnly upright on his heels, his face emaciated, his grey hair reflecting the dim light of a candle that stood on the inner lid of a ration can. Lt Colonel Tanaka, his staff officer sat face to face with him, also on a mat. Two lonely shadows were cast on the dirty wet canvass.
>
> 'I'm not going back, not a step! Are you going back, Tanaka? How can we abandon this position, after all the blood the soldiers have shed and the hardships they have endured? I cannot give such an order.'

SAVING PORT MORESBY

Major General Horii grasped his samurai sword that lay beside him, and drawing a little closer to the staff officer, added in a biting tone: 'I will not retreat an inch. I'd rather disguise myself as a native of these mountains and stay here!'

The staff officer was silent, watching the burning wick of the candle as though to avoid the commander's eyes, when a rustling sound was heard in the thicket outside, and the signal squad commander came in with another wireless message. It was an order from the Area Army Commander at Rabaul instructing the Horii detachment to withdraw completely from the Owen Stanleys and concentrate on the coast of Buna. This message was immediately followed by a similar order that came directly from the Imperial Headquarters in Tokyo. It was now beyond doubt that the order had not been issued by the Commander of the Area Army on his own discretion but had been authorised by the Emperor himself. His Majesty's order had to be obeyed. It is true there was a strong body of opinion among the hot-headed battalion commanders advocating a desperate single-handed thrust into Port Moresby. But staff-officer Tanaka remained cool and reasoned with them saying it was a suicidal action even if everything went well except the supply of food, which was in a hopeless condition.

The night was far advanced. It had begun to drizzle, softly. The headquarters was in confusion sending out messages to the front positions instructing them to make preparations for immediate withdrawal. The stout adjutant dashed back and forth through the rain between the headquarters and the signal unit.

As I passed by the signalmen's tent, where Lt Colonel Tanaka was talking over the telephone with front-line commanders, I heard a deep angry voice ringing through the receiver: 'It's all very well for a staff officer like you to say that, but how can we artillerymen leave our guns? No, we can't!'

CHAPTER 38

'I know – I know how you feel,' said Lt Colonel Tanaka in his calm, soothing tone, though I noticed a touch of suppressed excitement in his voice. 'But it can't be helped, if you are short of hands. Guns are valuable, of course, but soldiers are more valuable. See? Leave your guns and ammunition and all that. It can't be helped now. We must take every living soldier with us – every living soldier, sick or wounded.'

I groped down the dark slope with Sato, my fellow correspondent, and came back to our hut. Photographer Katayama and the liaison men were waiting anxiously for me. I told them briefly that the detachment had been ordered by the Imperial Headquarters to leave the position that night and go back. We all felt relieved, for uneasiness and despair about the campaign had been gnawing at our hearts for some time. We decided to start at once. We did not like to accompany the main body of the troops which might be thrown into disorder while marching back along the narrow mountain path. Our rice ration, too, was running low. We were afraid of being starved on the way before we got to the coast. So, we equipped ourselves as lightly as possible with a camping outfit and what little food we had about us and set out on our way back.[17]

The unknown leader of the signals unit with *No. 2 Infantry Section, No. 1 Company, I/144th Regiment* was now busy ensuring that orders were being sent and received, even so, he made no comments in the second-last entry in his diary for 24 September: 'The line which connects the No. 3 Bn HQ with the No. 2 Bn HQ was damaged by artillery fire in the night but was repaired by No. 1 Sec'.[18] That night, the staff officer, Lieutenant Colonel Tanaka Toyonari, remained busy issuing the relevant orders, and when required was forced to argue the point: 'It would be suicidal action even if everything started well. Supply of food is in a hopeless state and our troops are too weakened for any sustained action. We don't know how large an enemy force we would face, but reports suggest they're well reinforced. It's more than likely we are greatly outnumbered. In that case, it would be pointless suicide'.[19] Regardless, they had their orders that came directly from the Imperial Headquarters in Japan, and all prepared to withdraw northwards.

Somewhere around Ioribaiwa, 1st Lieutenant Hirano Kogoro, commanding *No.3 Platoon, No. 1 Company, I/144th Regiment*, recorded in his diary just before midnight on hearing the news: 'Like a bolt from the blue received order to withdraw at 1130. It left us momentarily in a daze. It is truly regrettable to retreat from our present position after pressing so close to MORESBY at the cost of enormous sacrifices and casualties. Departed at 0730. At 1700, reached the area where the equipment of the Coy was left'.[20]

Among these men of the Japanese of *I/144th Regiment*, also getting the news was 27-year-old Sergeant Imanishi Sadashige, with *No. 2 Company*. He was a veteran of the war against China, and on returning home, he became a schoolteacher, but he was soon back in uniform with the *144th Regiment* and was among the first to invade Gona Beach on 21 July 1941. He recorded in his diary: 'We never knew how to retreat because we had never done it before. The order was to "change the marching direction", but we knew what it really meant. I thought the enemy would launch a counter-offensive once we started retreating, therefore we must escape quickly. Otherwise, we could not resist the enemy with our strength and number of troops. I thought it would be very difficult for the weakened soldiers around me'.[21]

The sergeant now told the men in his platoon to throw away their grenade launchers, as they had no grenades left, and his men did not have the strength to carry the launchers back across the mountains. Sergeant Imanishi recalled, however: 'The men would not throw them away, no matter how much I told them to do so, so I removed the fuses, grabbed their launchers from their hands and threw them into the jungle. The only thing I thought of was my men's lives. I

CHAPTER 38

couldn't let them die in vain, so I only thought how to make them survive. Sergeants and soldiers never think about the whole military operation'.[22]

Close by was Lieutenant Nakahashi Kokichi, with the *55th Mountain Artillery*, who recorded in his diary on hearing the news: 'The order came like a blot from the blue causing an overflowing … of emotion, which could not be suppressed; it was compounded by feelings of anger, sorrow and frustration. The purpose, the dreams and the desires of the officers and soldiers of the South Seas Force had vanished in an instant'.[23]

* * *

The war diary of the 2/14th Battalion records the welcomed news that Lieutenant Maurice 'Mokka' Treacy and his party had finally made their way to the Australian lines. They also got word of Captain Sydney Buckler's party: '0715 hrs, news received that Lieut. TREACY had reported into our B Echelon, also that news had been received of Capt. BUCKLER and party …. News received from the frontlines, that a slight push forward by the 2/25 Bn had resulted in a short clash with the enemy, while our artillery continued to harass enemy positions'.[24] Finally, the men were informed they would be moved out of the line down to Port Moresby on 26 September. The diarist recorded it was a damp night, and quiet except for a bombing raid against Port Moresby with some shelling by Australian artillery. Captains Phillip Rhoden and Clifford Thompson, along with Lieutenant Stan Bisset contacted Lieutenant Colonel Albert Caro at Uberi and received word regarding the relief of the battalion.

Later, Brigadier Arnold Potts, accompanied by 32-year-old Captain Ken Murdoch, from Melbourne, arrived. They informed Rhoden that lieutenants Alan McGavin and Jack Clements, who had been with the stretcher party of the 2/14th Battalion, had reported in safely with Cooper's men, and that the latest news regarding Captain Buckler indicated that he was now making his way south.[25]

SAVING PORT MORESBY

The commander of the 36th Militia Battalion, Lieutenant Colonel Isaachsen, was informed that his battalion was to take over from the remnants of the 2/14th and 2/16th battalions by the 26 September. That day, the militiamen, who had spent most of their time on wharf duties, headed into the Owen Stanleys for the first time.[26] The battalion history records the advance into the mountains and the system of supply at the end of the Kokoda Track, which was still partially reliant on pack animals:

> Down in the valley where it passed through Uberi village, the Kokoda Track was knee-deep in mud and slush. About twenty men of a light horse troop were in charge of the mules and horses which struggled into the mountains with supplies from Owers' Corner. These supplies were brought forward over a corduroy road by motor vehicle to Owers' Corner. A 'flying fox' was in course of construction down the long precipitous slope from Owers' Corner, but at this stage the zig-zag descent was negotiated by these amazing pack animals. It has been found that the mules and the small brumbies from Bootless Inlet were better fitted for this purpose than the plantation horses whose bigger feet were unsuited to the narrow mountain tracks. In a lather of sweat and down to their hocks in mud, these animals struggled forward in single file under crippling loads of ammunition and other supplies approaching 200 pounds in weight. It was cruel to see these poor wretches being goaded forward twenty feet at a time through the slush, after which they would be allowed to pause on unsteady legs for a couple of minutes before being urged forward again. Their bellies contracted and expanded like huge bellows as they tried to control their breathing in the thin mountain air. One small grey mare, particularly, looked as though her sides would burst at any moment. It was not only soldiers who suffered on the Kokoda Track.[27]

CHAPTER 38

That day, Brigadier Arnold Potts visited the 2/16th Battalion, accompanied by Major General Arthur Allen. The senior officers were finally informed that the battalion would be relieved by the militiamen of the 36th Battalion the next day. The battalion diarist recorded underwhelmingly and briefly: '0800 D Coy patrol leave to relieve C Coy patrol. 1200 Brig AW Potts DSO MC visited Bn at UBERI. 1230 Div Comd visits Ben area. 1300 Received movement order for 25 Sept'.[28]

* * *

Meanwhile, the rescue party – led by warrant officers Ronald Preece and Hylton Jarrett, who had been sent out to bring in Corporal John Burns, Private Alfred Zanker and the wounded – had been ambushed by the Japanese near Nauro and forced to withdraw. It was now that another party led by Sergeant Arthur Koehne, a 28-year-old police constable from Adelaide, with the 2/27th Battalion, was sent out to bring in the stranded party.[29]

While travelling along the northern banks of the Naoro River, the sick and wounded, who were being looked after by the dedicated Burns and Zanker, were feeling very much the worse for their experiences. Indeed, the gravely wounded Corporal Leonard Williams, a 41-year-old clerk from Linden Park in South Australia, succumbed to his wounds that day; Burns recorded in his report:

> After spending most of the night on the go, Friday dawns with a blazing hot sun and millions of flies. Again, I spent the night with Cpl Williams and at 0800 hrs he had a drink and at 0810 hrs we found him dead. We immediately dug a grave approx. three feet deep by means of a tin hat and a machete. It was a hard hour's work. At 0930 we buried him with just a little prayer. The event cast a gloom over the whole of the party. We were all very thrilled when the sun set on this sad day, it gave us relief from the millions of flies and the heat. The party was anything but happy.[30]

SAVING PORT MORESBY

* * *

By 24 September, the strained relationship between Blamey and Rowell was obvious to all. The Australian Commander-in-Chief asked Major General Samuel Burston, a 54-year-old physician from Adelaide, and now head of the Army's medical services (who was also an old friend of Rowell's) to convince Rowell to accept Blamey's request to act as his deputy, but it fell on deaf ears. Rowell was convinced that Blamey had come up to Port Moresby to steal his glory now that the Japanese advance had been halted and Australian offensive operations were commencing, regardless, Rowell had grown to loathe Blamey.[31]

39
'THE ORDER TO RETREAT HAD CRUSHED THE SPIRIT OF THE TROOPS'

On the 25 September, Captain Thorlough O'Bryen and his men of 'B' Company, 2/25th Battalion continued to push hard in support of lieutenants Stanley Steel and Bruce Walker, who were still holding the old Brigade Headquarters' position just behind the main ridgeline of Ioribaiwa on the Grassy Plateau. O'Bryen decided on an early morning frontal assault to push back the Japanese. This assault was led by Lieutenant Archibald Barnett, a 23-year-old clerk from Lismore in New South Wales, and 28-year-old Lieutenant Howard Strachan, from Rockhampton in Queensland, and the men of 11 Platoon, and a section from 10 Platoon. They attacked along the track; in support was 12 Platoon with the Headquarters Company to their rear. O'Bryen recalled: 'Coy HQ joined Lieut. Strachan when party arrived at high ridge area where casualties were being received. Pls were endeavouring to outflank enemy, but we could see him moving around both flanks and a mortar being moved to our right flank. I therefore decided to withdraw to high ground in rear … and hold this area … immediately in rear. I wish to record Lieut. BARNETT'S skilful handling of his party in overcoming enemy fwd posns'.[1] Barnett provided a more detailed account of the fighting in his report:

> At 0500 hrs., Cpl [James] MITCHELL moved fwd with four men to cut and remove tins strung on sig wire, stretching to within 25 yds of enemy posn on high ground. Cpl MITCHELL reported job done at 0520 hrs., and I moved fwd with a section

of eight men under Cpl [Thomas] HOUSE. We moved quietly and slowly. One sig wire had been missed, but almost the complete section was through before we were heard. Almost simultaneously Cpl HOUSE and enemy LMG opened fire. Enemy threw a grenade. Section went to ground. We worked up on the left flank, throwing grenades which seemed to be well placed. I sighted one enemy was appeared to be wearing a helmet and fired. We closed in, but the enemy had evacuated so quickly that I thought he had gone over the side. He left all his equipment.

According to arrangements made, I ordered Lieut. STRACHAN through with his section under Sgt [David] THORBURN. The remainder of 11 Pl., under Sgt [Thor] HARTMAN, did not seem to be following, so I sent a runner back to bring him on and tell 12 Pl. to follow. I suspected a ruse to come in behind me. When almost at the old Bde HQ posn., Lieut. STRACHAN reported a barricade of saplings across the track. I put him to ground behind it and moved half of my section up on the left with Sgt HARTMAN and 2 Brens. to cover us, whilst I endeavoured to push 12 Pl. under Lieut. [James] HOWES up the left of the position.

We reached a posn approx. 20 yds from Lieut. STRACHEN'S left, and over a slight ridge, when very heavy fire came from above us, just over the barricade. Lieut. HOWES was stopped, and I managed to get to a position where I could see a log sangar which appeared to be the main position. We tried grenades, some of which seemed to be effective. The fire continued. They returned only 2 grenades on our side, which were ineffective, but they were more numerous on the right.

Lieut. STRACHAN endeavoured to outflank, but sustained casualties. A similar effort on Lieut. HOWES' flank was unsuccessful owing to the steepness of the climb.

CHAPTER 39

The only way open to attack was straight over the barricade, and, as it was a climb to there and there was direct fire on it, I decided to confer with the OC, who had arrived. From him I got the information that a mortar was seen moving to our right flank and LMG fire already appeared to come from the right. It was decided to retire to the high ground taken and consolidate. I had no casualties; Lieut. STRACHEN had four.[2]

The brigade diarist recorded: '055 hrs: 2/25 Bn reports quiet night. B Coy attacked first light – no report. 0745 hrs: JAP driven off small knoll on main track 300 yds south of IORIBAIWA by 2/25 Bn. Offensive action continued'.[3] With this, 'A' Company advanced to the Kunai Patch, just 800 metres behind the Grassy Plateau in support of 'B' Company.[4]

* * *

All indications are that these were the Japanese of *No. 9 Company, III/144th Regiment*, as Lieutenant Horibe, acting commander of *No. 6 Company, II/144th Regiment*, recorded in his diary that they were forced to stand-to at around 6 am, as about 50 Australians attacked the neighbouring *No. 9 Company* and 'precautions were strictly enforced'.[5] Soon after, two days rations were issued, consisting of less than half a litre of rice per man. Another patrol was sent forward and returned to report that there was no change to the situation in front of their sector.

* * *

By 10 am, Major General Arthur Allen and Brigadier Kenneth Eather had decided at a commander's conference that all battalions would move forward the next day to take Ioribaiwa Ridge. The 2/33rd Battalion would advance on the right, the 2/31st on the left, and the men of the 2/25th Battalion would continue their advance in the centre. Soon after, Allen returned to divisional headquarters at Port Moresby.[6]

Meanwhile, Captain O'Bryen's CO, Lieutenant Colonel Charles Withy, with the 2/25th Battalion, who had not been present at the commander's conference, was busy reorganising his position to assist O'Bryen and 'B' Company in maintaining his pressure against the Japanese. The company had by now retired to the high ground a few hundred metres to the rear to seek cover from accurate Japanese mortar fire. Soon after, word arrived from 'D' Company that they would be advancing to reinforce 'B' Company; they were expected to be in position by 10 am. By 11.30 am, the Battalion Headquarters had moved forward to the newly occupied position held by 'D' Company, and just before 4 pm, the men of 'A' Company had advanced north of the Kunai Patch on the main track, where a subsidiary track branched off to the right leading towards Ponoon.[7]

Meanwhile, Lieutenant Colonel Alfred Buttrose was preparing for his men of the 2/33rd Battalion to move forward to take up a defensive position in the area near Major Moore's former Salvation Army post. From there, they were to push on the next day to reoccupy the right flank of Ioribaiwa Ridge. That night, they took up a bivouac near the aid post along the main track.[8] At this time, Captain Trevor Clowes and his men of 'D' Company, were still out on their patrol towards Ioribaiwa Ridge; they soon struck the Japanese along its southern slopes close to the top of the ridgeline east of the main track and Ioribaiwa village, as recalled by Sergeant William Crooks who would later author the battalion history:

> At dawn on 25 September, the rains now gone and all tired and weary from an uncomfortable, wet and sleepless night, the patrol again set off slowly climbing up the jungle-clad precipitous slopes [of Ioribaiwa Ridge]. At 1100 hours the patrol broke out into a native garden of banana, pawpaw and taro. It was a small area perhaps fifty yards wide by forty deep, in the centre of which stood a small native store hut. The garden had been dug up and the patrol saw the first of the footprints of the Japanese

CHAPTER 39

boot with its separated big toe piece. Captain Clowes sent A and B patrol groups, drawn from 16 and 17 platoons, and which he himself led, to the right of the garden, and C Group, from 18 Platoon, under Lieutenant Richard Cox to the left.

It was 1200 hours when C group set off west, Lieutenant Cox leading, with 7, 8, and 9 Sections following in that order. Bringing up in the rear was platoon Sergeant [William] Crooks. The group moved well-spaced out across an open patch of taro, over the two-foot-high pig fence, and on to a well-marked native track that led into the jungle. At 1210 hours, the patrol was moving swiftly, all thinking they had plenty of ground to cover before they would come up with the enemy. Some were still hurriedly eating the last of their biscuits, and others checking their weapons, when suddenly blasts of machinegun fire opened up from directly in front, and from the right. Two grenades burst in the trees above them. There was instant disorganisation. All were temporarily concerned only with the thought of self-preservation. The patrol disintegrated into the bush on either side of the track. Some men began to run back. In perhaps two minutes all were scattered about the jungle. The fire of the Jap M.G.s kept up, cutting twigs and branches from the trees. It was fortunate all five leaders had been in action before, Lieutenant Cox and his three section leaders had all served in the same platoon in Syria. Two of them, Frank Smith and Wally Mackay had at different periods ably commanded the same sections. In a few moments the leaders had recovered from their initial surprise. The first to move was Lieutenant Cox who, although momentarily surprised when the fire opened up, had turned to organize his platoon, only to see the men clearing into the jungle in all directions. However, as he made his way a little to the rear, he heard the section leaders and the platoon sergeant hurriedly reorganising everybody into a defensive position.

By that time Jap riflemen had taken up the fight, and the heavier and slower sound of the Jap M.M.G., or the 'Woodpecker', as it was called, could also be heard. All seemed to be very close – so close that the sound of bolts being cocked could be heard. Jap voices were clearly distinguishable, smoke could be seen, and the smell of cordite hung in the heavy atmosphere of the jungle. Lieutenant Cox realised that it would be suicidal to continue now that the Jap knew the patrol was there, and quickly gave the signal to retire – a direction which was efficiently carried out by leapfrogging in sections. In ten minutes, the group was back in the garden position hurriedly taking up fire positions along the pig fence. Miraculously no one had been hit. Nobody had seen a single Jap, nor seen any movement, nor had anyone fired a shot. The men were a little upset and shamefaced at the moments of near-panic. The leaders were all angry at the outcome and were getting rid of their anger by more careful siting of positions and insisting on more speedy reactions to their orders. The Jap fire now had become continuous. It had kept up throughout the withdrawal and was now clearing the tops of the kunai and foliage in the garden patch. Bullets were kicking up spurts of dirt along a lower taro patch fifteen feet beyond the pig fence. Shortly it came in through the hut putting everybody to the floor, but not before one bullet had rendered the 108-set useless. Lieutenant Cox ordered all to watch their various fronts and decided that all would remain until Captain Clowes' return.

It will be recalled Captain Clowes had set out, leading A and B groups, at the same time as C group left. Moving through the garden they had also struck a track that led north but Captain Clowes, seeing so many Jap footmarks in the patches of mud, decided to move into the bush, and was just clearing the track as the fire was heard on C group. This brought him temporarily to a halt. None of the fire was falling near him so he decided to

CHAPTER 39

push on. Slowly and cautiously now, since the presence of the other group was now known to the enemy, the patrol began to climb up to what appeared to be a crest. After ten minutes of crawling, walking and swinging from tree-to-tree over the last precipitous stretch, the leading men – the OC, Corporal [Gordon] Goldsborough, and Private Peter Barr – emerged onto a false crest and looked up, perhaps twenty or thirty yards over ground that had apparently been cleared, to a barricade of logs and bushes that had been laid out like a barbed wire fence. Here Captain Clowes signalled a halt, and then with the two men moved further to the right and up again until, beyond the barricade they could clearly see a single file of Japs, some forty yards away. The Japs were lining up for what appeared to be a meal parade. They were all laughing and chattering away, with one man dishing out ladles of something. Quickly Captain Clowes took in the scene and decided it would be foolhardy to attempt to seize a prisoner in broad daylight. He signalled to the others to withdraw. Quietly the three rejoined the base and with the group now altogether made their way back, reaching the outskirts of the main base in the garden area at 1330 hours. Having heard the firing throughout the whole of his patrol and having summed up the situation he quietly gave the order to withdraw from the position.

The patrol descended for half-an-hour and concentrated well out of sight of Jap positions on the ridge. The patrol then struck west through the bush, heading for the main track. Their route took them down one open kunai spur and through more gardens which all showed Jap tracks and evidence of where the enemy had dug for food. By 1700 hours the patrol had reached the main track, where a temporary halt was made in the middle of 2/25th Battalion, which was deployed around the old kunai patch on the main ridge. Captain Clowes told Lt Col Withy of his findings and at the same time gave his patrol report over the line back to Brigadier Eather at brigade HQs on Imita. At 1800 hours after an hour's jolting descent down the main track a halt was called, and all bivouacked for the night just off the track.[9]

* * *

Meanwhile, Lieutenant Colonel Colin Dunbar, commanding the 2/31st Battalion was still on Imita Ridge, covering the left flank and now sent out a 50-man patrol led by Captain Leslie Hurrell, a 28-year-old agriculturist from Cobargo in New South Wales, with his 2i/c 30-year-old Lieutenant Christopher Ryland, from Hughenden in Queensland. Hurrell commanded two rifle platoons assembled from a number of specialist troops, including those from the carrier platoon and mortarmen – riflemen were required in the mountains. They were to move out to reconnoitrer a way for the battalion to approach and attack the Japanese right flank on Ioribaiwa Ridge. Soon after, these men heard machinegun fire in the direction of Ioribaiwa, indicating that heavy fighting was taking place along the ridge. Undoubtedly this was the Japanese ambush against Lieutenant Cox's men, and the fighting by the men of the 2/25th Battalion near Ioribaiwa village.[10]

The 3rd Battalion war diary also records the sound of the fighting up ahead: 'Mixed rifle fire and MG also grenades heard since 0545 and continuing, apparently from South of IORIBAIWA. 1900 hrs artillery fire'.[11]

* * *

That day, Japanese Major General Horii Tomitarô arrived at Nauro and gave a speech to the assembled troops: 'For more than 20 days … every unit forced its way through deep forests and ravines and climbed over scores of high peaks in pursuit of the enemy. Traversing mud more than knee deep, clambering up steep precipices, bearing uncomplainingly the heavy weight of artillery ammunition, our

CHAPTER 39

men overcame the shortage of our supplies, and we succeeded in surmounting the Stanley Range. No pen or word can depict adequately the magnitude of the hardships suffered. In proving your ability in surmounting all these impossible difficulties, you should feel compassion in your hearts towards the many victims who fell in battle during that period'.[12]

Still trying to maintain communications between elements of the *144th Regiment* was the unknown leader of the signals unit with *No. 2 Infantry Section, No. 1 Company, I/144th Regiment*. That day, he recorded the last entry in his diary: 'The line was damaged again by artillery fire around 1100 and was repaired for the second time by the No. 1 Sec'.[13] The diary was recovered by Australian troops a month later south of Oivi on the northern lowland fetid swamps, likely taken from his dead body.

Meanwhile, Lieutenant Colonel Tsukamoto Hatsuo, commanding the *I/144th Regiment* had given orders for his men to withdraw from Ioribaiwa. The battalion commander had been wounded at Myola from Allied bombing but had gone forward with his men. He was now issuing orders for his men to retire towards Menari. At midday, he sent a message to the *South Seas Force* commander, Major General Horii, that they had withdrawn successfully from their positions and that the ability to get their wounded back to Menari was 'not favourable', and that the battalion would try to hold the Australians for a few days in the Nauro area; ending with a request for rations to be sent at once.[14]

Among Tsukamoto men was 1st Lieutenant Hirano Kogoro, commanding *No.3 Platoon, No. 1 Company, I/144th Regiment*. He recorded in his diary that the men of the *South Seas Force* were retreating and the men's physical and mental condition had rapidly declined. He recorded in his diary for 25 September that men had strived on believing that they would soon be in Port Moresby; however, now they were in full flight back north over the mountains. The young Japanese officer recorded despondently: 'Departed at 0700. Many stragglers were dropped behind'.[15] With the *I/144th Regiment* was 2nd

SAVING PORT MORESBY

Lieutenant Onogawa, *No. 2 Platoon, No. 3 Company*, who had rushed forward from Myola after getting word that they were to rejoin the regiment as the offensive to take Port Moresby would commence on 20 September. Now he received orders that he was to take up a defensive position somewhere between Isurava and Kokoda Plateau, while the rest of the regiment would take up positions along the northern slopes of the Owen Stanleys, and the men of the *41st Regiment* would defend Kokoda itself.[16]

Now heading north was Private Teruoka Akira and his comrades of the *Tanaka Unit, 55th Division Medical Unit*. They had earlier received orders to evacuate their aid post and head back over the mountains to Eora Creek with the sick and wounded in the first instance. Teruoka and the rest of the men were already on the track heading north when behind them they could hear artillery and mortar rounds exploding in what was their forward positions along Ioribaiwa Ridge. Most men were now aware of the general withdraw.

Indeed, the men and *No. 2 Company, Takasago Formosan Naval Volunteers* heard it themselves from Captain Uyama, who was part of Major General Horii's staff. He informed them that the assault against Port Moresby had been cancelled, and all patients were being moved to Eora Creek to the rear, and they were to assist in helping move them.[17]

Meanwhile, Japanese war correspondents Okada and Sato Toshio, along with their photographer, Katayama, and their small party were making their way back to Buna having left their rudimentary hut on Ioribaiwa Ridge the previous night. Okoda later recalled: 'As the day dawned, we saw the troops in the rear making preparations for the retreat. We hurried on past them. We were exhausted and hungry. It was dreadful to think of the long journey before us. But we walked on driven by a strong desire to get away as soon as possible from the ill-fated campaign and from the terrible scenes of battle in the mountains. The order to retreat had crushed the spirit of the troops which had been kept up through sheer pride. For a time, the soldiers

CHAPTER 39

remained stupefied among the rocks on the mountain side. Then they began to move, and once in retreat, they fled for dear life'.[18]

* * *

Just south of Uberi, the men of the 2/14th Battalion were preparing to finally be relieved: 'During the morning a recce was made for an area suitable for a bivouac for the relieving Coy of the 36 Bn, guides were provided and Coy arrived at 1200 hrs. The Pls of the relieving Coy were addressed by Capt. RHODEN, Capt. THOMPSON and Lieut. BISSET, regarding Jap tactics'.[19] At just before 4 pm, Lieutenant Ken Evans set out on another patrol with twelve men to Omaritana, while Lieutenant John Crameri's patrol reported in at 4.30 pm, after having been out for two days.[20] The battalion war dairy of the 2/16th Battalion merely records for 25 September: '0845 A C Coy move out to relieve D Coy patrol. 1130 Div Comd visits CO & talks to members of Bn. 1200 Forward coys of 36 Bn arrive in UBERI to relieve us'.[21]

* * *

Captain Bert Kienzle at Dump No. 66 had spent the last few days with a fever but was now ordered to make his way to the headquarters of the Australian 7th Division at Bisiatabu in the base area. He recorded that his responsibilities were to be expanded: 'I was advised that I was now attached to this HQ and to give all assistance regarding carrier organisation, tracks, intelligence reports received from natives regarding enemy movements etc. Met Maj. General Allen'.[22] The next day, he reported to ANGAU headquarters where he received the personal thanks of his CO, Major General Basil Morris, for the outstanding service that he and his men had performed over the last three months.[23]

* * *

That day, on returning to Port Moresby, Honner Force was broken up and the men returned to their respective units. Major Harry Harcourt, commanding 2/6th Independent Company, who had worked closely with Lieutenant Colonel Ralph Honner, later wrote in his report regarding Honner Force: 'The operations of Honner Force engaged a large portion of the Coy for part of the month and although this effort was abortive it was not because of lack of fitness of the Coy's detachment. At the end of the month, it was possible to plan for an attempt to be made against the enemy L of C and a strong patrol was standing by to leave for this task when the month ended'.[24]

* * *

Further north and east, Corporal John Burns and Private Alfred Zanker were still awaiting rescue for the sick and wounded of the 2/27th Battalion. It had been a week since they had been left behind; Burns recorded:

> The 25th arrived and only after what seemed to be the longest and hardest night of my life. Zanker and I managed about two hours sleep during the night but the poor lads on the stretchers couldn't go off [to sleep]. Pte [Thomas] Burke of the 2/16th Bn had taken a definite step towards the end at this stage, and he lapsed into semi-consciousness. He was in a bad way. There was nothing we could do for him except a dose of morphia to put him out of agony every now and then. It took Zanker and I all our time to hold him on his stretcher when he started throwing himself around. During the morning we washed and shaved all the lads again and re-made a couple of stretchers for them. About midday the weather became so threatening, and together with the fly menace I decided to shift the lads around under the native huts. I realised I was taking a big risk of our movement being observed but the lads were becoming too weak to allow them to be saturated too often and the sun was too strong out in the open for them. The change seemed to do them a lot of good.[25]

CHAPTER 39

Meanwhile, just north-west of Brigade Hill, at Enivilogo village, it had been days since the promised return of the Papuan carriers to Lieutenant Robert McIlroy and his small party of three men – one was suffering from a severely injured knee and the other was wounded in the Japanese ambush. Sometime that day, two lost men from 'A' Company, 2/14th Battalion, who had been lost since the battle for Isurava, stumbled across McIlroy and his men.[26]

* * *

That morning, General Blamey went to Milne Bay, and without notifying Lieutenant General Rowell, he gave orders to Major General Cyril Clowes, commanding Australian forces there to dispatch a force to Wanigela. Blamey recalled: 'On the second day after my arrival, I visited Milne Bay accompanied by Brigadier General Walker, US Forces, temporarily commanding the air forces in New Guinea. From our discussions there it appeared that we might develop close cooperation between the air and land forces and overcome the difficulties of movement to some extent by the transport of land forces and supplies by air. This necessitated the finding of landing grounds. An examination of the landing ground at Wanigela Mission, approximately 100 miles [160 kilometres] north-east [*sic* north-west] of Milne Bay, made it appear possible that the first step might be made in this way, and it was ultimately decided to fly the 2/10th Battalion, A.I.F., to Wanigela from Milne Bay.[27]

In his diary, Lieutenant Colonel Fenton recalled Rowell's first open clash with Blamey: 'Saw General Rowell who said he had had a terrific argument with C-in-C who was determined to stay. Rowell said he told C-in-C quite frankly what he thought of him and was surprised that the C-in-C had put up with it. "Any real man would have sacked me on the spot". C-in-C had threatened him with a retire list. "I thought I might get out of it with a drop in rank"'.[28]

It was becoming increasingly evident to Rowell and his staff that they were effectively being supplanted by Blamey, and a clear rapture between the two generals had been established. This is further suggested by Major General Charles Willoughby, who wrote on visiting Rowell that day at his headquarters: 'Scattered tentage, some attempts at camouflage; on the whole an efficient mobile campaign headquarters .... Rowell is of dignified but amiable personality. The staff regards him with respect and affection; that feeling prevails through other command echelons .... There is hesitancy to change personnel in all units; there is evidently a considerable camaraderie, that has developed in the Middle East'.[29]

40
'TWO DAYS WERE SPENT RAFTING DOWN THE RIVER AMID BEAUTIFUL SCENERY'

On 26 September, the battalions commanded by lieutenant colonels Alfred Buttrose and Colin Dunbar manoeuvred to their new positions, while Lieutenant Colonel Charles Withy and his men of the 2/25th Battalion kept the Japanese to his front busy.[1]

The men of the 2/33rd Battalion had spent the night around the former Salvation Army Red Shield post and moved out early that morning towards Dump No. 44 and Ioribaiwa Ridge beyond. A sergeant and five of his militiamen from the 3rd Battalion led the way. It would not be long before the 2/33rd battalion would come across their mates of 'D' Company now located just south of Dump No. 44. As recorded in the battalion war diary, Captain Trevor Clowes reported to Buttrose: 'Enemy in barricaded posns right flank of GAP[.] D Coy also reported that enemy has control of track where Bde HQ was formerly established'.[2] Sergeant William Crooks with 'D' Company recalled how they were denied their first hot meal in three days:

> At dawn on 26 September, Captain Clowes had the patrol moving again, passing for a cold breakfast at Camp 44, and continuing the march again at 0900 hours. At noon a halt was made on the flats along the Ua-ale [*sic* Ua-Ule] creek with the order that it would be for two hours, and cooking was permissible. The patrol prepared for its first hot meal since breakfast at Imita on

the 23rd. Just as the fires got going the voice of the CO, Lt Col Buttrose was heard yelling: 'Get those fires out D Company. Saddle up and join on the tail of the battalion. We're moving up on Ioribaiwa.' This was tough, but nobody's fault – an advance is an advance. However, by the time the remainder of the battalion had cleared through the area all in D Company had managed to brew up a mug of tea and joined on to the tail end of the battalion at 1345 hours for the weary climb up the ridge again. For the fifty men in the D Company patrol this would be the third climb up this dreaded ridge.[3]

Also joining their advance was a three-inch mortar team. By 6 pm, the Battalion Headquarters and men of 'C' and 'D' companies had taken up a position to the rear of the 2/25th Battalion, near the Kunai Patch area. Meanwhile, the men of 'A' and 'B' companies had moved east along the narrow subsidiary track from the Kunai Patch area leading towards Ponoon. As darkness fell, these men took up a position below the southern and eastern slopes of Ioribaiwa Ridge, and from here they would launch an attack to take the heights around Ponoon as soon as the rest of the battalion arrived.[4]

That night, having ordered the 2/31st and 2/33rd battalions to join the 2/25th Battalion in their advance, Brigadier Kenneth Eather recorded in his diary: 'Things moving very well. Jap on the defensive. Does no patrolling whatever that we can discover …. Tubby [General Allen] rings me at 2200 hours. Seems pleased with way things are shaping up'.[5]

* * *

Earlier, the men of the 2/31st Battalion had advanced from Imita Ridge towards Ioribaiwa. Captain Leslie Hurrell and Lieutenant Christopher Ryland (commanding the two composite platoons assembled from the carrier platoon and mortarmen), who had been sent out the previous day by Lieutenant Colonel Colin Dunbar to reconnoitrer a route to

CHAPTER 40

the Japanese right flank, had spent the night at Dump No. 44 just below Ioribaiwa Ridge. At first light, Hurrell and his 50-man patrol moved west towards the Spotter's Hut area, using the track that passed through the 'Large Grass Patch', south-west of the hut area. The rest of the battalion on reaching Dump No. 44 spent the night in the general area awaiting word from Hurrell.[6]

Meanwhile, during the night and early morning hours, the men of the 2/25th Battalion had held their forward position near the Grassy Plateau and the Kunai Patch. The Japanese fired into their positions, but the Australians withheld their fire, keen not to give their precise positions away. The men of 'C' Company moved forward to take up a position below 'A' Company, while 'B' Company, reporting two Japanese attempted to advance down the track – both were killed, soon after 'B' Company changed positions with 'D' Company. It was now that Captain Peter Blundell, a 33-year-old grazier from Stanthorpe in Queensland, commanding 'D' Company was forced to retire due to illness. Lieutenant William Crombie, a 28-year-old station overseer from Longreach in Queensland, took over command of the company.

During this time, patrols from 'B' Company reported a Japanese light machine gun was firing from a gap on the right, while combined fire from a light and medium machine gun was observed to the left. With this, the Australian artillery fired into the Japanese positions, and soon after, about a dozen Japanese were seen hurriedly in the distance retiring over the ridgeline and presumably down its northern slopes.[7]

The diarist of the 3rd Battalion recorded that at 1.15 am, artillery, 'fire and bursts of MG fire and three explosions from vicinity Kunai Patch. 1055 – 1135 hrs continuous artillery fire. Five rounds observed slightly right of village; remainder appeared to fall beyond ridge. 1105 – 1109 hrs continued bursts of MG fire from vicinity village'.[8] Sergeant Colin Kennedy recalled their time on Imita Ridge: 'With time to relax, everyone resumed an interest in the outside world. Those few who received news cuttings of the progress of the war elsewhere were very popular'. Also on the ridge, 37-year-old Lance Corporal Roland Cattley, from Kelvin Grove in Queensland, recalled a similar sentiment: 'We have absolutely nothing to read and nobody knows what is going on at home unless it is in letters'.[9]

At noon, Brigadier Walter Smith, commanding the Australian 14th Infantry Brigade, made his way forward to confer with Brigadier Eather; both men moved to a forward observation post to get a better understanding of the proposed advance.[10] The 25th Brigade war diary records: 'Logs used in JAP barricades were sawn, not chopped. JAPS using signal wire with tins attached as night alarms. Enemy using amn with approx. same calibre as .303. 1810 hrs: 2/25 Bn still in contact, 2/33 Bn moving out to right flank from Post 44, 2/31 now located Post 44 preparatory to moving to left flank on 27 Sep. General patrols and supporting fire during day'.[11]

* * *

At dawn, Lieutenant Sakamoto Atsushi, commanding *No. 2 Machinegun Company, II/144th Regiment*, was already awake and preparing his men for their withdrawal from their position on Ioribaiwa Ridge. The sick and wounded were assembled at 8.30 am: 'The dead were buried', he scribbled in his diary sometime that night: 'It is truly regrettable having to leave ... the bodies of our comrades and the ground that we won so dearly. Sleep peacefully my friends. Farewell. We shall meet again in Heaven'.[12]

Sometime that day, 1st Lieutenant Horibe, acting commander of *No. 6 Company, II/144th Regiment*, recorded in his diary the effects of the Australian artillery over the last few days: 'At IORIBAIWA. Shelled on 22, 23, 24 and 26 Sep – 3 wounded as result'.[13] He soon after received his orders to vacate his position and fall back to Nauro; he left one platoon behind to cover their withdraw – these men would follow as soon as the rest of the company had made good their escape. Horibe and his men spent all that night marching north to reach Nauro, arriving at the village during the early morning hours of the

CHAPTER 40

next day. His blocking platoon on the ridge would join them there; his company strength was now just 51 men.[14]

That day, Lieutenant Nakahashi Kokichi and his gunners of the *55th Mountain Artillery* had arrived at Menari in the pouring rain. The lieutenant ordered the few remaining shells be fired from their mountain gun into the Australian lines. After firing the last of their ammunition, the gunners rolled the gun down a slope, where it crashed into the thick jungle undergrowth and likely remains to this day. The young lieutenant later wrote: 'This at least gave an uplift to our downcast spirits and to those of our dead comrades …. It was compounded by feelings of anger, sorrow and frustration. The purpose, the dreams and the desires of the officers and soldiers of the *South Seas Force* had vanished in an instant'.[15]

Meanwhile, 1st Lieutenant Hirano Kogoro, commanding *No.3 Platoon, No. 1 Company, I/144th Regiment* had watched as some of his men fell out of the line. He waited and hoped that some might yet reappear that day. He recorded in his diary: 'Sgt NISHIKAWA and the rest departed at 0800. I remained to wait for the stragglers. By 1700, all of them had finally been assembled. Under the moonlight, marched on to overtake the BUTAI'.[16]

Early that morning, Private Teruoka Akira, with the *Tanaka Unit, 55th Division Medical Unit*, awoke to find himself alone – somehow the men of his unit had left without him. He now headed down the track after them, accompanied by an Indigenous carrier who was carrying his equipment. It had been raining, but he noted in his diary that he was not cold, as the night before he had a blanket. Even so, his illness had returned with a vengeance, and he was feeling extremely weak. Later that day, he went in search of water and collapsed near a creek; he spent the night there. This was Private Teruoka Akira's last diary entry. His diary was later recovered by Australian troops, likely from his body somewhere near the creek.[17]

* * *

Finally, the men of the 21st Brigade were relieved, falling back to the bivouac areas in the Port Moresby area. Major Abert Moore had been told the day before that his mates of the 2/14th and 2/16th battalions would be moving down the line that day, so he and Corporal Jock Inglis were waiting for them: 'I arose early, had our stove smoking as the dawn broke across the ranges, and baked about 300 scones, plus jam and apple tarts. Along the table, we set out [a] scone, slice of tart and slice of fruit cake for each man, [and] also a cup of coffee. I estimated that we served 250 men of the two units'.[18]

For the men of the 2/14th Battalion, it had been a month to the day since they had reinforced the men of the 39th Battalion while the battle for Isurava commenced.[19] The battalion war diary records that at 7 am, a platoon from 'D' Company, 36th Battalion left to contact Lieutenant Ken Evans' patrol that had gone out the previous day. They were to tell him that he was to report to Newton's Dump (just south of Owers' Corner) where he would contact the remainder of the battalion later in the day. At 10 am, Lieutenant David Rainey's standing patrol on the high ground north of Uberi was relieved by another platoon from 'D' Company of the 36th Battalion. Just after midday, the battalion moved to Newton's Dump. From there they were transported to Manurinumu, then transported to Koitaki where rear details were established. They finally arrived at the base area at around 4.30 pm and were given a hot meal. The diarist also recorded: 'A beer issue, unit canteen and pictures were available to troops during the evening. The officer's mess was opened for the evening meal'.[20]

The month before, lead elements of the 2/16th Battalion had also begun to arrive to defend Alola, just south of Isurava, from Japanese flanking movements that were conducted on 26 August. Now they were finally out of the line. The battalion war diary records sparingly: '1000 Tps take part in demonstration of Jap tactics for 36 Bn. 1515 Bn left UBERI area. 1800 Bn arrived at Bivouac area KOITAKI'.[21]

CHAPTER 40

The militiamen of the 36th Battalion now took over the positions held by the men of the 2/14th and 2/16th battalions. They were located near Uberi with a three-inch mortar detachment. These men settled in for the night as recorded in the battalion's published history: 'The wet and gloomy atmosphere of the mountain jungle, with its very tall trees, its steaming heat by day and biting cold at night, was again in sharp contrast with Koitaki area. At night, to greet the soaking rain, a million jungle noises rose from the countless insects which inhabited the decaying mass of vegetation which carpeted the ground. Preparing for the defence of their areas, the companies adopted tactics similar to those used by the enemy, such as erecting obstructions in front of weapon pits by driving into the hillside's thin poles with sharpened ends projecting, and by creating false turns in jungle tracks to lead the unwary into ambush'.[22]

* * *

Captain Sydney Buckler and his party of 39 men were still pushing south from Sengai No. 2. Buckler decided to push on ahead, leaving Lieutenant Charles Butler in control of the party. He could move quicker alone; he was in a hurry to get help for his stricken men left in the care of Private Thomas Fletcher.[23]

Meanwhile, his junior officer who he had sent south weeks before, Lieutenant Maurice Treacy, having rested for three days at village of Dorobisolo, began his journey to Port Moresby with the wounded Private Frederick Rockliffe and Private William Avent. Soon after arriving at the township, Treacy concluded his report about a week later: 'After resting for three days, during which time many good meals were taken, the party continued SOUTH eventually reaching the KEMP-WELCH [River] and natives in a nearby village-built rafts for the party. Two days were spent rafting down the river amid beautiful scenery. Eventually the party reached Port Moresby, travelling by truck via RIGO. Pte Rockliffe was there evacuated to hospital. The following day 2nd October, the party reported back to the unit after an absence of 32 days. News that Capt. Buckler's party was safe was confirmed much to the relief of the members of the party'.[24] This would turn out not to be entirely true.

* * *

Further north, Corporal John Burns and Private Alfred Zanker had yet another restless night and morning looking after the wounded who had been left behind by the main party of the 2/27th Battalion; Burns recorded for that day:

> After a night when no one slept again, Saturday arrived. Zanker and I were run off our feet during the night, the boys had developed diarrhea and we would no sooner lift one lad back and the next man would call for us. How we kept going night after night and day after day was a miracle. I think all our many prayers were answered. Burke was gradually getting worse, he was going through a living hell, this was his second day of unconsciousness. A native from the nearby tribe visited us during the afternoon and I asked him if he knew his way to ITIKI and he said 'Yes' so I told him to come back the following morning and I would send the strongest of walking cases on. They had to take a chance whichever course they undertook – stay in the village or a break for safety – and it meant less movement around the stretchers, so I sent them off. Heavy rains were the order of the day and we had to put the stretchers three feet off the ground away from the water. Our dry firewood supply was very limited [and] we boiled very little yams and concentrated on roasting a supply for the party moving out on the morrow.[25]

* * *

Lieutenant General Sydney Rowell – on hearing of Geneal Blamey's orders to his subordinate Major General Cyril Clowes (at Milne Bay)

CHAPTER 40

to send the 2/10th Battalion to Wanigela, he cabled Blamey outraged that he had been completely blindsided in the decision-making process. He clearly felt that he should have at the very least been consulted in any decisions regarding the use of his force:

1. By the terms of para 3 of LHQ Operation Instruction No. 30 dated 9 Aug 42, GOC 1 Aust Corps, which appointment I was holding at the time, was designated Commander 'New Guinea Force'.

2. Para 4 of the same instruction authorised me to exercise operational control over all military forces constituting New Guinea Force at that time and such other troops as may have been subsequently assigned thereto.

3. Your arrival at Moresby to operate as Commander-in-Chief from here will inevitably vary my authority and it is submitted that the position needs to be defined so that all concerned will know what variation is contemplated. This applies particularly to Allied formation commanders now exercising command under my orders as well as the joint service aspect of Navy, Army and Air cooperation. In this connection I instance your verbal orders to Commander Milne Force at Milne Bay on 25 Sep 42 regarding the despatch of a detachment to Wanigela.

4. The question of the exercise of powers of the Authorised Person under National Security (Emergency Control) Regulations will also need to be determined. By the terms of para 18 of LHQ Operation Instruction No. 30, the powers of the Authorised Person are vested in me. The wording of the regulation is as follows: 'The Senior Officer of the Military Forces for the time being present and having the operational command of the Military Forces serving in that part'. It would now appear that this authority should pass

to you and that fresh delegations be made to replace those already issued to me.[26]

Blamey did not reply directly but, later in the day, issued the following directive to Rowell:

> I have been directed by the Prime Minister and the Commander-in-Chief, South West Pacific Area, to take control of the forces in the New Guinea area.
>
> For the present it does not appear necessary or desirable to set up an additional headquarters staff. Therefore, I propose to exercise command through yourself and the present staff.
>
> I will be glad if you will direct that arrangements are made to furnish me promptly with all tactical and other information and alterations in the functions, allocations, dispositions, and location of troops.
>
> I will be glad if you will ensure that all messages and information for Headquarters Allied Land Forces or Headquarters Australian Military Forces are submitted to me, including Situation Reports, before dispatch.
>
> In the event of my absence from Headquarters, where it is apparent that such absence would cause undue delay in furnishing such information, you will forward it direct at the same time taking such action as is necessary to ensure that I am kept fully informed.
>
> The above applies also to matters of Administration of any importance .... [27]

The news must have hit Rowell and his staff like a bomb shell. Things again came to a head when Blamey alighted his plane returning from Milne Bay with a suspected hangover. Rowell and others noted that a crate of scotch was also unload from the aircraft. To Rowell and

CHAPTER 40

others, nothing had changed from their days when fighting the Vichy French, Italians and Germans.[28] Lieutenant Colonel Fenton recorded a conversation he had with Rowell soon after: 'He said, C-in-C was holding on. "Wish there was a good air raid. I think he'd bolt for home." Rowell seemed more depressed than previously .... I said it was essential that Wilmot should leave for the mainland that night. Wilmot was prepared to risk everything to help, and if it didn't go immediately the C-in-C could hold him up indefinitely. Rowell asked to see Wilmot and I rang him'.[29] Lieutenant General Rowell and ABC journalist Chester Wilmot had become good friends during the fighting in Syria and Greece, each supporting the other as best they could. Fenton was present during the discussion between general and journalist:

> Returned to Rowell to hear Chester cross-examining him about various aspects of the campaign. Chester was masterly. Rowell told the story of what happened in Greece when the retreat began; how Blamey wouldn't go forward himself because he was afraid and wouldn't let Rowell go until he insisted on going. Blamey made the Divisional Commanders come back sixty miles to get their orders. He was so pathetic trying to give those orders that Rowell had to step in and do it for him. Then when the time came to travel down to the plane, Blamey, almost sick with anxiety to get away, ordered Rowell to be one of his party. Rowell demurred saying he considered his duty to remain with the troops. Blamey then said it was an order, and Rowell, after repeating his protest, obeyed. For the sixth place, Blamey took his son, Tom, and Rowell described the scene under the olive trees when [Brigadier William] Bridgeford burst out: 'The old bastard is taking his son with him', in a voice that everyone heard, including Blamey. Then, said Rowell, Blamey damned him in his report on Greece saying he hadn't sufficient reserves of nervous energy to cope with the campaign.[30]

Fenton organised for Wilmot to be sent back to Australia that night; he was to do as much as he could to get Australian politicians, including the Prime Minister, along with the Australian public, to back Rowell. Wilmot would arrive too late to affect the outcome and would find himself branded persona non grata by Blamey and his staff and had his journalist accreditation soon after revoked. Indeed, Blamey, the great hater, threatened to force Wilmot into the Army where a 'latrine unit was waiting'.[31] Like Parer, Wilmot found himself covering someone else's war, working with the BBC. He would land with British airborne troops in Normandy on D-Day – 6 June 1944 – and would go on to report the final downfall of Nazi Germany.

* * *

Meanwhile in the United States the American public were being informed that their troops had come to the rescue at the last-minute saving Port Morseby from the Japanese at the end of the Kokoda Track. American journalist, Jules Archer, who was in New Guinea at the time, later wrote to but the record straight: 'Ironically, the American press jumped to the conclusion that it was the U.S. Infantry that had prevented the enemy conquest of Port Morseby. Hanson Baldwin, writing in *The New York Times*, declared, "American soldiers were rushed into action and were instrumental in saving the day". Up to that point, however, the only troops engaged in fighting across the Owen Stanleys had been the Aussies'.[32]

41
'DID I TELL YOU I'M TRYING TO COLLECT 3 DECORATIONS OF THE HIGHEST?'

The 27 September was another quiet day for Brigadier Kenneth Eather and his men, and as darkness fell, the Japanese positions were still under a heavy artillery bombardment with lieutenant colonels Alfred Buttrose and Colin Dunbar closing in on Ioribaiwa Ridge from the flanks. Meanwhile, Lieutenant Colonel Charlies Withy's men of 'B' Company, 2/25th Battalion moved forward to relieve the men of 'D' Company. These men pressed slowly forward and penetrated the Japanese defensive works, which blocked the Kokoda Track. The Japanese were falling back, avoiding direct contact with the Australians. A combined patrol from 'A' and 'D' companies, 2/25th Battalion reported a Japanese rearguard behind a barricade on the track. Soon after, a detachment of the battalion's mortars moved in between the positions of 'C' and 'A' companies in the grassy plateau area.[1]

Eather now proposed to launch an all-out attack to take and occupy the whole of Ioribaiwa Ridge the next day. In preparation, just before 3 pm, another Australian mortar detachment occupied the Kunai Patch on the main track. The men of the 2/1st Pioneer Battalion now also left Imita Ridge to move up and repair this part of the track.[2] Still at Dump No. 44 was Dunbar and his men of the 2/31st Battalion. The battalion war diary records that word soon arrived from Captain Leslie Hurrell that he had identified a large Kunai Patch area to the west, which gave 'access to Jap right flank. On this information Bn moved to it. Water was difficult to obtain. Meanwhile shelling of ridge was carried out by 14 AUST FD REFT'.[3]

To the east that morning, Buttrose and the remainder of the 2/33rd Battalion had advanced east along the subsidiary track from the Kunai Patch area to join the men of 'A' and 'B' companies' positions immediately below the southern slopes near Ponoon village. They would attack the ridgeline at first light the next morning, as recalled by Sergeant William Crooks: 'Throughout the afternoon of the 26th and all day of the 27th, in rain that started at noon on both days and persisted through both nights, the battalion climbed and manoeuvred into its forming place. Major Cotton had brought forward one M.M.G., and one three-inch mortar, both of which were put into a position beside B Company to which was also attached an artillery observer. Throughout the 27th line was laid from the main track to all companies, the hardworking and dedicated signallers uncomplaining carrying the hundred-pound reels [50 kilograms], with their own gear and weapons, up and down that broken and mountainous terrain'.[4]

Buttrose's plan to capture the Ponoon area was for the men of 'C' Company to attack on the left, with 'B' Company on the right. Supporting each respectively would be 'D' and 'A' companies. These men were now close to the area where the men of 18 Platoon had been ambushed just days before. Sergeant Crooks recalled: 'During the night our guns harassed the whole ridge, and at dawn, methodically swept from the right to the left'.[5] He also recalled the conditions along the track and in the mountains at this point:

> It was a place of perversities and heartbreak. There were the long arduous and mind deadening climbs, which at times reduced us to silent tears. Then the steep bone-jarring descents that tortured already overstrained muscles and lungs, as we tripped over hidden roots and fell in the ever-present slimy mud. There was always the hated and consistent heavy rain that came out of the low-lying clouds every afternoon and pelted down through

CHAPTER 41

the nights. We all knew the lack of warm food and drink; the biting weight of forty and fifty pounds [around 20 kilograms] of equipment on those impossible slopes; our clothes were never removed and were forever wet from rain or sweat. Nights were broken by duty, rain and cold. There was the loss of dignity in excreting before the eyes of your fellows. There was the pitiful sight of the wounded, deathly pale and wordless on their makeshift hessian stretchers, surrounded by their native carriers. The lonely resting places of strangers and close friends buried beside the track through this dark and tumbled prehistoric land, got nothing but a passing look. The pathetic hoarding of a few spoonfuls of sugar gathered from a burst airdrop, and the secretive whispers to one's section mates who would share this wonderful prize, told how primitive had become our urges. The writing of letters to one's loved ones on toilet paper, with no envelopes to hide the sad things. We became used to, but we could still wonder at the miracle of them being carried down the track, and in fact duly delivered to suburbia or bush homes in far-off Australia.[6]

* * *

Still south of Kagi was Japanese 1st Lieutenant Hirano Kogoro, commanding *No.3 Platoon, No. 1 Company, I/144th Regiment*. He was trying to catch up with the rest of the battalion and recorded in his diary for 27 September: '0430. Reached a ravine and rested. Sent out men to dig taros. I set out alone to join the main force'.[7]

Meanwhile, Major Koiwai Mitsuo, commanding the *II/41st Regiment*, was with his men somewhere further north; the major recalled after the war: ' Our bodies were completely fatigued, so climbing even the smallest hill required a great effort. The majority of members of the unit were carrying wounded on stretchers. My battalion's 7th Company was meant to be protecting the rear, but they were carrying casualties from the field hospital, as were members of the force's wireless radio unit … counting each step, grasping for each breath. Despite some encouragement to struggle through pain, many were stooped over, their eyes filled with tears and without even the strength to urge themselves on. My heart was filled with sorrow'.[8]

Somewhere further down the track was Private Nishimura Kokichi, with *No. 3 Platoon, No. 5 Company, II/144th Regiment.* Unknown to him, he still had a machinegun bullet buried in his shoulder from the fighting at Mission Ridge. He had spent time recovering in an aid post and had been part of the walking wounded advancing with the rest of his battalion. A hospital bed would be waiting for him at Port Moresby, or so he had been told. Now he was part of the long line of unattached wounded making their way back over the mountains. Even though his shoulder had been strapped and bandaged, every step sent a painful jolt to his wound. What concerned him most, however, was not his physical pain, but the trauma of being the sole survivor of his section. He would eventually stumble into Kokoda, before making his way to Buna.[9]

Lieutenant Sakamoto Atsushi, commanding *No. 2 Machinegun Company, II/144th Regiment* recorded in his diary that night: 'Men are searching in the moonlight for food. Sickness increased'.[10]

* * *

At this time, the 2i/c of the 3rd Battalion, Major Robert Grantham, left for the rear area to arrange for the remainder of the battalion that had been left at Port Moresby to move forward, including the remainder of the machinegun company and mortar teams. At 7.30 pm, the battalion CO Lieutenant Colonel Allan Cameron received orders from Brigadier Eather that he was to prepare his men to move forward by 11 am the next day. A patrol sent out earlier had discovered a new track an hour's distance on the right flank. This track was believed to have been cut by Australian engineers two months before while trying to discover a mule track to Ioribaiwa.[11]

CHAPTER 41

* * *

Throughout the day, reports had come into Eather's headquarters that the Japanese had vacated the top of the ridgeline and were now occupying the reverse, northern slopes of the ridge, likely to avoid the fire from the Australian artillery. Indeed, the 25th Brigade diarist recorded: 'Arty smashed barricades in parts on right flank (2/33 Bn)'.[12] Even so, all expected a Japanese counterattack once the Australians occupied the ridgeline.[13]

At 7.30 pm, Eather held a brigade conference with all battalion commanders, using a telephone hook-up. The brigade war diary records: 'Hook-up was highly commended by Brig EATHER; he was able to speak clearly with each CO and the COs were able to converse with each other – all COs in the picture as to the other's activities. The following instns were issued by Brig EATHER – At first light the 2/33rd Battalion was to advance and take the ridgeline around Ponoon'. As this attack was made, the men of the 2/25th Battalion were to push forward to take the centre of the ridge around the village area. Meanwhile, to the left, the men of the 2/31st would take the area around the Spotter's Hut area. On achieving their objectives, all battalions were to consolidate their positions. Each was to maintain contact with the Japanese by pushing out fighting patrols. The 3rd Battalion would then advance and concentrate as a reserve on reverse slope of the ridge. The Australian 25-pounder guns where to support each attack as defined in the brigade orders: 'Targets – (i) IORIBAIWA, (ii) SPOTTERS HUT, (iii) Reverse [forward] slope of ridge'.[14]

* * *

The previous day – after having made sure the exhausted veterans of the 2/14th and 2/16th battalions had been well fed as they passed through his Red Shield post – Major Albert Moore had headed back to the base area to speed up the forward movement of supplies to his station. Having reached the Port Moresby area, it would be a very busy couple of days, as recorded by Moore: 'A number of Salvation Army officers were arriving on the island for Red Shield work among the forces and, as I had been appointed the senior man, it meant that I would need to start organising work with them, though most arrived attached to units. I gathered the Red Shield representatives together and we had a conference. There was a request from the 2/14th that we erect a marquee in their area, so I secured one from Moresby and set about getting tennis tables, writing tables and games of various sorts. Soon the centre was in full operation and available to any who cared to use it. The whole scene was fast-changing and new demands were being made on us to meet the needs of the men resting and regrouping'.[15] Among these new arrivals was Salvationist, Captain John McCabe, attached to the 16th Brigade, who recalled at the time: 'Our Brigade, which included the 2/1st, 2/2nd and 2/3rd Australian Infantry Battalions, were moved up and re-equiped with jungle greens, because we were wearing khaki at that particular time, and we were given American gaiters'.[16] Within weeks, these men would first confront the Japanese at Templeton's Crossing.

* * *

That day, Corporal John Burns and Private Alfred Zanker were getting desperate. They sent five of the strongest walking cases with a native who had promised to guide them to Itiki that morning. Burns scribbled a note and handed it to one of the men to deliver to the first Australian he came across, which was to be forwarded to their commanding officer; it read:

27 Sept 42

CO
2/27 Bn AIF.

We have now been stranded here for nine days and there is still no sign of relief. When volunteered to stay with the wounded I was told help would be back in a week (I'm not concerned over

CHAPTER 41

my personal wants, it is the medical supplies that is of primary importance) we have no dressings whatsoever left and the lad's wounds badly need fresh dressings. I have decided to let the stronger men move off in the morning and try to reach safety, if there has been no party sent out to help us. I ask for a patrol to be sent out, if possible, with medical supplies and rations and smokes. I have been feeding the lads on hot mashed yams and pawpaw of an evening plus a few roasted yams hot, and during the day we live on cold roasted yam and green pawpaw.

I regret to report that Cpl L.P. Williams died at 0810 hrs on 24th Sept 42. We buried him on the side of the hill NE of the village we are occupying. Pte Burke is very low, but the rest of the stretchers are on the improve.

Trusting in the future and looking forward to a tin of bully beef.

(signed) John H. Burns A/Cpl. [17]

The fact that Corporal Burns, Private Zanker and their men were all looking forward to a tin of bully beef is testimony to their suffering and near starvation and dwindling repetitive diet. Burns went on later to record in his report:

On the 27th the native guide arrived at approx. 0800 hours, and I saw the lads off on their way to what we prayed would be safety. Corporal [Albert] Riches, Private [Leonard] Riches, Private [Cecil] Tee, Private [Stanley] Ashby and Private [Thomas] Corfield comprised the party. We gave the natives 10/- for his troubles. Private Burke kept Zanker and me busy all night and Private [Robert] Martlew was also weakening. We boiled a little water and gave Burke a sponge over, put my spare shirt on him and put him in a new stretcher. We then gave him a needle and he slept in peace – a real blessing for everyone. Heavy rains again worried us during the day and our

SAVING PORT MORESBY

Air Force decided it would be nice to strafe the track about 200 yards south of us. I don't know if they had a target or not. It certainly didn't give us the nicest of feelings when our own boys were coming so close.[18]

* * *

Still in Port Moresby was Brigadier Arnold Potts, who was keen to again command the 21st Brigade in the field. He wrote to Doreen providing some details of the fighting along the track, including: 'Later on I can tell you some quite interesting yarns and the last was produced nothing tougher in soldiers than this one. Lads shot through legs, or knee or foot or both legs possibly or high body wounds, smashed arms and shoulders; grinning like Cheshire cats and refusing point blank to be carried (and that meant 7 or 8 days sheer agony on appalling tracks), the stock phrase being "We'll make it" followed up with a grin, "And won't the little B------ pay for this". Oh well, we extracted our price, and it was a high one, and gained experience. I'd better get back to work again. But just to mark the end of a savage, but on the whole satisfactory period, I'd like a fortnight with just you …. Did I tell you I'm trying to collect 3 decorations of the highest [VC]. Here's hoping – this entre nous of course'.[19] Of the three, only Private Bruce Kingsbury, a 24-year-old estate agent from West Preston in Victoria, would be awarded a VC for his actions during the battle for Isurava.[20]

* * *

Also in Port Moresby, Rowell and Blamey agreed to certain arrangements – a compromise had been reached, or so Rowell had been led to believe. Later that day, Blamey went behind Rowell's back and complained that NGF was not making enough effort in providing him with prompt and accurate information.[21]

42
'THE POSN ON IORIBAIWA RIDGE NOW CONSOLIDATED'

On 28 September, the attack against Ioribaiwa Ridge went forward as planned. On the right Lieutenant Colonel Alfred Buttrose and his men of the 2/33rd launched their attack against the ridgeline around Ponoon. Sergeant William Crooks recalled: 'At 0900 hours, H-hour for attack, C and B companies moved off in extended line, with about two paces between men, each company with two platoons forward and one in reserve following up. By 1000 hours both companies were on top of Ioribaiwa Ridge, both reporting over the telephones lines that not a Jap had been sighted. At BHQ, which had been set up in a clearing in the centre of the battalion and on the axis of attack, this news was pleasant and surprising. Other than the noise of the exploding [Australian] shells, no other firing had been heard, and some had doubts as to whether the two forward companies were in the true positions'.[1]

Indeed, Buttrose was so concerned that he rushed forward with his intelligence officer, Lieutenant Robert Howland, and signal officer, William Weale, when reports arrived that his men had occupied the ridgeline and had encountered no enemy troops. On arriving at the forward position, he found that his men had indeed occupied their objective, and no Japanese were to be found. Now as planned, Captain Lindsay Miller, commanding 'C' Company, and Captain Thomas Archer, leading 'B' Company, sent out patrols to maintain contact with the enemy. By 1 pm, the battalion had consolidated its position along the eastern ridgeline, with 'C' and 'B' companies pushed down the northern slopes. Later, their respective patrols returned with news that they could not locate any enemy troops, but they had come across considerable amounts of abandoned ammunition and equipment.[2]

The men of 'D' Company, who now occupied the ridgeline, had not forgotten about Lieutenant John Mosley and Corporal Harold Bradford – they had been wounded in the fighting during 15 September. Captain Trevor Clowes sent out Lieutenant Ted Logan and his men of 17 Platoon to the area of the fighting with orders to try and locate them. Sergeant Crooks recalled: 'By 1600 hours Lieutenant Mosley had been found. He lay unburied. And near naked, his clothes[,] equipment and weapons gone. Lieutenant Moseley was lying where he had been hit; Corporal Bradford was found half-an-hour's march further on the same day by a patrol from A Company together with another Australian, and probably from 3rd Battalion. It appeared both must have been captured after being wounded because Jap positions were in the area, and none of them was [sic] the position attacked by 16 Platoon. Both were buried where they were found. Lieutenant Logan continued on, reporting to 2/25th Battalion before commencing the return journey back along this melancholy route. The night of the 28th was another night of high wind and driving rain. None enjoyed sleep, as none had time to erect shelters, the day having closed while the last of the reconnaissance patrols was rejoining'.[3] The battalion diarist recorded sometime that night: 'Bodies of Lieut. MOSELEY and Cpl BRADFORD [were] found and buried at [the] scene of action on 15 Sept'.[4]

* * *

In the centre, the men of the 2/25th Battalion had also launched their attack, focusing on the area around Ioribaiwa village. The men of 'B' Company on the left and 'C' Company on the right were supported respectively by their mates of 'A' and 'D' companies. In the centre of their position was the Headquarters Company and personnel. The

CHAPTER 42

men of 'D' Company were pushed out along the main track from the ridgeline.[5] The battalion war diary records that by 2 pm, the men of 'A' Company were on the high ground left of Ioribaiwa village, with 'C' Company on the right helping to cover the track. These men were soon occupying Ioribaiwa village itself, as the Japanese had gone. Among the first to enter the village was 30-year-old Captain Oswald O'Grady, from Brisbane in Queensland, leading 'A' Company. He saw a Japanese Rising Sun flag fluttering from a pole and brought it down with a burst from his Tommy gun; he quickly folded the flag and placed it in his pack. He would carry the flag all the way through the fighting to the Gona beachhead.[6]

By 4 pm, the 2/25th Battalion Headquarters had moved to the ridge above the old position held by the 21st Brigade Headquarters. All observed that the Japanese had apparently been intent on holding their position here, as dugouts and barricades were constructed along the whole ridgeline. They now lay empty of men, but equipment was strewn all over the place. The enemy had withdrawn in haste. The men of the battalion now began to collect the Japanese equipment (as well as looking for souvenirs); all searched for anything of intelligence value. At this point, a strong patrol led by 29-year-old Lieutenant Robert Hazard, from Toowoomba in Queensland, with 'C' Company, moved forward from Ioribaiwa towards Nauro. The battalion war diary concludes: 'No enemy seen or heard during advance on IORIBAIWA. All posns, evacuated by enemy before our attack'.[7]

* * *

Meanwhile, to the left, the men of the 2/31st Battalion launched their attack around the Spotter's Hut area. Covering the extreme left flank of the attack were the men of 'D' Company, and to their right was 'B' Company, then the Headquarters Company, 'C' Company, with the men of 'A' Company covering the battalion's right flank. They too occupied their objects facing no opposition. As ordered by brigade the night before, several patrols were pushed forward to contact the retreating Japanese, and they came back reporting they had encountered no Japanese. The battalion diarist recorded: 'No sign of enemy, which was very surprising as we had expected to find him in strength. Strong Jap positions with commanding fields of fire were in evidence. At least two coys must have occupied this hill. Found some of our missing dead and buried them. Contacted 2/25 Aust Inf Bn at IORIBAIWA village. Found quantity of Jap equipment. Shells for field gun, amn for SA (small arms), mortars and grenades. Look out towers were sighted with ladders in taller trees. Many Jap graves'.[8]

The stretcher-bearer with the battalion, Private Geoffrey Hamlyn-Harris, recalled searching the now-vacated Japanese defensive positions on the ridge:

> Some of us were permitted to visit the deserted lines of the enemy and came back with interesting souvenirs. There was a Japanese flag amongst these which was considered a great prize by its finder, a sheet of white silk with a red spot in the middle, covered all over with Japanese writing, apparently the history of the battalion to which it belonged. There were cute wallets, handbooks, shells from their mountain guns, rifle cartridges, tin helmets, camouflage nets, and many other odds and ends. What interested me most at the time were some ash boxes, collapsible bamboo affairs for holding the ashes of the dead, which later could be sent back to Japan to the soldier's people. It was plain that the Jappo had underestimated the number of those who would be killed. Recent battle grounds which we crossed later were ghastly with their decomposing bodies in masses and heaps covered by liquid mud, whilst these useless little contraptions were in our kits as souvenirs or being mailed home as curiosities![9]

*\ *\ *

CHAPTER 42

Soon word was coming into brigade from all battalions that there was no opposition. The Japanese had abandoned their forward positions and much of their equipment. By the end of the day, the three battalions were in occupation of the entire Ioribaiwa area, with the militiamen of the 3rd Battalion joining them the next day. The expected Japanese counterattack did not eventuate; indeed, all indications were that the enemy was retiring in haste. The brigade war diary records at 4.30 pm: 'Congratulatory message received by Brig EATHER from Maj. Gen. ALLEN, on reoccupation of IORIBAIWA. Brig gave interview to various newspaper representatives (LONDON reps) who somehow managed to contact us, the BM (Maj. LARKIN) gave them the general situation'.[10] It was also recorded two hours later: 'The posn on IORIBAIWA RIDGE now consolidated – No sign of enemy – withdrawn north. Strong patrols pushed fwd along tracks to gain contact. Quantities of JAP stores and eqpt left on IORIBAIWA RIDGE'.[11] The tables had finally turned – the supply conditions that had so troubled the Australians now strangled the Japanese advance as they pushed south away from their supply base, while the Australians got ever closer to theirs. But none who had fought the Japanese doubted that much desperate fighting still lay ahead, even though the threat to Port Moresby had been lifted.

* * *

With the forward Australian troops, journalist George Johnston recorded the long-awaited start of the Australian offensive to push the Japanese back across the track all the way to their beaches at Gona and Buna; and from there, into the sea:

> Blasted by the shells of our 25-pounders firing across Ua-ule [*sic* Ua-Ule] Valley, the mountain ridge and village of Ioribaiwa, limit of the Japanese advance towards Port Moresby, was captured today by the A.I.F. and our troops are pushing

ahead through a heavy rainstorm towards Nauro Creek, Menari, Efogi, and the rest of the scattered villages that line the Kokoda Trail.

The Japs offered little resistance, although they had built up a high timber palisade across the top of the ridge in front of an involved system of weapon pits and trenches. The 25-pounders blew great holes in the palisade and the Australians went in with the bayonets and grenades. The Japs didn't wait for anymore. They scuttled northward through the jungle, abandoning a stack of unburied dead, a great dump of equipment and ammunition, and leaving to us the steep ridge down which we retreated only a couple of weeks ago with the Japs rolling stones and grenades down on us and plastering our rearguard with fierce mortar and machinegun fire.

The Japs have left a lot of graves on Ioribaiwa Ridge and trampled in the mud between the bodies of the dead is an elaborate shirt of scarlet silk with a black dragon embroidered on it. Most of the corpses are emaciated .... The evidence scattered everywhere along the track and through the jungle is that this Japanese army was at the point of starvation and riddled with scrub typhus and dysentery. The stench of the dead and the rotting vegetation and the fetid mud is almost overpowering. One of our doctors carried out a couple of autopsies. Many of the Japs, he said, had died because hunger had forced them to eat the poisonous fruits and roots of the jungle ....

Our progress will profit by these grim reminders of an advance that went too fast. The Australians are pushing ahead very cautiously, building up store dumps and medical posts as they go, taking meticulous care about sanitation and hygiene, advancing in three prongs that are exploring every side-track and clearing every yard of jungle as they go.

CHAPTER 42

In this dense terrain of mattered vegetation and half-hidden native pad-pads and steep gorges there are the ever-present threats of ambush, counter-infiltration and outflanking. But the Australians are climbing slowly and grimly up the southern flanks of the Owen Stanleys with knowledge that the only Japanese behind their thrusting spearheads are dead ones.[12]

* * *

Having returned to Dump No. 66, Captain Bert Kienzle was organising carriers and supplies to be moved forward with the advance, as documented in his report: 'Investigated carrier positions at Uberi and "66" and reallocated ANGAU personnel and carriers to suit existing demands. I noted a large number of carriers sick at Bisiatabu [and] also restlessness owing to desertions which were taking place daily. This was due to several factors, pressure of the enemy, heavy work continuously, and proximity of the Rigo district and coastal natives'.[13] Also still at Dump No. 66 was Captain 'Doc' Vernon, who scribbled in his diary the conditions at the camp, which had grown to a significant field hospital:

> Life at Camp 66 was drab and consisted mostly of evacuating cases of dysentery to Bisiatabu and doing the best we could with those too ill to be moved; but the results as regards [to] mortality were good, and we could really feel we were pulling serious cases through. I often walked into Uberi where WO Maxwell had a fairly well-equipped hospital when we discussed ways of dealing with bombing casualties, Uberi being another seemingly easy target …. In addition to contending to heavy sickness, everyone was a bit down in the mouth about our reverses. Some were inclined to be pessimistic, but personally I never felt anything but complete confidence that the Japs would be staying where they were or on the Moresby side of the ridge beyond Uberi. It might be taken for granted that out defences there would be able

to deal with an army that had had to cross the Owen Stanley Range without proper military roads and with a makeshift supply column. Any big battle there should have been entirely our affair.[14]

* * *

Now falling back with the rearguard of the Japanese force, were mostly the veterans and survivors of the *144th Regiment*, who had been fighting the Australians since the first day of the Japanese invasion three months before (21 July 1942). At midday, advanced elements of the regiment had arrived at Efogi. The men of the regiment had begun their march at 4.30 am, with the Regimental Headquarters leading the way. With them was their CO Colonel Kusunose Masao, likely still suffering from malaria and still being carried on a stretcher. Close behind was the signals unit, with the men of the *II/144th Regiment* bringing up the rear.[15]

The men of 1st Lieutenant Hirano Kogoro's *No. 3 Platoon, No. 1 Company, I/144th Regiment* arrived at Efogi at around 3 pm without their commander. Hirano had stayed behind to bring forward several stragglers but was now hurrying to catch up with the rest of the battalion; his stragglers would have to find their own way to the battalion. At Efogi, his men were provided with a small ration of rice, less than half a litre, but the men were happy to receive seven cigarettes per man.[16] Even so, they were soon on the march again and finally reached Kagi just before dark. It was now that Hirano arrived at Kagi, recording in his diary: 'Rejoined the Battalion at Kagi. Reported to Capt. FUJISAKI'.[17]

Not far from Hirano was Japanese war correspondent Okada Seizo, who would be one of the few survivors of the fighting in Papua. He recalled in his unpublished manuscript of his experiences during the Kokoda Campaign, and that the confusion and a strong desire to survive had now taken root among the retreating Japanese soldiers of Horii's *South Seas Force*:

CHAPTER 42

None of them had ever thought that a Japanese soldier would turn his back on the enemy. But they were actually bearing a retreat! There was no denying that. As soon as they realised the truth, they were seized with an instinctive desire to live, neither history nor education had any meaning to them now. Discipline was completely forgotten. Each tried for his life to flee faster than his comrades .... During short respites on the way we often repeated the popular Japanese saying: 'Where there is life there is hope'. The Australian forces did not stir for two days after the Horii detachment had begun to retreat, probably because they did not know exactly what had happened at our front positions .... Our party, and small independent units went down without bothering about anything; but the headquarters unit was constantly delayed by the sick and wounded whom they had to pick up on the way. At each key point there was a platoon of the Tokoyama Engineer Unit, who waited for the last of our men to pass and blew up the cliffs or cut off the log-bridges to delay the Australians in pursuit.[18]

* * *

Meanwhile, to the north, Captain Sydney Buckler was crossing the Owen Stanleys from Sengai. He was desperately trying to get help to Private Thomas Fletcher and his wounded men back at Sengai No. 2 near the Japanese garrison at Gorari, and he was still forward of the main escape party. On 28 September, Buckler came across American troops of the 126th Regiment at Oidobi Rest House. After sharing a plate of bully beef, the Americans guided the sick and exhausted Australian officer to Dorobisolo, half-a-days march away.[19]

* * *

At midnight, in the Papuan village near the northern banks of the Nauro River, Corporal John Burns reported that Private Thomas Burke, a 25-year-old farmer from Hyden in Western Australia, died.

Burns and Private Alfred Zanker buried him beside Corporal Leonard Williams; Burns recorded in his report:

> The 28th and we were a sad party once more, our pal Pte Burke passed away at midnight and at dawn Zanker and I dug a grave with our only implements, the tin hat and machete. We buried him beside Cpl Williams. We then washed and shaved the lads and made them as comfortable as possible. The flies were at us in a big way again. The native boy who took our boys off on their return arrived back during the morning and brought seven sticks of trade tobacco, it was like a Xmas dinner for the boys, the only smiles I had seen in days appeared.[20]

* * *

Back in the base area of Port Moresby, officers of the 2/14th Battalion were assigned to spend some days with incoming Australian battalions and the US 126th Infantry Regiment, 32nd Division, to inform them of the tactics used by the Japanese. The diarist of the battalion recorded that at 8.30 am, Lieutenant Alan McGavin and six men left for the American regiment, and at 9 am, Lieutenant Stan Bisset and Lieutenant John Thurgood, with three NCOs, were detached to the 2/2nd Battalion, and another two NCOs were also detached to the 2/1st Battalion; all these groups were to instruct on lessons learnt during the campaign against the Japanese. The duration of the detachment was to be for three or four days. Lieutenant Norman Cairns was seconded to the 21st Brigade and appointed as the adjutant. At noon, the 2/14th, 2/16th, 2/27th battalion Bren carrier platoons came under the command of the 16th Brigade.[21]

* * *

While the Australians of Maroubra Force were advancing north and the men of the 16th Brigade were preparing to advance into the Owen Stanleys, the internal infighting within the Australian high command

CHAPTER 42

had reached its climax. At 9 am, Blamey informed Rowell that he was to be relieved of his command. To rub salt into the wound, Blamey backdated his command of all troops in New Guinea to 24 September, indicating he had taken over command of the Australian forces just as they launched their first large scale offensive operations against the Japanese on the Kokoda Track. Not to be outdone, MacArthur would later claim he had arrived in Papua at this time to command the Australian advance. The American warlord, however, would not arrive in New Guinea until 6 November – over a month later – and then he never visited the front lines.[22]

Rowell was informed by Blamey that he had telegraphed an 'adverse report' on him to both the Australian Prime Minister, John Curtin, and General MacArthur, providing no further details. Blamey had written nothing short of a character assassination of Rowell to the Australian Prime Minister:

> I would like to say that the personal animus displayed towards me was most unexpected …. In regard to his [Rowell's] … claim that I had failed to safeguard his interests in accepting the direction of yourself and the C-in-C, SWPA, I informed him perfectly frankly of the exact incidence of events which led me to come to New Guinea, and there appears to be no ground for any resentment or objections on his part. It seemed to me when I received your directions and those of the C-in-C, SWPA that it behoved me to carry out those instructions, and there can be no doubt that when the consequent instructions were given to General Rowell, it was his duty also to carry them out without question, cheerfully and cooperatively. I endeavoured to induce him to see this point of view, but his resentment was too deep.
>
> I informed him that I did not propose to make any alteration in the method of command, and I would do nothing that would derogate from his authority. He asserted his intention of refusing to accept the situation and remain in New Guinea. I

pointed out that such an attitude would be unacceptable to any Government and that it would certainly mean his retirement from the Forces .... This was the substance of my interview with him on the day of my arrival on the 23rd September.[23]

Soon after, Blamey recommended Rowell's early retirement and that he revert to his pre-war rank of colonel. Prime Minister Curtain dispassionately observed that Blamey was indeed, 'a great hater'.[24]

In his letter to MacArthur, Blamey requested that Lieutenant General Edmund Herring, a 52-year-old barrister from Melbourne in Victoria, be sent to Port Moresby to replace Rowell and take command of what was now called the Advanced New Guinea Force.[25] Historian David Horner, in his seminal book *Crisis of Command*, wrote: 'Blamey had chosen his man carefully and he knew that Herring would be completely loyal. It is true that this was to be his first operational command of a formation, for as CRA [Commander, Royal Artillery] he had never had to bear the responsibility of independent decisions, but for a while at least, Blamey would be at his shoulder to offer advice and encouragement'.[26] Or to put it as Rowell had rightfully feared, he would be a figurehead and Blamey would call the shots.

Rowell left New Guinea on the same day he was dismissed; he had not been provided with Blamey's allegations against him, as such he was given no opportunity to reply. Before leaving for Australia that day, Rowell visited Major General George Vasey, Deputy Chief of the General Staff at his headguards. Vasey scribbled a note to his wife, which he asked Rowell to deliver: 'He is to go – in fact he takes this letter with him – relieved of his command. Purely on personal grounds too – not for any failure on his part to command here. The feeling between he and Albert [Blamey] is really only known to themselves and this is the result. As I said in my previous letter Syd took a very poor view of Albert's arrival and conditions have got steadily worse, until this morning Syd was informed, he was to go home .... I find it difficult to assess Syd's action. He may have been precipitate. I feel

CHAPTER 42

that without their previous associations, the situation should not have become intolerable so quickly. God only knows who will follow Syd. I trust it is not our friend G.B. [Gordon Bennett]'.[27]

Just before leaving, Rowell also wrote a farewell letter to his good friend Major General Arthur 'Tubby' Allen:

Dear Tubby,

Events moved rapidly to a crisis this morning. I had accepted the situation and was prepared to bite hard for the time being, however, difficult it might have been. However, the C-in-C himself decided this morning that the position was untenable and that I was the cause of it through my temperament. The result now is that I am ordered home, to what job I do not know, if any.

I have tried not to let personal matters get on top [of me] in this difficult situation, but I am not able to go beyond a certain point in eating dirt.

I'm desperately sorry to have to go away at this juncture. It almost seems to me like a retreat, and I knew my detractors will make it so. However, my friends know the true position and I can only hope that I can do something to remove the root causes that make such actions as these possible.

I'm grateful to you for all the help you have given me. We have had a long association together and I felt we would see this war out in company. Perhaps we will yet, but one can never tell.[28]

At the time, Vasey made it clear who he sided with when he wrote to Lieutenant General Sydney Rowell the very next day: 'I can only express my bitter disappointment from both the personal and service points of view. I had looked forward to having a lot of fun with these Japs for I think you and I make a very good team. From the service point of view, I think it is disastrous. The very best luck in your next battle. I still believe that honesty is the best policy and gets recognition in the end'.[29] Within hours of Rowell leaving Port Moresby, Brigadier Kenneth Eather signalled: 'Ioribaiwa ridge occupied .... Strong offensive patrols sent out both flanks and centre'.[30]

* * *

Meanwhile, Brigadier Arnold Potts was back in command of his brigade in the field and issued a Special Order of the Day to the men of the 21st Brigade, with reference to the militiamen of the 39th Battalion:

> AUSTRALIAN MILITAREY FORCE
> SPECIAL ORDER OF THE DAY
> BY BRIGADIER A W POTTS, DSO, MC
> COMD 21 AUST INF BDE
>
> HQ 21 Aust Inf Bde
> In the Field
> 28 Sep 42

The Bde Comd desires to congratulate members of the 21 Aust Inf Bde and attached troops for the skill and gallantry they displayed in the fighting at ISURAVA and ABUARI and the subsequent stubborn withdrawal to IORIBAIWA.

Although heavily outnumbered by the Japs it is confidently asserted that enemy casualties were far in excess of those suffered by our forces and this was entirely due to the courage and teamwork of the units and incidentally to the sound training adopted by the unit commanders and the standard of fitness attained.

On behalf of the Bde, regret is expressed for our comrades who were killed, wounded or missing and it is urged that all ranks endeavour in the future [to] use the experience so hardly gained;

CHAPTER 42

and so enabled us to extract a full payment for those we lost. This opportunity is taken to congratulate our comrades of the 39 Bn on their gallant stand at KOKODA, DENIKI and ISURAVA.

The Bde Comd also wishes to thank all ranks for their loyalty, cheerfulness, and courageous cooperation and to express the pride he feels in commanding such a team of fighters.[31]

Brigadier Potts likely had no idea that Blamey was already set on removing him from active command. Major General Allen, however, must have suspected his own tenure was hanging by a thread, given Blamey's record. At the first sign of trouble, Blamey would almost certainly use Allen as a sacrificial scapegoat – Allen had good reasons to be concerned.

43

'LEFT A GOOD DEAL OF STUFF HERE'

The morning of 29 September opened with a heavy downpour. Covering the right were the men of the 2/33rd Battalion. A four-day fighting patrol was to be led by Captain Trevor Clowes and his men of 'D' Company. They had continually been out on patrol forward of the battalion since entering the Owen Stanleys. Why they were again being called upon to lead the way must have made some men question the decision. From the Ponoon area, these men were soon advancing north-east along the old police track. At noon, the captain used the 108 radio set to report to Lieutenant Colonel Alfred Buttrose that they had not observed any Japanese troops and were now pushing towards Nauro, which lay directly north of their current position.[1] With this patrol was Sergeant William Crooks, who recalled:

> At dawn on 29 September brigade orders to send out a fifty-man patrol from the battalion were received. Its missions were to try and contact Japs; find another track to Nauro; and to try and find any men of the 2/27th Battalion. This patrol inexplicably became D Company's task. Whether it was because two company commanders were now feeling sick and feverish; or that Captain Clowes had exemplified all that was best in a dedicated regular soldier is a matter for conjecture. Sufficient to say the men of D Company were staggered at the order. For the third time they would be the vanguard, and doing their second four-day patrol, where no others, other than HQ Company back at Imita, had done any. The order caused much anger

CHAPTER 43

and discontent. In the exhausting activity of those days, it was natural to detail for duty a company that had been reserve in a recent attack, but as the attack had proved [a] walk-over, a little thought could have made D Company a happier band of men that day and perhaps better for what was to come later at Gorari. A fairer course would have been to select any other than D. In the event there was a hurried search for rations for the four days patrol. Only one day of hard rations could be found and the other three days were made up by calling in all emergency rations held in reserve at brigade.

By 0900 hours the patrol moved out by the same track as 18 Platoon's patrol on 15 September. At the creek crossing below the ridge, Captain Clowes again detailed the groups for the jobs ahead: 18 Platoon would take a three-hour lead, under Sergeant Crooks (Lieutenant Cox remaining at Ioribaiwa with Lieutenant Copp and the nucleus of the company); 16 Platoon, under Sergeant Audsley and 17 Platoon, under Lieutenant Logan, would follow on behind to clear the side-tracks and be available to form a firm base if needed. At 1100 hours the patrol set off up out of the creek, following the well-defined track that led up from Ponoon. Usually, volunteers were called for the jobs of point scouts, as invariably it called for men of strong nerves and keen intelligence. Never was there any dearth of men volunteering. All knew that the chances of survival if any enemy delaying parties or posts were struck would be slim; the first shots would certainly be aimed at the scouts. Their only chance of not being hit would be bad shooting.[2]

Meanwhile, in the centre of the line along the ridge, another patrol was being organised. At 9 am, Captain Colin Andrews, 22-year-old assistant agriculturist from Crow's Nest in Queensland, with the 2/25th Battalion led his platoon out on a five-day patrol. They would reconnoitre the main Kokoda Track – their objective was to also make contact with the retreating Japanese. By 11.30 am, he reported via wireless that smoke was observed on the high ground to his left. Two hours later, he reported having crossed Ofi Creek but still had made no contact with the enemy. He reported to the battalion: 'Contacted Lieutenant Hazard's patrol which was on a hill across a creek, reporting no trace of enemy except for a small amount of equipment seen on track'.[3] Not long after, Lieutenant Robert Hazard's patrol returned from the Nauro area, having also failed to contact the enemy; Hazard also noted that Japanese equipment had been jettisoned along the track.[4] Meanwhile, still on Ioribaiwa Ridge, Lieutenant Colonel Charles Withy presented a situation report to brigade:

> Weather: Very hot during the day and rather cold during the night – rains heavily each day, with winds ever so slight.
>
> Conditions: Track is well worn and covered with a heavy coat of mud. The Track is 'stepped' with jungle timber at its leading edge, is held in place with bush pegs in the difficult places and is generally in a state of despair. Pioneer Battalion working to repair the worst areas.
>
> Food: Usually brought by native carriers, mostly hard rations, occasional ration of butter and jam to supplement bully beef and hard biscuits.
>
> Men: Showing signs of stress through unusual exercise in climbing mountains, where steep descent requires the building of rough steps. Scrub itch and diarrhoea are prevalent.[5]

Also still on Ioribaiwa Ridge but further west covering the left flank around the Spotter's Hut area were the men of the 2/31st Battalion. Another strong 50-man patrol was sent out that morning led by Major Ewan Robson, a 36-year-old solicitor from Sydney in New South Wales, and 24-year-old Lieutenant John Mules from Auchenflower in Queensland. They were to conduct a four-day patrol to reconnoitrer a

CHAPTER 43

track that was found running parallel and west of the main track from Ioribaiwa to Nauro, known as the 'Old Kokoda Track'. Reporting soon came back from this patrol that the way forward was clear.[6]

* * *

By midday, the Australian 3rd Battalion, commanded by Lieutenant Colonel Allan Cameron, had joined Brigadier Kenneth Eather's 25th Brigade's advance to Ioribaiwa Ridge. The battalion war diary records: '0800 Bn moved forward in following order A Coy, HQ Coy, Bn HQ, B Coy, C Coy, D Coy ... 1220 hrs defence positions occupied on reverse slope, after liaison with WITHY. Plan of dispositions is being submitted'.[7]

That night, the 25th Brigade diarist concluded the situation: 'Bde HQ opened 1500 hrs 300 yds SW of IORIBAIWA .... Bde in posn IORIBAIWA RIDGE. 2/33 Bn right flank, 2/31 Bn left flank, 2/25 Bn and Bde HQ in centre, 3rd Bn in centre in reserve. 2/1 Pnr Bn responsible for defence and improvement of main track vicinity IMITA RIDGE. 2/6 Aust Fd Coy now under command 25 Bde, advised to make recces of JAP defences on IORIBAIWA RIDGE with dimensional sketches'.[8]

Later that day, a tired by satisfied Brigadier Kenneth Eather recorded in his diary: 'Frightful trudge up and down hills .... Sent patrols forward to find Jap. Seems to have gone back past Nauro in a hurry. Left a good deal of stuff here .... Very tiring on men. After such a march must have a day's rest'.[9]

* * *

Meanwhile, the Japanese troops were ordered by their commander, Lieutenant General Tomitarô Horii, that a defensive line was to be stablished at Eora Creek. Major Horie Tadashi commanding the *II/144th Regiment* was now informed that his battalion was designated the *Stanley Detachment*. The battalion commander issued his orders: '1200 hours Received Horie Operation order no. 122 stating that

Major Horie as Stanley Detachment Commander will fortify the position at Eora as a base for future offensive'.[10]

Having finally caught up with his battalion at Kagi the day before, 1st Lieutenant Hirano Kogoro, commanding *No. 3 Platoon, No. 1 Company, I/144th Regiment*, recorded in his diary for 29 September: 'Capt. FUJISAKI left with the Staff Officer(s). 1st Lt. MAEDA took charge of the Battalion consisting of No. 1 Coy and the Battalion HQ. Reached the top of a bamboo-thicket hill where 40 of our men assisted No. 5 Coy [*II/144th Regiment*] in construction work [just south of Templeton's Crossing]. At 1620, camped near a river. Spent unpleasant night under pouring rain'.[11]

Close by would have been 2nd Lieutenant Onogawa, *No. 2 Platoon, No. 3 Company, I/144th Regiment*, who recoded in his diary that the entire force 'seems to be constantly retreating. Rations will be distributed at Kagi. Medical, signals and mountain artillery units are descending the mountains'.[12] Early that morning, the men of the regiment began moving north. They would march throughout the day and night, not bivouacking for the night until 1 am the next morning.

* * *

Captain Sydney Buckler that day finally made his way to Dorobisolo. There was Lieutenant William Nichols and his men of 'D' Patrol, 2/6th Independent Company. Nichols and his men had a week earlier found Lieutenant Maurice Treacy and his men of the 2/14th Battalion; they were tasked with searching out any Japanese infiltration parties within the Jaure area, about 30 kilometre north-east from Dorobisolo, as the crow flies. They were then at Dorobisolo to resupply. Nichols recalled that Buckler walked into their camp alone, informing him that he had come ahead of his party of one officer and 37 other ranks to find help and supplies. Private Les Gosden, a 21-year-old turner and fitter from Carlton in Victoria, with 'D' Patrol recalled: 'Captain Buckler of the 2/14th Battalion came out of the jungle after having followed some tracks made by the "Jaure" patrol and was lucky that the guard on the

CHAPTER 43

base did not shoot him'.[13] Buckler was sent onto Rigo, just east of Port Moresby, while Nichols arranged for a party to go towards Jaure to bring in Buckler's men.[14]

* * *

Corporal John Burns, Private Alfred Zanker and their wounded mates were no doubt praying that the small party of the fittest walking wounded, being led by 28-year-old Corporal Albert Riches, from Wayville in South Australia, were making good progress on their way south for help. Even so, it was difficult to keep the men's spirit's up, not to mention their own; Burns recorded:

> Tuesday 29th finds the planes bombing and strafing just opposite us over the river, they were most anxious to get at something as they returned again about 11 o'clock. We kept ourselves well out of sight. About 1400 hrs a native policeman arrived, and I asked him would he try and obtain some food and dressings from the nearest outpost and he said he would and off he went. The boy's spirits were falling fast at this stage, Zanker and I found it more difficult to instil any confidence in them. We read the New Testament as much as possible and had general discussions on different subjects to try and keep their minds occupied. The strain was beginning to tell on us two lads too, we were becoming very weary.[15]

44

'... THE WAY LAY OPEN FOR OUR TROOPS TO ADVANCE'

The morning of 30 September was described as being extremely hot.[1] The 25th Brigade war diary records that at 7.30 am: 'Quiet night. Filth left by JAPS being cleaned up. Patrols still pushing towards NAURO – still no contact with enemy. Recces in progress fwd of IMITA RIDGE as to possibilities of mule track'.[2]

Covering the right flank, the men of the 2/33rd Battalion at Ponoon were continuing to dig-in. No word had been received from Captain Trevor Clowes since midday the previous day; the captain and his men were on their second day of a four-day-long patrol.[3] Sergeant William Crooks, who was leading the forward platoon in their eastern approach towards Nauro along the old police track, recalled the disturbing fact that their newly furnished emergency rations that were to provide food for four of their five days were not only useless, as they required cooking, but also because men to the rear had tempered with them:

> For the next two days the patrol pushed on, remaining always on the track that continued generally north-east. Two large villages were found, searched and passed by. 18 Platoon remained in the lead, each section taking two hour turns in front. By mutual agreement, the scouting was shared by six men – Privates Jock Proudfoot and Billy Musgrave in 7 Section; Johnny Condon and Alex Breakwell in 8; and 'Tibby' Price and Teddy Café in 9 Section. The track route was fairly easy going, following around

CHAPTER 44

ridges and spur lines. Small native gardens dotted the valleys and glens. At no time was there any sign seen of Japs or even of their tracks. No natives were sighted and both villages were deserted. After the first day [29 September] the hard rations had been consumed, so the men began to open the emergency rations, to discover that in every case they had been tampered with. The flat new issue tins had all been opened and all the contents less the dehydrated vegetable powder had been taken out and the waterproof insulting tape replaced around the tin. More than thirty of the tins were empty and the contents replaced by dirt to make weight. The new ration, making its first appearance in this campaign, was designed for one day's meals for one man. It consisted of four ounces of dehydrated vegetable and meat, which necessitated cooking before eating; a four-ounce compressed mixed fruit plug; two hard pack biscuits; a tea and sugar tablet; a small tin of meat past or jam, and two P.K. chewing gum tablets. Nobody had thought that these were useless on forward protracted patrols over four days. Cooking, by creating smoke, was suicidal. These ration errors revealed in this campaign were not repeated in later campaigns. The rations then came in solder-sealed tins, were of a better variety, and contained methylated wax cubes for smokeless fires. For D Company this was an unhappy patrol. For the next three days they existed by eating the cold soaked dehydrated ration for breakfast, with nothing for lunch or tea.

On the second day, just as light was failing, a sign on the track, consisting of a vine tied in a reef knot, and holding an arrow pointing down the track, indicated that perhaps some of our men or some of 2/27th Battalion had passed this way. Below the track at this point, the country fell away sharply into a valley, the floor of which was perhaps 1000 feet down. Part way down into huts could be seen in the centre of a garden area. From far below came the muffled roar of falling water. At that moment the rains came down, and a heavy impenetrable mist came swirling up from the valley. At the same moment a runner arrived at 18 Platoon from the OC, only an hour or so behind, with orders to halt and make camp. By dark the company were all together and sat the night on this high spur that had ended suddenly at a cliff edge. Lightning and thunder kept up all through the night. Hungry, wet and weary, we discussed man's more sordid weakness which allowed him to rat emergency rations. After two days of marching, we had still no sign of Nauro, the finding of which would perhaps cancel a return march and maybe provide some food.

That night Captain Clowes decided in view of the food situation and the complete lack of communication with BHQ on the 108 set since the previous night, he would abandon the patrol and return next day.[4]

Still on Ioribaiwa Ridge were the men of the 2/25th Battalion. That morning they had discovered soil that had recently been disturbed. On digging, the men soon found several artillery shells for a mountain gun, as well as other equipment. Of significance, however, they also discovered several Japanese documents along with what appeared to be textbooks. These were immediately collected and sent to brigade for scrutiny by intelligence officers.[5]

Meanwhile, the forward patrol led by Captain Andrews had approached Nauro from the Kokoda Track to find that it had been abandoned by the Japanese. The battalion war diary records: '0815 Report from Capt. ANDREWS' patrol – Quiet night. Reached high ground overlooking Nauro. Posns had been prepared but evacuated by enemy'.[6] Indeed, by noon, he and his men were in the village as recorded by the battalion diarist: 'Report from Capt. ANDREWS – reached NAURO at 1130. Found one sick Jap. Left two men and pushed on ½ mile to [Naoro] river'.[7] It was now that the ill Lieutenant

CHAPTER 44

Colonel Charles Withy was forced to hand over command to Major Richard Marson, a 38-year-old dairy farmer from Toogoolawah in Queensland, due to illness.[8]

The 25th Brigade war diarist recorded for noon: 'Considerable amount of JAP material and eqpt abandoned on IORIBAIWA RIDGE. JAPS appear to have left in a hurry. Articles incl quantity of eqpt, steel helmets, rifles, bayonets, belts and pouches, water decn [de-sanitation] outfits, clothing 192 rds 75mm amn (mountaingun shells), mortar shells, hand grenades and MMG amn, medical supplies, documents, diaries, notebooks, textbooks, letters, postcards, identification discs, badges [of] rank – all were fwded to HQ 7 Aust Div …. It is hoped that much valuable infm may be gained from the captured articles. Much of the material was carefully buried by the JAPS before withdrawing and more is being unearthed. They seemed to have retreated as lightly as possible'.[9] It concludes: 'All quiet to 1300 hrs. Offensive patrols still pushing towards NAURO. Lt Col WITHY evacuated with severe skin disease. 2/25 Bn patrol arrived NAURO 1130 hrs. No contact with the enemy – one sick JAP was captured. Patrol proceeding beyond. Ordered to send sick JAP to Bde HQ immediately and to tie him to a stretcher if necessary'.[10]

On the left flank, the men of the 2/31st Battalion around the Spotter's Hut area – which was approximately one kilometre northwest of Ioribaiwa village – were also continuing to dig-in. For the first time in weeks, the men could build fires and cook their own rations, as the Japanese seemed to have vanished. Major Robson's patrol that had been sent out the previous day to reconnoitrer the 'Old Kokoda Track'. Word arrived from Robson early that morning via radio that the track was rough going and diverged to the west away from the main track. He also reported seeing 'many traces of our troops, webbing, tins and blankets'.[11] Robson and his patrol would not return to the battalion until 2 October. He later recorded in his report for 30 September that he followed the track going down north of a prominent feature into the valley and then veering east and coming to a single hut (1st Hut).

The area was covered in kunai grass. Soon after, he and his men passed the '2nd Hut' about one kilometre north and followed for a time a track leading west.[12] The 2/31st Battalion war diary also records: 'After 1 1/4 hrs track petered out, returned to 2nd Hut. Thence by track east, passed water (small trickle); bad surface to 3rd hut – time two hrs. Three recce patrols sent out, one around side of feature to north; one to south and third to investigate smoke to west. Latter patrol contacted a native impressed for bearing by Japs. Was very frightened; refused any infm except that he pointed to high feature to north and said – "Japs gone over hill". Other patrols failed to locate main track to NAURO or parallel track'.[13]

The war diary of the 3rd Battalion records for 30 September: 'Quiet night. Two shots fired vicinity D Coy … thought to be accidental. Area was found to be in dangerously filthy condition; this is being rectified now and further medical inspection of area will follow'.[14]

* * *

Moving north towards Isurava, Japanese 1st Lieutenant Hirano Kogoro, commanding *No.3 Platoon, No. 1 Company, I/144th Regiment*, recorded in his diary: 'Ordered 41 men to carry wireless sets from the hill to the provision storehouse area. At 1000, as soon as the men on detail returned, marched off through the rain and mud. At 1600, billeted and ate handful of rice which a Corporal had somehow managed to find for me. Met Acting Regimental Commander, Lt Col TSUKAMOTO'.[15]

Meanwhile, just south at Eora Creek, 2nd Lieutenant Ebuchi, *No. 2 Platoon, 9 Company, III/144th Regiment*, was informed by Captain Uyama, who was part of Major General Horii's staff, that when his battalion arrived, they were to push onto the Kokoda Plateau. The young lieutenant waited anxiously for the battalion. He now witnessed the *South Seas Force* Headquarters, Regimental Headquarters, and headquarters of the *I/144th Regiment* and *II/144th Regiment* pass through. He was informed the men of the *II/144th Regiment* were

CHAPTER 44

rushing forward to obey Horii's orders to begin constructing defensive works around Eora Creek Village, just north of Templeton's Crossing, and to be prepared to defend this position from the advancing Australians.[16]

* * *

Captain Bert Kienzle with ANGAU was still in Port Moresby with the headquarters staff of the Australian 7th Division. He records in his official report: 'Gave information regarding tracks, also interrogated natives who had been held by the enemy. They advised shortage of food with the Japs'.[17]

Meanwhile, his good friend Captain Geoffrey 'Doc' Vernon was preparing to leave Dump No. 66 as the Australian advance progressed into the Owen Stanleys. He recorded in his report for ANGAU that he issued orders to Warrant Officer Maxwell to advance from Uberi: 'Towards the end of the month when it appeared likely we should move forward again, I instructed him to move forward to 66 and handed the hospital there over to him. Before leaving, 66 had been rebuilt and was now looked upon as a model camp'.[18] He recorded in his personal diary: 'In the last days of September the tide of war did turn in our favour. Japs were driven out of Ioribaiwa, mercilessly bombed along the whole mountain route, and the way lay open for our troops to advance'.[19]

* * *

Just a few kilometres east of Captain Andrew's patrol, Corporal John Burns and Private Alfred Zanker with the wounded of the 2/27th Battalion were still waiting for help to arrive. They had been stranded almost two weeks before, and morale was an increasing problem; the young corporal recorded: 'The 30th Sep finds the boys' morale and spirits very low, we shave and sponge the lads again, but the smell [gangrene] was getting too strong to stay beside them too long. If only we could have had some dressings for them. There was no sign of our police friend but one of his tribe visited us and said he had gone to find meat for us and that made us a lot brighter. Heavy rains in the afternoon almost flooded us out. Private [Robert] Martlew took a turn for the worse and we had to watch him very closely. The flies were worrying Cpl Roy McGregor very much and we spent the hours of daylight taking it in turns to keep them away from him. The yam stock was beginning to run low, so we were praying for help more than ever'.[20]

* * *

Lieutenant Robert McIlroy and his small party were still likely somewhere near the village of Enivilogo, with the two men from 'A' Company who had recently joined them. The young lieutenant recorded for that day: 'During this period the wounded man had improved considerably but unfortunately the injured man got weaker and on 30 Sept he died. With the assistance of the mission natives a proper burial was carried out and the remainder of the party guided by the natives moved on later in the afternoon'.[21]

* * *

On the other side of the Owen Stanleys at the Japanese beachhead at Sanananda, Father James Benson from the former Gona Mission was about to accompany another Japanese patrol. Captain Nikaiama with the military police, instructed NCO Furokoa and interpreter Susuki to take Benson with them as they were to reconnoitre the Soputa area. Benson wrote after the war:

> I was asked to go inland, again with Furokoa and Susuki, to Soputa. It seemed that some of the natives there had drifted back into their village, and Nikaiama wanted me to go and tell them they must clean and repair the roads in their district. I did not see any harm in doing this – for if the natives refused to keep the roads in order willingly the Japanese could easily have made them do so by force.

CHAPTER 44

We went to several villages around Soputa; in one was a signals station; in another company headquarters; and in the last we visited we found a farriers' camp, where there were three blacksmiths' forges set up in the bush; all of them were shoeing horses. Furokoa talked a great deal at each of these places, and I began to wonder if gossip was the real purpose of our journey. After lunch we walked along tracks to the west of Soputa, away from the river and into the country of coffee plantations. Here we found a large field hospital, and Furokoa talked all afternoon to various doctors, attendants, and patients as we wandered round the wards.

At sundown we went into one of the doctor's tents, where space was made for us on the blanketed floor; soon candles were lit, and Furokoa settled down to a good gossip with the two doctors and their four assistants. I was kept squatting near the entrance of the tent. After a little while one of the doctors who spoke excellent English came across to me. He had done a post graduate course at one of the London medical schools, and he had been on the staff of two London hospitals. He seemed very friendly, and we both enjoyed an evening of reminiscences. But, as it became late, I began to feel a little apprehensive. A medical assistant had come into the tent; he squatted beside me and started fiddling about with the boxes stacked along the wall of the tent. From time to time the doctor broke off our conversation to give his assistant directions in Japanese; and I presently noticed that the man was preparing two hypodermic syringes; he seemed to be taking a great deal of time about it. At last, after further talk with the doctor, the assistant bowed himself out of the tent, much to my relief taking his syringes with him. I must have sighed my thankfulness aloud, for the doctor asked me what was the matter. I explained my fear that the injections were for me – 'to put me to sleep so that I couldn't run away!' The doctor rolled on his back on the rugs, kicking his feet in the air, cackling in uncontrollable laughter. When he recovered control, he slapped me on the back saying: 'Our wounded soldiers need all the drugs we can get. When we decide to get rid of you, we will do it some other way!'

After the late evening meal – which was better than I had had for many a long day – cigarettes were passed round, and we talked of many things. My doctor and I went touring together, up through Bedford, to Huntingdon, Godmanchester, St. Ives and Cambridge. Then he told me of the beauties of Japan, and we ended up by discussing Buddhism and Shinto. The other doctor was a tall man, very well built, and wearing pince-nez glasses; he rather reminded me of Leon Blum. He spoke fair English and talked of the Shinto beliefs for a long time. Eventually we lay down to sleep, and after the big doctor had seen me saying my prayers, he said: 'I, that speak to you, am a Shinto priest'.

So I lay down that night alongside the Shinto priest, and until we dropped off to sleep the others were chaffing him in a gentle, friendly way; much as a young curate might have to meet the playful gibes of your 'man of the world' or the blushing youthful padre the wisecracks of a ward room; only the Shinto priest was neither youthful nor blushing, and appeared quite as worldly as anyone else.

There was a small flickering light throughout the night, a floating wick in oil; and it cast strange shadows on the walls of the tent. From where I lay at the foot of the tent pole, it appeared like a small portable shrine burning away above my head. It was a strange night.[22]

* * *

Back in the Port Moresby area, the militiamen of the 39th Battalion were already back in action conducting patrols to ferret out any

CHAPTER 44

potential Japanese infiltrating parties. That day, Major General Arthur Allen visited the battalion as recorded in the battalion war diary: 'Maj. Gen. A.S. ALLEN (Cond 7 Aust Div) called at Bn HQ expressly to ask the CO to convey to all ranks his congratulations on the splendid work they did in the unit's first actions, some of which he considered outstanding'.[23] Indeed, it would be the gallant militiamen of the 39th Battalion, led by Lieutenant Colonel Ralph Honner, who would in mid-December 1942 finally break through the Japanese beachhead defences at Gona and capture the position.

However, the Australians of the 16th and 25th brigades would first have to advance northward along the Kokoda Track, along with their mates of the 3rd Battalion. These men would face stubborn resistance by the Japanese *144th Regiment*, *41st Regiment* and their supporting elements now digging-in at Templeton's Crossing, Eora Creek and soon after, Oivi and Gorari, only then could the battle for the Japanese beachheads finally commence.

EPILOGUE

Still lost was Lieutenant Robert McIlroy and his small party of the 2/14th Battalion. They continued to head south in search of the Australian frontlines. In early October, they moved back towards the Seventh Day Adventist Mission station, with the two Papuan converts who had stayed with them for weeks, caring for these men. The ordeal of these men was about to come to an end – finally. McIlroy recorded for his report: '1-2 Oct: During this period, we moved with all the people from the mission who had been waiting for us since 17 Sept. The mission party consisted of 1 old man, 1 young man, 3 small boys (one of whom spoke English and acted as interpreter throughout), 7 women and 5 small girls'.[1]

The next day, these men were still making their way south. He recalled resting 'for the Sabbath (the natives being Seventh Day Adventists). They decided that to facilitate movement, the old man and 3 young lads would take us on [the] next day. As none of [them were] sure of the location of the Jap forces, they decided to lead us to JAIL GARDENS (near LALOKI), and because of the youngsters, [they] would keep away from any tracks in case of clashes with the enemy'.[2]

Next morning – 4 October – they had moved out of Jail Gardens, leaving the remainder of the mission Papuans behind. They quickly crossed the a tributary of the Brown River in search of the Australian frontline positions. As they headed south, however, the bulk of the Australian frontline troops were now north and east of their position.

Lieutenant McIlroy and his men spent the following day travelling along a creek bed. Their rations consisted of just one taro, one small fish (about 10 centimetres long) and some Papuan cabbage.[3]

EPILOGUE

Finally, the one-month ordeal of Lieutenant McIlroy and his lost section of the 2/14th Battalion was coming to an end. He wrote in his report to the battalion: '6 Oct: Continued following the creek bed for an hour then struck across country reaching the Brown River approx. 1000 hrs and after following the river down for about 2 hrs came across an old garden. We were picking some green pawpaw to cook for lunch when a patrol from an [2/6th] Independent Coy came up the track, they guided us back to their camp area where we severely liberated their supplies of tobacco and cigarettes [and] also ate far more than was good for us'.[4]

They were quickly taken to the Australian 14 Field Ambulance and from there to Port Moresby. Days later, while recovering in hospital in the base area, McIlroy learnt of the fate of several members of his patrol who had become separated from him. Of the main party he had lost contact with on 10 September, seven had joined another isolated party under Sergeant William Irwin, these men made it back to the Australian lines on 21 September. McIlroy was also informed that of the four men who had attempted to push forward to Menari before the afternoon rains, only two had so far shown up. He was informed that Privates Alexander Roberts and Lionel Smith had made their way to Ioribaiwa on 14 September where they rejoined the battalion; he and his men would later learn that Corporal Lewis Waller and that Private Albert Bell had been killed at the Japanese ambush on 10 September.[5]

Finally, he was informed that Corporal Stewart Gedye and his party consisting of privates Douglas Brown, Wilbur Cahill and Alexander Matschoss, who had been sent forward to bring back carriers, had been attacked by local Papuans; only Gedye and Brown would survive.[6] Lieutenant Robert McIlroy concluded in his report: 'Note: 2 of the 4 men that went ahead on 10 Sep have arrived in, but they have no knowledge of the other two. Of the main body (separated 10 Sep) several have arrived back, but others have not yet been traced. Of the 4 men who went out ahead (16 Sep) 2 got through on 3 Oct and as far as can be ascertained the other 2 were killed by hostile natives. Fortunately, I was [later able] upon my return to make satisfactory repayment to the mission natives for their assistance'.[7]

* * *

Meanwhile, the other lost party led by Corporal John Burns, Private Alfred Zanker and their sick and wounded men of the 2/27th Battalion were by 1 October still on the northern banks of the Naoro River just east of the main Kokoda Track. They were waiting for the help promised the day before by the Papuan constable. Burns later wrote:

> Thursday morning the policeman turned up with three tins of army emergency ration. He told us the Japs had taken all the food from the dump. He also told us that two policeman and 25 bearers would arrive at ten the following morning to take us away. I thought he had rounded up a party for us. After the policeman's departure I occupied their time [sick and wounded] by giving a talk on the finer points of baking [Burns was a baker by trade] and then we had a discussion on the merits of the different books of the New Testament. It all helped to pass away the long hours. At 1600 hrs Zanker and I cut up enough yams for us stronger lads and flavoured it with a little M&V ration; for the weaker lads I made some M&V broth. It was grand to have a different flavour to just yams. It usually used to take us an hour for all the lads to have a meal, five of the lads had to be fed by us as they couldn't move, but this meal took me almost two hours.[8]

The next day, to their great relieve, an Australian patrol arrived led by Medical Officer Captain Robert Wilkinson, a 31-year-old physician from Rose Park in South Australia, who was with the 2/4th Australian Field Ambulance. Also among these men was Warrant Officer Ronald Preece and a Papuan carrier party attached to ANGAU. Preece would

EPILOGUE

later deservedly be awarded the Military Medal for his devotion and bravery and dedication throughout the Kokoda Campaign. Soon they collected the wounded, and all were heading south-east to Ta-ai over rough country where at this village the track turned in a south-west direction to Nigabaifa then to Jawarere before heading directly west to Itiki.[9] They took this route to avoid any congestion on the main Kokoda Track with the Australian advance northwards from Ioribaiwa. Burns later wrote in his report:

> Friday 2nd October comes after we had been kept busy looking after Cpl McGregor and Pte Martlew. Our first job at dawn was to go down and fill the bottles and then gave all those who could eat them a roast yam for breakfast. At 0930 Zanker and I were making a new stretcher for Pte Martlew and in runs our police friend and he was most excited. At first, I thought he meant the Japs were coming and I ordered the men who could walk to hide in a clump of thick undergrowth nearby while Zanker and I stood by the stretchers. Soon afterwards I saw a line of troops moving across the garden but couldn't identify them until I recognised Sgt [Arthur] Koehne and I knew all was well. It was a grand sight; you have no idea how the boy's spirits came to life again. The Doctor – Capt. Wilkinson of the 24 Fd Amb immediately set to and gave attention to the lads, while Sgt Koehne boiled a billy of beef tea, the tin of bully we had been longing for, we still were denied – the Doc wouldn't play. At 1210 hrs the first stretcher moved off and we were on our long trek to safety. It wasn't until 1730 hrs that a halt was called. We had covered a lot of ground, crossing the Nauro [sic] River approx. seven times, the last three hours being a very steady climb. For the evening meal the stretchers had a cup of beef tea and a little rice with Nestles condensed milk. Zanker and I managed a tin of salmon and some rice. It was like a Xmas dinner at home.[10]

The next day, the walking wounded that Burns had sent south on 27 September to get help reached the Australian lines. Later that day, a copy of the note that Burns had scribbled requesting urgent help for his wounded arrived at the headquarters of the 2/27th Battalion. On the bottom of this typed copy of Burns' original note is scrawled in handwriting: 'Above note received by Bn on 3rd Oct 42 brought in by 2 of the coys left with the stretchers. They had also brought with them 4 of the sick and wounded who could walk – hence the 7 days to do the journey. By this time relief had reached the stretchers and they were already on their way in. Altogether 4 different patrols were sent out with rations. Medical supplies and bearers to locate stretcher party, but it was not until 2nd October that Capt. Wilkinson found them'.[11]

Corporal Burns continues his narrative of their long journey back to the Australian lines; they would not arrive back in the base area for another week:

> At seven o'clock on the third we were on the move again after another meal of rice. We climbed steadily all day. From the top of the mountains late in the afternoon we secured the most magnificent of all views. At 1800 hrs we bedded down for the night and had a light stew and boiled rice for tea. The Doc then plays a real Mandrake trick and produces a block of chocolate for all of us.
>
> On the fourth we left at seven after a breakfast of stew. The track was over terrific country until seven at night when we reached the rendezvous. It was dark when we arrived there so we all just bedded down without any tea. Cpl McGregor was feeling the strain at this stage.
>
> The fifth was a much easier day and after about six hours marching, we reached a creek where we bathed the wounded and re-dressed their wounds. It brightened them up a lot. Leaving the creek at 1700 hrs we made the village of Tahi

EPILOGUE

[*sic* Ta-ai] by 1800 hrs. Rations were very scarce at this point, the native bearers had a hungry night like the rest of us. Cpl McGregor was very weak at this stage and Doc Wilkinson decided to stop behind with him for a day or two to strengthen him up.

At 0630 on the 6th, we left for Nigabyfa [*sic* Nigabaifa] and it was a hard day, three separate ranges by 1600 hrs. The 36th Militia Bn had a beautiful stew awaiting us and it was very welcome. The bearers were also given a big meal. We washed the lads here again in preparation for the last day of our trek.

After a good meal on the 7th, we were away at 0600 hrs. It was raining but everyone forgot all about the rain, the hospital was all they were thinking of. After a trip which took a lot out of us because of the slippery nature of the track we reached Jawarere at 1130 hrs and here the 2/4 Fd Amb had a meal and chocolate awaiting us. There was also a big meal for the 'hero bearers'. At 1300 hrs we left for Subatani. The pace was very fast, the bearers found the open road much to their liking, in fact they almost ran Zanker and I off our feet. At 1700 hrs we sighted Subatani. I can't describe how we all felt and to meet a few of our pals was even better. At 1730 hrs I handed the wounded over to Capt. Ryan of 2/4 Fd Amb and they were given a hot bath and a pair of pyjamas and then on the ambulance to hospital.

When the party with Cpl McGregor arrived[xxv] in on the evening of the 9th, it brought to a close a chapter of hardships which few wounded have ever had to endure previously.

No tribute can be paid which is too high for those native bearers, for without them it would have been impossible to have reached safety in time to save the lives of two of our men. And to their

---

xxv Tragically, after surviving the journey, Corporal Roy McGregor, a 23-year-old cellar hand from Berrie in South Australia, died from his wounds three days later.

master WOII Ron Preece of NGF. His name will be a byword to the boys of the stretchers for a long time.

SX4557 John H Burns Cpl.
HQ Coy 2/27 Aust Inf Bn.[12]

Corporal John Burns and Private Alfred Zanker were Mentioned in Dispatches for their dedication to these men.[13]

* * *

Meanwhile, to the east, Lieutenant William Nichols' party from 'D' Patrol, 2/6th Independent Company finally contacted the lost men from Captain Sydney Buckler's party, who were helped by the commandos on their way south and had finally reached Dorobisolo on 6 October. With safety at last in sight, one of the 'walking' wounded, Private Albert King, a 23-year-old labourer from Fitzroy in Victoria, died at Dorobisolo. The ordeal for the rest of the party, however, had not ended, as another two days of walking lay ahead before they would be floated down the Kemp Welsh River by raft to the American camp at Kalikodobu, just east of Rigo. Among the walking wounded was Sergeant John Gwillim, who with his men had been ambushed just south of Isurava Rest House, and who had fallen back to warn Buckler that the Japanese had cut the track, forcing them all to descend into Eora Creek Valley. The official historian of the Kokoda campaign observed: 'It was strange that the wild circle of their wanderings should have brought Buckler's main party and Treacy and his two men to the same point of exit [Dorobisolo] from the Papuan mountains within such a short time of one another, after they had set out in opposite directions.[14]

Twenty-two-year-old Private Keith Taylor from Melbourne with 'D' Patrol recalled he stayed at Arapara, which was about ten kilometres south of Dorobisolo, to 'lead the 2/14 troops to the food depot and assist in their evacuation by raft down the Kemp Welsh River. These rafts were made up of two long logs, which had a high buoyancy, and

EPILOGUE

then bound together with vines to make a stable raft. A native on the rear end of the main logs assisted in steering the rafts which headed for Kalikodobu'.[15]

Meanwhile, Buckler, having finally arrived at Port Moresby, quickly organised a flight to be flown over Sengai Village to drop medical supplies and food to the sick and wounded. Indeed, he insisted on going on the flight himself. The plane circled the village but found no sign of Private Thomas Fletcher or the seven sick and wounded. They tried again the next day, dropping two packages, but again nothing.

Private Thomas Fletcher and the sick and wounded men, including Corporal John Metson, who had refused to be carried and crawled on his hands and knees across the Owen Stanleys had already been discovered and murdered by a Japanese patrol.[16] On 4 October, they shot Fletcher and bayoneted the sick and wounded where they lay.

ACKNOWLEDGEMENTS

I am indebted to those mentioned in this book who wrote down and/or recorded their experiences. Without these critical historical records this narrative could never have been told in any meaningful detail. It has always been important for this author to tell the story using the words of those that were there – it brings the narrative into a whole new light as many voices are heard as opposed to the monotone of a single writer. Like all researchers, I am also indebted to these individuals for unselfishly donating their precious documents, writings and 'curios' to numerous research institutions for others to study. This also applies to relatives who have provided similar valuable records. For those copyright holders I was unable to locate or who did not reply to my request to use quoted material in this book, I trust that the material quoted meets with your approval.

I would also like to thank the 39th Battalion Association for allowing me to quote from Vic Austin's book *To Kokoda and Beyond: the story of the 39th Battalion 1921-1943*; 'Doc' Geoffrey Vernon's diary, published by the Association; and stories from the diggers themselves published in the battalion's journal, *The Good Guts*. The author did not have access to the journals themselves but has used Carl Johnson's book *Blood over Mud* which has reproduced these stories first published in the battalion's journal to tell the story of the digger's in their own words.

I would also like to thank the following authors/publishers for allowing me to quote from the following books: Much thanks to Lex McAulay for allowing me to quote from his ground breaking book, *Blood and Iron: the battle for Kokoda 1942* (Hutchinson Australia); a big thanks to Robyn Kienzle for allowing me to quote from her excellent book, *The Architect of Kokoda: Bert Kienzle – the man who*

ACKNOWLEDGMENTS

made the Kokoda Trail, (Hachette, Sydney); I would like to thank the Australian War Memorial (AWM) Research Center for their assistance during my visits and also thank the AWM for granting permission for me to quote from the official historian of the Kokoda Campaign, Dudley McCarthy, and his book, *South-West Pacific Area – First Year Kokoda to Wau* as well as permission to quote from their extensive collection of diaries, letters and official papers held in their care. Gratitude to the Department of Veterans Affairs and the University of New South Wales for enabling me to quote from the extensive collection of interviews of Australian veterans of the campaign as part of the *Australians at War Film Archive (AWFA)*. I would like to especially thank William Grayden for allowing me to quote from his captivating account of his time during the Kokoda Campaign as published in *Kokoda Lieutenant: The Triumph of the 21st Brigade*, as well as his publisher Peter Bridge OAM of Hesperian Press – a truly remarkable publication. The same again applies to Hesperian Press for allowing me to quote from William Cousens' fascinating account of his time on the track and at Gona entitled: *Kokoda and Gona New Guinea 1942*.

At Big Skey I would like to thank my publisher Denny Neave for believing in this book, when all too often others incorrectly stated there was nothing more to be said about the Kokoda Campaign – we still have much to learn. A similar thanks to my editor XXX who took on the manuscript and quickly and efficiently had it ready for publication; similarly, to XXX for drafting the maps from the authors initial appalling rough sketches and amendments. It goes without saying any remaining errors are those of the author alone.

My biggest thanks are to my three wonderful and special children, Emma, Anita and Lloyd, and my precious little granddaughter Naomi.

BIBLIOGRAPHY

Australian War Memorial, Canberra

Personal Records

AWM PR00297 Nakahashi, *War History of the Force which was Sent to the South Seas*, (translated by Lt F.C. Jorgensen); given to Alf Salmon, 39th Battalion at a *144th Regiment* reunion meeting in Tokyo, 26 August 1972.

AWM MSS0732 Okada, Seizo – *Lost Troops*, unpublished manuscript.

AWM PR00787 Vernon, Captain Geoffrey – diary (original of AWM 54 253/5/8).

AWM 52 8/3/14 September 1942: *Report by Lieut Treacy for War Diary*.

AWM 52 8/3/14 September 1942: *Report of VX 39721 Lieut R.M. McIlroy on movement of patrol sent out from EORA CK 1 Sep 1942*.

AWM 52 8/3/27 September 1942 – Appendix G: *The History of the Stretchers by Cpl J H Burns*.

AWM 3DRL 2381 2/7 – Notes on the operations of Maroubra Force.

AWM 3DRL 2381 3/7 – Notes on the operations of Maroubra Force.

AWM 3DRL/4005 Private Records Private Akiyoshi Hisaeda.

AWM 419 14/2 OPS *Report by Captain S.H. Buckler*, 12 October 1942.

Unit War Diaries

AWM 52 1/5/14 7th Australian Division General Staff Branch War Diary: September 1942.

AWM 52 1/10/1 004 ANGAU Headquarters War Diary: September 1942.

AWM 52 8/2/21 21st Brigade War Diary: August to October 1942.

AWM 52 8/2/25 25th Brigade War Diary: August to October 1942.

AWM 52 8/3/14 2/14th Battalion War Diary: September 1942.

AWM 52 8/3/16 2/16th Battalion War Diary: August to October 1942.

AWM 52 8/3/25 2/25th Battalion War Dairy: September 1942 to December 1942.

AWM 52 8/3/27 2/27th Battalion War Diary: September 1942.

AWM 52 8/3/31 2/31st Battalion War Diary: August 1942 to March 1943.

AWM 52 8/3/33 2/33rd Battalion War Diary: September to December 1942.

BIBLIOGRAPHY

AWM 52 8/3/39 3rd Militia Battalion War Diary: September 1942 to April 1943.

AWM 52 8/3/78 39th Militia Battalion War Diary: July to December 1942.

AWM 52 8/3/78 39th Militia Battalion War Diary: July to December 1942 – Appendix.

AWM 52 8/3/91 55/53rd Militia Battalion War Diary: September to December 1942.

Written records, 1939–45 War

AWM 54 253/5/8 Vernon War Diary Part 1, July–November 1942.

AWM 54 577/6/8 Kienzle Report.

AWM 54 577/7/26 Notebook no. 32, 2nd Lieutenant Noda Hidetaka Dairy.

ATIS records of POW interrogations and captured documents

AWM 55 Current Translation 3/2 No. 87, Private Watanabe Toshio Diary.

AWM 55 Current Translation 3/2 No. 210, Lieutenant Nose Munekichi Diary.

AWM 55 Current Translation 3/2 No. 218, 2nd Lieutenant Hirano Kogoro Diary

AWM 55 Current Translation 3/2 No. 266 Diary of unknown leader of signals unit with No. 2

Infantry Section, No. 1 Company, 1/144th Regiment.

AWM 55 Enemy Publication No. 27, Field Log of *Sakigawa Tai*.

AWM 55 Enemy Publication No. 28, Intelligence Reports Issued by *Yazawa Butai HQ*.

AWM 55 Enemy Publications 5/3 No. 33, File of Nankai Shitai Orders 16 August – 15 October.

AWM 55 Enemy Publications 5/4 No. 39, File of Nankai Shitai Orders 2 October – 7 November.

University of New South Wales (UNSW)
Australians at War Film Archive (AWFA)

Baldwin, Raymond: Archive number 1214, interviewed 24 November 2003.

Bisset, Stanley: Archive number 1223, interviewed 22 December 2003.

Corbett, John: Archive number 1303, interviewed 9 December 2003.

Guest, William: Archive number 588, interviewed 13 August 2003.

Johns, Robert: Archive number 1195, interviewed 26 November 2003.

Osbourne, Allen: Archive number 1319, interviewed 3 February 2002.

Phelan, Kenneth: Archive number 509, interviewed 17 June 2003.
Rhoden, Philip: Archive number 2549, interviewed 5 June 2000.
Sambell, Eric: Archive number 2231, interviewed 5 June 2000.
Smith, Lionel: Archive number 1386, interviewed 16 January 2004.
Thompson, Robert: Archive number 1594, interviewed 5 March 2004.
Tongs, Bede: Archive number 1151, interviewed 23 April 2002.

National Archives Australia

Most basic biographical details of Australian military personnel discussed are checked against the on-line electronic files of the NAA using identified army SX, V or VX serials; most serial numbers taken from Johnson, 2006 or McCarthy, 1959.

Official Histories

Bullard, S. (2007) *Japanese Army Operations in the South Pacific Area: New Britain and Papua Campaigns, 1942–43*, Australian War Memorial (English translation of the Japanese Official History).

Gillison, D. (1962) Royal Australian Air Force 1939–1942, *Australia in the War of 1939–1945*, (Air Series) Australian War Memorial, Canberra.

McCarthy, D. (1959) South-West Pacific Area First Year, Kokoda to Wau, *Australia in the War of 1939–1945*, (Army Series) Australian War Memorial, Canberra.

Milner, S. (1955) The War in the Pacific: Victory in Papua, *United States Army in World War II*, Center of Military History United States Army, Washington, D.C.

Tanaka, K. (1980) *Operations of the Imperial Japanese Armed Forces in the Papua New Guinea Theatre during World War II*, Japan Papua New Guinea Goodwill Society, Japan.

Walker, A.S. (1957) The Island Campaigns *Australia in the War of 1939–1945*, (Medical Series) Australian War Memorial, Canberra.

Published Unit History

Austin, V. (2007) *To Kokoda and Beyond: the story of the 39th Battalion 1921–1943*, Australian Military History Publications, Australia.

Biggs, S. & Biggs, L. (1967) *Ike's Marines: 36th Australian Infantry Battalion, 1939–1945 – The Story of an Australian Infantry Battalion and its part in the War against Japan*, The 36th Battalion Association (2003 edition).

BIBLIOGRAPHY

Budden, F.M. (1987) *The Chocos: The Story of the Militia Infantry Battalions in the South West Pacific Area 1941–1945*, Summer Hill, NSW.

Budden, F.M. (1973) *That Mob: The Story of the 55/53rd Australian Infantry Battalion, A.I.F.*, Summer Hill, NSW.

Burns, J. (1960) *The Brown and the Blue Diamond at War: The Story of the 2/27th Battalion A.I.F.*, 2/27th Battalion Ex-servicemen's Association, Adelaide.

Byrnes, G.M. (1989) *Green Shadows: A History of the Papuan Infantry Battalion*, Queensland Corrective Services Commission

Crooks, W. (1971) *The Footsoldiers: The Story of the 2/33rd Australian Infantry Battalion, A.I.F., in the War of 1939–45*, Printcraft Press (Second edition 2020 Everbest Printing Co. Ltd).

Draydon, A.W. (2000) *Men of Courage: A History of 2/25th Australian Infantry Battalion, 1940–45*, published by the 2/25th Australian Infantry Battalion Association, Loftus.

Johnson, C. (2006) *Mud Over Blood: the 39th Infantry Battalion 1941–43 – Kokoda to Gona*, History House, Victoria.

Kennedy, C. (1992) *Port Morseby to Gona Beach: 3rd Australian Infantry Battalion 1942*, self-published, Canberra.

Laffin, J. (1994) *Forever Forward: The History of the 2/31st Australian Infantry Battalion. 2nd AIF 1940–45*, published by the 2/31st Australian Infantry Battalion Association, NSW Branch.

Perrin, A.E. (1990) *The Private War of the Spotters: A History of the New Guinea Air Warning Wireless Company February 1942 – April 1945*, NGAWW Publication Committee, Victoria.

Powell, A. (2003) *The Third Force: ANGAU's New Guinea War, 1942–46*, Oxford University Press, Melbourne.

Russell, W.B. (1948) *The Second Fourteenth Battalion: A history of an Australian infantry battalion in the Second World War*, published by the 2/14th Battalion Association, Melbourne (1986 edition).

Trigellis-Smith, S. (1992) *The Purple Devils: A History of the 2/6 Australian Commando Squadron formerly the 2/6 Australian Independent Company 1942–1946*, 2/6 Commando Squadron Association, Melbourne.

Uren, M. (1959) *A Thousand Men at War: A History of the 2/16th Australian Infantry Battalion, AIF*, William Heinemann (Australian Military History Press, Loftus, Australia, 2009 edition).

Academic Dissertations

Harwood, J.M. (1979) *The Australian Militia in New Guinea 1940–1945*, MA dissertation, Macquarie University

Williams, P. (2008) *The Kokoda Campaign, July–November 1942, An Analysis*, Ph.D. dissertation, Charles Darwin University

Books/Book Chapters

Alexander, J.H. (2001) *Edson's Raiders: The 1st Marine Raider Battalion in World War II*, Naval Institute Press, Annapolis, Maryland.

Anderson, N. (2014) *To Kokoda, Australian Army Campaign Series No. 14*. Australian Army History Unit, Canberra.

Anderson, N. (2018) The Battle of Milne Bay, 1942, *Australian Army Campaign Series No. 24*. Australian Army History Unit, Canberra.

Archer, J. (1984) *Jungle Fighters, A G.I. War Correspondent's Experiences in the New Guinea Campaign*, Julian Messner, New York.

Benson, J. (1957) *Prisoner's Base and Home Again: The Story of a Missionary POW*, Robert Hale, London.

Bergerud, E. (1996) *Touched with Fire: the land war in the South Pacific*, Penguin Books, New York.

Blake, G. (2019) *Jungle Cavalry: Australian Independent Companies and Commandos 1941–1945*, Helion & Company, Warwick, United Kingdom.

Borneman, W.R. (2016) *MacArthur at War: World War II in the Pacific*, Little Brown and Company, New York.

Bradley, P. (2012) *Hell's Battlefield: the Australians in New Guinea in World War II*, Allen & Unwin, Sydney.

Braga, S. (2004) *Kokoda Commander: A Life of Major-General 'Tubby' Allen*, Oxford University Press, Melbourne.

Brune, P. (1991) *Those Rugged Bloody Heroes: from the Kokoda Trail to Gona Beach, 1942*, Allen & Unwin, Sydney.

Brune, P. (1994) *Gona's Gone! The Battle for the Beach-head, 1942*, Allen & Unwin, Sydney.

Brune, P. (2003) *A Bastard of a Place: The Australians in Papua: Kokoda, Milne Bay, Gona, Buna and Sanananda*, Allen & Unwin, Sydney.

Brune, P. (2007) *Ralph Honner: Kokoda Hero*, Allen & Unwin, Sydney.

Cameron, D.W. (2020) *The Battles for Kokoda Plateau: Three weeks of hell defending the gateway to the Owen Stanleys*, Allen & Unwin, Sydney.

BIBLIOGRAPHY

Cameron, D.W. (2022) *The Battle for Isurava: Fighting on the Kokoda Track in the heart of the Owen Stanleys*, Big Sky Publishing.

Campbell, J. (2007) *The Ghost Mountain Boys: Their Epic March and the Terrifying Battle for New Guinea – The Forgotten War of the South Pacific*, Crown Publishers, New York.

Chan, G. (2003) *War on Our Doorstep: Diaries of Australians at the frontline in 1942*, Hardie Grant Books, South Yarra.

Collie, C. & Marutani, H. (2009) *The Path of Infinite Sorrow: the Japanese on the Kokoda Track*, Allen & Unwin, Sydney.

Cooper, A. (2014) *Kokoda Air Strikes: Allied Air Forces in New Guinea, 1942*, New South Books, Sydney.

Cops, G. (un-dated) *My Early Papuan Diary, V58666 – VX117679 George Albert COPS, 39th Aust, Infantry Battalion*, self-published.

Cousins, W.F. (2010). *Kokoda and Gona New Guinea 1942*, Hesperian Press, Western Australia.

Cox, L. (2020) *Cuppa Tea, Digger: Salvos serving in World War Two*, Salvo Publishing, Canberra.

Dawson, J. (2003) *Kokoda Survivor*, self-published

Day, D. (2003) *The Politics of War: Australia at War, 1939–45: From Churchill to MacArthur*, HarperCollins Publishers, Sydney.

Dornan, P. (1999) *The Silent Men: Syria to Kokoda and to Gona*, Allen & Unwin, Sydney.

Drea, E.J. (2017) Making Japanese Soldiers, in James, K. (ed.) *Kokoda: beyond the legend*, Cambridge University Press, Melbourne, pp.188–205.

Duffy, J.P. (2016) *War at the End of the World: Douglas MacArthur and the Forgotten Fight for New Guinea, 1942–1945*, Nal Caliber, New York.

Eather, S. (2003) *Desert Sands, Jungle Lands: A Biography of Major General Ken Eather*, Allen & Unwin, Sydney.

Edgar, B. (2010) *Warrior of Kokoda: A Biography of Brigadier Arnold Potts*, Australian Military History Press, Loftus, Australia.

Fitzsimons, P. (2004) *Kokoda*, Hodder, Sydney.

Frank, R.B. (1990) *Guadalcanal: the definitive account of the landmark battle*, Penguin Books, USA.

Friedman, K.I. (2007) *Morning of the Rising Sun: the heroic story of the battle for Guadalcanal*, Friedman, USA.

Gailey, H. (2000) *MacArthur Strikes Back: Decision at Buna: New Guinea, 1942–43*, Presidio, USA.

Gailey, H. (2004) *MacArthur's Victory: The War in New Guinea. 1943–1944*, Presidio, USA.

Gallaway, J. (2000) *The Odd Couple: Blamey and MacArthur at War*, University of Queensland Press, Brisbane.

Gamble, B. (2018) *Kangaroo Squadron: American courage in the darkest days of World War II*, Da Capo Press, New York.

Grayden, W. (2015) *Kokoda Lieutenant: the triumph of the 21st Brigade – recollections of an A.I.F., platoon commander, 1942*, Hesperian Press, Carlisle, Western Australia.

Hall, T. (1981) *New Guinea, 1942–44*, Methuen, Australia.

Ham, P. (2004) *Kokoda*, HarperCollins, Sydney.

Hamlyn-Harris, G. (1993) *Through Mud and Blood to Victory*, Fast Books, Glebe, NSW.

Happell, C. (2008) *The Bone Man of Kokoda: the extraordinary story of Kokichi Nishimura and the Kokoda Track*, Pan Macmillan Australia

Hawthorne, S. (2003) *The Kokoda Trail: a history*, Central Queensland University Press Hayashi, S. & Cox, A.D. (1959) *KŌGUN: The Japanese Army in the Pacific War*, The Marine Corps Association, Quantico, USA.

Henderson, J. (1992) *Onward Boy Soldiers: The Battle for Milne Bay, 1942*, University of Western Australia Press, Perth.

Hogan, T, Hogan, N, Hogan B. (1992) *From Grabben Cullen to Kokoda*, self-published, Goulburn, NSW.

Horner, D.M. (1978) *Criss of Command: Australian Generalship and the Japanese Threat, 1941–1943*, Australian National University Press

Horner, D.M. (1992) *General Vesay's War*, Melbourne University Press

Horner, D.M. (1995) *The Gunners: A History of Australian Artillery*, Allen & Unwin, Sydney.

Horner, D.M. (1996) *Inside the War Cabinet: Directing Australia's War Effort 1939–45*, Allen & Unwin, Sydney.

Horner, D.M. (1998) *Blamey: Commander-in-Chief*, Allen & Unwin, Sydney.

Hull, W. (1995) *Salvos with the Forces: Red Shield Services during World War Two*, The Salvation Army, Albert, Victoria.

James, A. (2011) *Stan Bisset Kokoda Wallaby: The Rugby International Who Became a Kokoda Hero*, Allen & Unwin, Sydney.

James, B. (2008) *Field Guide to the Kokoda Track: an historical guide to the lost battlefields*, Tower Books, Sydney.

Johnston, G.H. (1944) *New Guinea Diary*, Angus & Robertson, Sydney.

BIBLIOGRAPHY

Johnston, M. (1996) *At the Frontline: Experiences of Australian soldiers in World War II*, Cambridge University Press, UK.

Johnston, M. (2000) *Fighting the Enemy: Australian soldiers and their adversaries in World War II*, Cambridge University Press, UK.

Johnston, M. (2015) *Stretcher-Bearers: Saving Australians from Gallipoli to Kokoda*, Cambridge University Press, Melbourne.

Keogh, E.G. (1965) *South-West Pacific 1941–45*, Grayflower Productions, Melbourne.

Kienzle, R. (2011) *The Architect of Kokoda: Bert Kienzle – the man who made the Kokoda Trail*, Hachette, Sydney.

Kienzle, R (2017) On the Trail of an Extraordinary Man, in James, K. (ed.) *Kokoda: beyond the legend*, Cambridge University Press, Melbourne, pp.238–51.

McAulay, L. (1991) *Blood and Iron: The Battle for Kokoda 1942*, Hutchinson Australia

McAulay, L. (1992) *To the Bitter End: The Japanese Defeat at Buna and Gona 1942–43*, Random House Australia

McDonald, N. (1994) *War Cameraman: The Story of Damien Parer*, A Lothian Books, Port Melbourne.

McDonald, N. (2004) *Chester Wilmot Reports: Broadcasts that Shaped World War II*, ABC Books, Sydney.

McDonald, N. & Brune, P. (1998) *200 Shots: Damien Parer, George Silk and the Australians at War in New Guinea*, Allen & Unwin, Sydney.

McKernan, M. (2006) *The Strength of a Nation: Six years of Australians fighting for the nation and defending the homefront in WWII*, Allen & Unwin, Sydney.

McLeod, J. (2019) *Shadows on the Track: Australia's Medical War in Papua 1942–43 (Kokoda – Milne Bay – The Beachhead Battles)*, Big Sky Publishing, Sydney.

Mayo, L. (1975) *Bloody Buna: the campaign that halted the Japanese invasion of Australia*, Australian National University Press, Canberra.

Paull, R. (1958) *Retreat from Kokoda*, Heinemann, Melbourne.

Pratten, G. (2009) *Australian Battalion Commanders in the Second World War*, Cambridge University Press, Melbourne.

Robertson, J. (1981) *Australia at War 1939–1945*, William Heinemann, Melbourne.

Rowell, S.F. (1974) *Full Circle*, Melbourne University Press, Victoria.

Smith, M.S. (2000) *Bloody Ridge: the battle that saved Guadalcanal*, Presidio Press, USA.

Steward, H.D. (1981) *Recollections of a Regimental Medical Officer*, Melbourne University Press, Victoria.

Sublett, F. (2000) *Kokoda to the Sea: a history of the 1942 campaign in Papua*, Slouch Hat Publications, McCrae, Australia.

Thomson, J. (2000) *Winning with Intelligence: A Biography of Brigadier John David Rogers, CBE, MC 1895–1978*, Australian Military History Press, Loftus, NSW.

Tomkins, D. & Hughes, B. (1969) *The Road to Gona*, Angus & Robertson, Sydney.

Tracey, R. (2017) Command Failures on the Kokoda Trail, in James, K (ed.) *Kokoda: beyond the legend*, Cambridge University Press, Melbourne, pp.112–31.

Veitch, M. (2019) *Turning Point: The Battle for Milne Bay 1942 – Japan's First Land Defeat in World War II*, Hachette, Sydney.

Wakeling, A. (2018) *Stern Justice: The Forgotten Story of Australia. Japan and the Pacific War Crimes Trials*, Penguin, Melbourne.

Williams, P. (2012) *The Kokoda Campaign, 1942: Myth and Reality*, Cambridge University Press, Melbourne.

Williams, P. (2017) Against overwhelming odds? Opposing strengths on the Kokoda Trail, in James, K. (ed.) *Kokoda: beyond the legend*, Cambridge University Press, Melbourne, pp.222–35.

White, O. (1945) *Green Amour*, Angus & Robertson, Sydney.

Yashida, H. (2017) Japanese Commander in Kokoda, in James, K. (ed.) *Kokoda: beyond the legend*, Cambridge University Press, Melbourne, pp.222–35.

Zobel, J.W. (2017) Victory at all costs: Douglas MacArthur's Papuan Campaign of 1942–43, in James, K. (ed.) *Kokoda: beyond the legend*, Cambridge University Press, Melbourne, pp.206–21.

Journals/Manuscripts/Newspapers

Grahamslaw, T. (1971) Missionaries Slaughtered in Grim Advance on Morseby, *Pacific Island Monthly* April 42(4): 71–75 & 117–123.

Grahamslaw, T. (unpublished manuscript) Recollections of ANGAU, https://www.pngaa.net/Library/RecollAngau.html.

Honner, R. (1967) This was the 39th, *Australian Army Journal*, July No. 218.

Reid, R (1997) The Simpson of the Kokoda Track: Major Albert Moore. Salvation Army, *Wartime: Official Magazine of the Australian War Memorial*, 1:26–29.

BIBLIOGRAPHY

Vernon, G. (1943) *A War Diary: The Owen Stanley Campaign July – November 1942 (the personal diary of Captain Geoffrey Hampden "Doc' Vernon MCD – AAMC)*, published by the 39th Battalion Association, Victoria.

ENDNOTES

Chapter 1
1. Bergerud, 1996; Duffy, 2016; Gallaway, 2000; McCarthy, 1959.
2. Quoted from McCarthy, 1959, p.225.
3. Quoted from McCarthy, 1959, p.225.
4. McCarthy, 1959.
5. Quoted from McCarthy, 1959, p.225.
6. McCarthy, 1959.
7. Quoted from McCarthy, 1959, p.226.
8. Quoted from McCarthy, 1959, p.226.
9. Anderson, 2018; Bergerud, 1996; Henderson, 1992; Horner, 1998; McCarthy, 1959; Veitch, 2019.
10. Horner, 1998; McCarthy, 1959.
11. Bergerud, 1996; Cailey, 2000; Campbell, 2007; Duffy, 2016; McCarthy, 1959.
12. Anderson, 2018; Bergerud, 1996; Henderson, 1992; McCarthy, 1959; Veitch, 2019.
13. Anderson, 2018; Bergerud, 1996; Henderson, 1992; McCarthy, 1959; Veitch, 2019.

Chapter 2
1. Bergerud, 1996; Cooper, 2014; Kienzle, 2011; McCarthy, 1959.
2. Gamble, 2018; Gillison, 1962.
3. Bergerud, 1996, p.139.
4. Cameron, 2022.
5. Quoted from Russell, 1948, p.156.
6. Quoted from Edgar, 2010, p.154.
7. Cameron, 2020, 2022; Perrin, 1990.
8. Perrin, 1990, p.73.
9. Cameron, 2022.
10. AWM 52 8/3/14.
11. Burns, 1960.
12. AWM 52 8/3/27.
13. McLeod, 2019; Walker, 1957.
14. Quoted from Brune, 1991, p.125.
15. Cameron, 2022; McLeod, 2019; Walker, 1957.
16. McLeod, 2019; Cameron, 2022; Walker, 1957.
17. McLeod, 2019; Cameron, 2022; Walker, 1957.
18. Quotes from Ham, p.211.
19. Quotes from Ham, p.211.
20. Quote from Johnston, 2015, p.227.
21. Cameron, 2022.

ENDNOTES

Chapter 3
1. Quoted from Braga, 2004, p.194.
2. Quoted from Braga, 2004, pp.194–95.
3. James, 2011; McCarthy, 1959; Russell, 1948; Uren, 1959.
4. Quote from James, 2011, p.192.
5. UNSW AWFA 1223.
6. AWM 52 8/3/16; Sublet, 2000; Uren, 1959.
7. Cameron, 2022; McCarthy, 1959.
8. AWM 52 8/3/16; McCarthy, 1959; Sublet, 2000; Uren, 1959.
9. AWM 52 8/3/16; Sublet, 2000; Uren, 1959.
10. Bullard, 2007, p.164.
11. Quoted from Russell, 1948, p.157.
12. Russell, 1948.
13. AWM 54 577/7/26.
14. AWM 52 8/3/27; see also Burns, 1960.
15. AWM 52 8/3/27.
16. Burns, 1960, p.112.
17. Quoted from Chan, 2003, pp.197–98.
18. AWM 52 8/3/27.
19. AWM 52 8/3/78.
20. Quoted from Ham, 2004, p.211.
21. Austin, 2007; Cameron, 2020.
22. Quoted from Austin, 2007, pp.173–74.
23. Cops Diary.
24. UNSW AWFA 509.
25. Budden, 1978; Cameron, 2022.
26. Brune, 2007; Cameron, 2020, 2022.
27. AWM 52 8/3/91.
28. Cameron, 2022.
29. Cameron, 2020, 2022.
30. AWM PR00787.
31. Russell, 1948, p.149.
32. AWM 52 8/3/14 Treacy Report.
33. AWM 52 8/3/14 McIlroy Report.
34. Johnston, 1944, pp.147–48.

Chapter 4
1. AWM 52 1/10/1 004; AWM 52 3/2/21; Sublet, 2000.
2. AWM 52 8/2/21.
3. UNSW AWFA 1594.
4. AWM 52 8/3/14.
5. AWM 52 8/3/27; Brune, 2007; Cameron, 2022; McCarthy, 1959.
6. AWM 54 577/7/26; Bullard, 2007; Williams, 2012.
7. AWM 55 Enemy Publication, 5/3 No. 33.
8. Bullard, 2007; Tanaka, 1980; Williams, 2012, 2017.
9. Bullard, 2007; Williams, 2008, 2012, 2017.

10. Quoted from Williams, 2008, 2012, pp.116–17.
11. Quoted from Williams, 2008, 2012, p.117.
12. Drea, 2017, p.204.
13. Quotes from Collie & Marutani, 2009, p.117.
14. AWM 52 8/3/16; Sublet, 2000; Uren, 1959.
15. Williams, 2008, 2012.
16. AWM 52 8/3/16.
17. AWM 52 8/2/21.
18. McCarthy, 1959; Russell, 1948.
19. AWM 52 8/3/14.
20. Burns, 1960.
21. AWM 52 8/3/27.
22. AWM 52 8/3/27; Burns, 1960.
23. Quoted from Brune, 2003, p.195.
24. AWM 52 8/2/21; Sublet, 2000.
25. Uren, 1939, p.142.
26. Williams, 2012.
27. Williams, 2012.
28. AWM 52 8/3/16; McCarthy, 1959; Sublet, 2000; Uren, 1959.
29. AWM 52 8/3/16.
30. AWM 52 8/2/21.
31. Quoted from Brune, 1991, p.144.
32. Cameron, 2022; McCarthy, 1959; Uren, 1959.
33. AWM 52 8/3/16; McCarthy, 1959; Powell, 2003; Uren, 1959.
34. AWM 52 8/3/16; Powell, 2003; Uren, 1959.
35. Uren, 1959, p.159.
36. Cousens, 2010, p.4.
37. Bullard, 2007, p.164.

Chapter 5

1. AWM 52 8/3/27; Brune, 2007; Burns, 1960.
2. AWM 52 8/3/27.
3. AWM 52 8/3/27.
4. AWM 52 8/3/27.
5. McCarthy, 1959.
6. AWM 52 8/2/21.
7. AWM 52 8/2/21.
8. AWM 52 8/3/91.
9. Quoted from Hull, 1995, p.159.
10. AWM PR00787.
11. Cameron, 2022.
12. AWM 419 14/2 Buckler Report.
13. AWM 52 8/3/14 Treacy Report.
14. AWM 52 8/3/14 Treacy Report.
15. AWM 52 8/3/14 McIlroy Report.

ENDNOTES

Chapter 6
1. AWM 52 8/3/16; Bullard, 2007; Williams, 2012.
2. Bullard, 2007, p.164.
3. Steward, 1983, p.121.
4. AWM 52 8/3/16.
5. AWM 52 8/3/16.
6. Quote from Brune, 2003, p.195.
7. Burns, 1960.
8. AWM 52 8/2/21; Cameron, 2020; McCarthy, 1959.
9. AWM 52 8/3/91; Budden, 1973, 1987.
10. AWM 52 8/2/21; AWM 52 8/3/27; AWM 52 8/3/91; Budden, 1973, 1987.
11. AWM 52 82/21.
12. McCarthy, 1959; Sublet, 2000.
13. AWM 52 8/3/27.
14. AWM 52 8/3/16.
15. AWM 52 8/3/14.
16. Quote from Brune, 1991, p.84.
17. Cousens, 2010, pp.4-5.
18. AWM 52 8/2/21.
19. AWM 52 8/2/21.
20. AWM 52 8/3/91; Budden, 1973, 1987.
21. AWM 52 8/3/78.
22. Cops Diary.
23. Austin, 2007, p.174.
24. Cameron, 2020, 2022.
25. UNSW AWFA 588.
26. AWM PR00787.
27. McAulay, 1991.
28. Quoted from McAulay, 1991, p.221.
29. AWM 52 8/3/78 39th Battalion Appendix.
30. AWM 54 577/7/26.

Chapter 7
1. McLeod, 2019; Walker, 1957.
2. Quoted from McLeod, 2019, p.134.
3. Quoted from McLeod, 2019, p.134.
4. AWM 52 8/3/39; Kennedy, 1992.
5. AWM 52 8/3/39.
6. AWM 52 8/2/39.
7. Kennedy, 1992, p.30.
8. Quote from Hull, 1995, p.159.
9. White, 1945, p.175.
10. Quote from Hull, 1995, pp.159–60.
11. AWM 52 8/3/14 Treacy Report.
12. AWM 52 8/3/14 McIlroy Report.
13. AWM 52 1/5/14.

14. Benson, 1957; Cameron, 2020; Tomkins & Hughes, 1969; Wakeling, 2018.
15. Benson, 1957; Cameron, 2022.
16. Benson, 1957, p.61.
17. Brune, 1994; Gailey, 2004; McAulay, 1992; McCarthy, 1959.

Chapter 8
1. Russell, 1948.
2. Quote from Edgar, 2010, p.162.
3. Quote from James, 2011.
4. Sublet, 2000, p.70.
5. AWM 52 8/3/16.
6. Russell, 1948, p.158.
7. Steward, 1983, pp.121–22.
8. UNSW AWFA 2549.
9. AWM 52 8/3/27; AWM 52 8/3/91.
10. AWM 52 8/2/21; Sublet, 2000.
11. AWM 52 1/10/1 004.
12. AWM 52 8/3/27.
13. AWM 52 8/2/21, Bell's patrol retired through Efogi (AWM 52 8/2/21).
14. AWM 52 8/2/21.
15. Quoted from Braga, 2004, p.196.
16. Quoted from Edgar, 2010, p.160.
17. Quoted from Edgar, 2010, p.160.
18. Quote from Bradley, 2012, p.61.
19. Quoted from Hall, 1981, pp.120–21.
20. UNSW AFWA 1214.
21. AWM 52 8/3/27.
22. AWM 52 8/3/16.
23. AWM 52 8/3/27; McCarthy, 1959.
24. AWM 52 8/2/21.
25. Russell, 1948, p.159.
26. Quote from Brune, 1991, p.154.
27. Grayden, 2015.
28. UNSW AWFA 1594.
29. Quote from Edgar, 2010, p.163.

Chapter 9
1. Bullard, 2007; Williams, 2008, 2012.
2. Bullard, 2007, p.165.
3. AWM 54 577/7/26.
4. AWM 55 Current Translation 3/2 No. 266.
5. AWM MSS0732.
6. McAulay, 1991.
7. Quote from Brune, 2007, p.189.
8. AWM 52 8/3/87.
9. Honner, 1967, p.37.

ENDNOTES

10. Cops diary.
11. Dawson, 2003, p.42.
12. Quoted from Austin, 2007, p.175.
13. Cameron, 2020.
14. Quoted from Austin, 2007, pp.176–77.
15. Cameron, 2020.
16. Quoted from Johnson, 2006, p.131.
17. AWM PR00787.
18. Hull, 1995, p.160.
19. AWM 52 8/2/21.
20. Kennedy, 1992, pp.30–31.
21. AWM 52 8/3/14.
22. AWM 52 8/3/14 McIlroy Report.
23. Quote from Horner, 1978, p.149.

Chapter 10

1. AWM 52 8/2/21; AWM 52 8/3/14; AWM 52 8/3/16; AWM 52 8/3/27; Burns, 1960; McCarthy, 1959; Russell, 1948; Uren, 1959; Williams, 2008, 2012.
2. Williams, 2012.
3. Tanaka, 1980, p.22.
4. AWM 52 8/3/16; Grayden, 2015; Sublet, 2000.
5. AWM 52 8/3/27.
6. AWM 52 8/2/21; Cameron, 2020, 2022; Grahamslaw, 1971, unpublished.
7. AWM 52 8/2/27; Burns, 1960; McCarthy, 1959; Williams, 2012.
8. Grayden, 2015, p.68.
9. Burns, 1960, p.116.
10. AWM 52 8/2/21.
11. AWM 52 8/3/27; Burns, 1960; Russell, 1948.
12. Cooper, 2014; Gillison, 1962.
13. Quoted from Cooper, 2014, p.457.
14. Steward, 1983, p.122.
15. AWM 52 8/3/27; Gillison, 1962.
16. Burns, 1960, p.115.
17. Tanaka, 1980, p.22.
18. Quoted from Bullard, 2007, p.165.
19. AWM 52 8/2/21.
20. Burns, 1960, p.115.
21. Burns, 1960.
22. AWM 52 8/3/14.
23. AWM 52 8/2/21.
24. AWM 52 8/2/21.
25. AWM 52 8/2/21; Gillison, 1962.
26. Russell, 1948.
27. AWM 52 8/3/27.
28. AWM 52 8/2/21; Grahamslaw, 1971, unpublished.
29. AWM 52 8/2/21.

30. AWM 52 8/3/27.
31. Burns, 1960, p.116.
32. UNSW AWFA 1195.
33. AWM 52 8/3/27.
34. AWM 52 8/2/21.
35. AWM 52 8/3/16.
36. AWM 52 8/3/27.
37. AWM 52 8/2/21; Burns, 1960.
38. AWM 52/2/21.
39. Burns, 1960, p.117.
40. Quote from Chan, 2003, p.199.

Chapter 11
1. Williams, 2008, 2012.
2. AWM 52 8/3/78.
3. AWM 55 Current Translation 3/2 No. 266.
4. Williams, 2008, 2012.
5. AWM 55 Enemy Publications, 5/3 No. 33.
6. AWM 54 577/7/26.
7. Williams, 2008, 2012.
8. AWM PR00787.
9. Vernon Diary, 1943, p.23.
10. AWM 52 1/10/1 004.
11. McLeod, 2019; Walker, 1957.
12. AWM 52 8/3/39.
13. Hull, 1995, p.160.
14. Quoted from Austin, 2007, p.175.
15. Johnston, 2015, p.229.
16. AWM 52 8/3/14 Treacy Report.
17. AWM 52 8/3/14 McIlroy Report.
18. Quoted from McCarthy, 1959, p.225.
19. Quoted from McCarthy, 1959, p.225.
20. Quoted from McCarthy, 1959, p.226.

Chapter 12
1. Quoted from Edgar, 2010, pp.165–66.
2. Quoted from Edgar, 2010, p.165.
3. AWM 52 8/2/21; Williams, 2008, 2012.
4. Burns, 1960; McCarthy, 1959.
5. Quote from Johnston, 2000, p.110.
6. AWM 52 8/3/27.
7. Quoted from Bradley, 2012, p.61.
8. Burns, 1960, p.117.
9. Happell, 2008; McCarthy, 1959; Williams, 2008, 2012.
10. Quoted from Williams, 2012, p.129.
11. Quoted from Ham, 2004, p.235.

ENDNOTES

12. Quote from Brune, 1991, p.156.
13. Collie & Marutani, 2009, p.121.
14. Cameron, 2022; Happell, 2008.
15. Paull, 1958, p.196.
16. Paull, 1958, p.196.
17. Quotes from Paull, 1958, p.197.
18. Quoted from James, 2011, p.203.
19. UNSW AWFA 1223.
20. AWM 52 8/3/16.
21. Steward, 1983, p.123.
22. Grayden, 2015, pp.69–71.

Chapter 13
1. UNSW AWFA 1303.
2. Williams, 2008, 2012.
3. AWM 52 3/2/21.
4. Quote from Paull, 1958, p.198.
5. AWM 52 8/2/21; AWM 52 8/3/16.
6. Cousens, 2010, p.5.
7. AWM 52 8/2/21; AWM 52 8/3/16
8. Burns, 1960, p.118.
9. Burns, 1960, p.118.
10. AWM 52 8/3/14.
11. AWM 52 8/3/27.
12. AWM 52 8/2/14.
13. Russell, 1948.
14. Sublet, 2000.
15. AWM 52 8/3/27.
16. McCarthy, 1959; Gillison, 1962.
17. AWM 52 8/3/27.
18. Burns, 1960; McCarthy, 1959.
19. AWM 52 8/3/27.
20. AWM 52 8/2/21.
21. AWM 52 8/2/21.
22. AWM 52 8/2/21.
23. AWM 52 8/3/16; Burns, 1960; McCarthy, 1959; Sublet, 2000.
24. AWM 52 8/3/16; Burns, 1960; McCarthy, 1959; Sublet, 2000; Williams, 2012.
25. AWM 52 8/3/16.
26. Braga, 2004.
27. AWM 52 8/3/27.
28. AWM 52 8/3/27.
29. Burns, 1960; McCarthy, 1959.
30. AWM 52 8/3/16; Williams, 2012.
31. Burns, 1960, p.118.
32. AWM 52 8/3/27.
33. Burns, 1960; McCarthy, 1959.

34. Quote from Brune, 1991, p.159.
35. Burns, 1960, p.118.
36. Sublet, 2000.
37. AWM 52 8/2/21.
38. AWM 52 8/2/21.
39. Russell, 1948, p.163.
40. AWM 52 8/3/78.
41. Williams, 2008, 2012.
42. AWM 55 Current Translation 3/2 No. 266.

Chapter 14
1. McCarthy, 1959; Steward, 1983; Sublet, 2000; Uren, 1958.
2. AWM 52 8/3/14.
3. AWM 52 8/3/16.
4. Grayden, 2015, p.72.
5. Grayden, 2015, pp.73–74.
6. Quote from James, 2011, p.206.
7. AWM 52 8/2/21; AWM 52 8/3/16.
8. Russell, 1948.
9. AWM 52 8/3/14.
10. Cameron, 2022; McAulay, 1991; McCarthy, 1959; Russell, 1948.
11. Quoted from Happell, 2008, p.50.
12. Quoted from Happell, 2008, p.52.
13. Happell, 2008.
14. Grayden, 2015, p.128.
15. Quoted from James, 2011, p.208.
16. Grayden, 2015, p.127.
17. Quoted from McCarthy, 1959, p.222.
18. McAulay, 1991; McCarthy, 1959; Russell, 1948.
19. Uren, 1959, p.149.
20. Grayden, 2015, pp.74–75.
21. Uren, 1959, p.148.
22. Uren, 1959.
23. Quoted from Edgar, 2010, p.168.
24. AWM 52 8/2/21; Graydon, 2015; McAulay, 1991; McCarthy, 1959; Russell, 1948.
25. Grayden, 2015, pp.75–76.
26. Uren, 1959, p.148.
27. Uren, 1959, p.148.
28. AWM 52 8/2/21; Uren, 1959.
29. Grayden, 2015, pp,76–78.
30. Uren, 1959, p.149.
31. AWM 52 8/3/16.
32. AWM 52 8/3/16.
33. AWM 52 8/2/21.
34. AWM 52 8/2/21.
35. Burns, 1960; McCarthy, 1959.

ENDNOTES

36. AWM 52 8/3/27; Williams, 2008, 2012.
37. UNSW AWFA 2231.
38. Burns, 1960; McAulay, 1991; McCarthy, 1959; Russell, 1948.
39. Grayden, 2015, p.78.
40. Quoted from Braga, 2004, p.198.
41. Quoted from Braga, 2004, p.198.

Chapter 15

1. Russell, 1948, p.165.
2. AWM 52 8/3/14; Burns, 1960; McAulay, 1991; McCarthy, 1959; Russell, 1948.
3. McCarthy, 1959, p.223.
4. AWM 52 8/3/16.
5. AWM 52 8/3/16.
6. Uren, 1959, p.150.
7. Steward, 1983, p.127.
8. AWM 52 8/3/27.
9. Cousens, 2010, p.5.
10. AWM 52 8/3/14.
11. Quote from Anderson, 2014, p.97.
12. Burns, 1960, p.120.
13. Burns, 1960, p.119.
14. Burns, 1960.
15. Quote from Brune, 1991, p.161.
16. AWM 52 8/3/27.
17. Burns, 1960.
18. Burns, 1960; McCarthy, 1959; Russell, 1948.
19. Quote from Johnston, 1996, p.45.
20. UNSW AWFA 1214.
21. Quote from Brune, 1991, p.182.
22. AWM 52 8/2/21; Brune, 1991; Russell, 1948.
23. Quoted from Braga, 2004, p.198.
24. Quoted from Braga, 2004, p.198.
25. Tanaka, 1980, p.22.
26. AWM 52 1/10/1 004.
27. AWM PR 00787.
28. Vernon Diary, 1943, pp.22–23.
29. AWM 52 8/3/39.
30. Kennedy, 1992, p.34.
31. Quote from Hull, 1995, p.160.
32. Cameron, 2022; Cox, 2020; McDonald, 1994; McDonald & Brune, 1998; Reid, 1997.
33. AWM 52 8/3/14 McIlroy Report.
34. AWM 52 8/3/14 McIlroy Report.
35. AWM 52 8/3/14 Treacy Report.
36. Quoted from Horner, 1978, p.151.
37. AWM 3DRL 2381 2/7.

Chapter 16
1. McCarthy, 1959; Williams, 2008, 2012.
2. Quoted from Happell, 2008, p.58.
3. Collie & Marutani, 2007; Happell, 2008; Williams, 2008, 2012.
4. Quoted from Ham, 2004, p.238.
5. Williams, 2008, 2012.
6. Milner, 1955; Williams, 2008, 2012.
7. AWM 52 8/3/78 39th Battalion Appendix.
8. AWM MSS0732.
9. AWM 52 8/ 2/21.
10. Grayden, 2015, p.83.
11. AWM 52 8/2/21.
12. Cousens, 2010, p.5.
13. Burns, 1960; McCarthy, 1959.
14. Burns, 1960; McCarthy, 1959.
15. AWM 52 8/3/27.
16. AWM 52 1/10/1 004.
17. Byrnes, 1989; Cameron, 2020.
18. AWM PR00787.
19. Vernon Diary, 1943, pp.23–24.

Chapter 17
1. AWM 52 8/2/21.
2. AWM 52 1/10/1 004.
3. AWM 52 8/3/14; AWM 52 8/3/16; AWM 52 8/3/27.
4. Steward, 1983, p.129.
5. AWM 52 8/3/16.
6. Paull, 1958, pp.297–08.
7. Grahamslaw, 1971, p.123.
8. AWM 55 Current Translation 3/2 No. 266.
9. McAulay, 1991.
10. AWM 55 Current Translation 3/2 No. 218.
11. Collie & Marutani, 2007.
12. AWM 52 8/3/14; McCarthy, 1959; Russell, 1948.
13. McCarthy, 1959; Russell, 1948.
14. AWM 52 8/2/21.
15. UNSW AWFA 1223.
16. Quoted from Paull, 1958, p.209.
17. Paull, 1958, p.209.
18. McLeod, 2019; Walker, 1957.
19. AWM 52 1/10/1 004.
20. AWM 52 8/3/27.
21. AWM 52 8/3/16.
22. Cousens, 2010, pp.5-6.
23. AWM 52 8/3/14; McCarthy, 1959; Russell, 1948.
24. Russell, 1948, pp.166–67.

ENDNOTES

25. Quote from Edgar, 2010, p.174.
26. Burns, 1960, pp.120–21.
27. Quoted from Brune, 1991, pp.166–67.
28. Burns, 1960; McCarthy, 1959.
29. Quoted from Brune, 1991, p.170.
30. Quoted from Brune, 1991, p.170.
31. Burns, 1960; McCarthy, 1959.
32. AWM 52 8/3/27.
33. Quote from Brune, 1991, p.167.
34. AWM 52 8/2/21; McCarthy, 1959; Sublet, 2000.
35. Quoted from Braga, 2004, pp.198–99.

Chapter 18
1. AWM. 52 8/3/39; Cameron, 2020; Kennedy, 1992; McCarthy, 1959.
2. Kennedy, 1992, pp.38–39.
3. Hogan *et al.*, 1992, pp.114–15.
4. AWM 52 8/3/39.
5. Quote from Hull, 1995, p.160.
6. Cameron, 2021; Johnson, 2006.
7. Quoted from Austin, 2007, p.176.
8. AWM 52 3/8/78.
9. AWM 52 1/5/14.
10. AWM 52 8/3/31.
11. AWM 52 1/5/14.
12. AWM 52 8/3/14 McIlroy Report.
13. AWM 52 8/3/14.
14. Horner, 1998; McCarthy, 1959.

Chapter 19
1. AWM 52 8/2/21; AWM 52 8/3/16; James, 2011; Russell, 1948; Uren, 1959.
2. AWM 52 8/3/14; AWM 52 8/2/21; Kennedy, 1992.
3. AWM 52/8/3/14; AWM 52 8/3/16; McCarthy, 1959; Russell, 1948; Uren, 1959.
4. AWM 3DRL 2381 3/7.
5. AWM 52/8/3/14; AWM 52 8/3/16; McCarthy, 1959; Russell, 1948; Sublet, 2000.
6. Trigellis-Smith, 1992.
7. Quote from Trigellis-Smith, 1992, pp.14–15.
8. AWM PR00787.
9. Vernon Diary, 1943, p.24.
10. AWM 52 1/10/1 004.
11. Quoted from Hull, 1995, p.161.
12. Paull, 1958, pp.216–17.
13. AWM 52 8/2/21; McCarthy, 1959; Trigellis-Smith, 1992.
14. AWM 52 8/3/14; AWM 52 8/3/16: Paull, 1958; Russell, 1948; Uren, 1959.
15. AWM 52 8/2/21.
16. AWM 52 8/2/21; McCarthy, 1959.
17. AWM 52 8/3/78 39th Battalion Appendix.

Chapter 20
1. AWM 52 8/3/25; AWM 52 8/3/31; AWM 52 8/3/33; Crooks, 1971.
2. Crooks, 1971, pp.144–46.
3. AWM 52 8/3/25; AWM 52 8/3/31; AWM 52 8/3/33; Crooks, 1971.
4. Paull, 1958, p.217.
5. Pratten, 2009.
6. Cooper, 2014; Crooks, 1971; Eather, 2003; McCarthy, 1959.
7. Borneman, 2017, pp.231–32.
8. AWM 52 8/3/27; Burns, 1960; McCarthy, 1959.
9. Quoted from Brune, 1991, p.171.
10. Quoted from Brune, 1991, p.171.
11. AWM 52 8/3/27; Burns, 1960; McCarthy, 1959.
12. AWM 52 8/3/27; Burns, 1960; McCarthy, 1959.
13. Quote from Brune, 1991, p.173.
14. Bradley, 2012; McCarthy, 1959.
15. AWM 52 8/3/14 Treacy Report.
16. AWM 55 Enemy Publications, 5/3 No. 33.
17. AWM 52 8/3/27; Burns, 1960.
18. UNSW AWFA 1386.
19. McCarthy, 1959; Russell, 1948.
20. AWM 52 8/3/14 McIlroy Report.
21. McCarthy, 1959.
22. AWM 55 Current Translation 3/2 No. 218.
23. Paull, 1958.
24. Benson, 1957.

Chapter 21
1. AWM 52 8/3/39; James, 2011; Kennedy, 1992; McCarthy, 1959.
2. Russell, 1948, pp.168–69.
3. AWM 52 8//3/14; Williams, 2012.
4. Uren, 1959, p.151.
5. AWM 52 8/3/16.
6. McCarthy, 1959; Russell; 1959.
7. AWM 52 8/2/21; AWM 52 8/3/16; McCarthy, 1959.
8. AWM 52 8/3/16.
9. AWM 52 8/3/14; AWM 52 8/3/16; AWM 52 8/3/39; McCarthy, 1959.
10. AWM 52 8/3/16.
11. AWM 52 8/2/21; AWM 52 8/3/39.
12. Kennedy, 1992; McCarthy, 1959; Trigellis-Smith, 1992.
13. AWM 52 8/2/21.
14. Quotes from Anderson, 2014, p.102.
15. AWM 52 8/2/21.
16. Blake, 2019, pp.110–11.
17. AWM PR00787.
18. Vernon Diary, 1943, p.24.
19. Quote from Hull, 1995, p.161.

ENDNOTES

20. AMM 55 Enemy Publications 5/3 No. 33.
21. AWM 55 Enemy Publications 5/3 No. 33.
22. AWM 55 Enemy Publications 5/3 No. 33.
23. AWM 52 8/3/78 39th Battalion Appendix.
24. AWM 55 Current Translation 3/2 No. 218.

Chapter 22
1. AWM 52 1/5/14.
2. Crooks, 1971, p.155.
3. AWM 52 8/2/25.
4. AWM 52 8/3/33; Crooks, 1971.
5. Crooks, 1971, p.156.
6. Quoted from Crooks, 1971, p.148.
7. Crooks, 1971, p.148.
8. AWM 52 8/3/33; Crooks, 1971.
9. AWM 52 8/3/23; AWM 52 8/3/31; AWM 52 8/3/33; Draydon, 2000; Eather, 2003; Laffin, 1994.
10. AWM 52 8/3/27.
11. AWM 52 8/3/27.
12. AWM 419 14/2 Buckler Report.
13. AWM 52 8/3/14 Treacy Report.
14. AWM 52 8/3/14 McIlroy Report.
15. Quoted from Horner, 1978, p.155.
16. Quoted from Edgar, 2010, p.175.

Chapter 23
1. AWM 52 8/3/39; Kennedy, 1992.
2. Kennedy, 1992, p.46.
3. AWM 52 8/3/16; Eather, 2003; Kennedy, 1992; McCarthy, 1959; Russell, 1948; Trigellis-Smith, 1992; Uren, 1959.
4. AWM 52 8/3/14.
5. AWM 52 1/10/1 004.
6. Crooks, 1971, p.156.
7. Crooks, 1971, p.158.
8. Crooks, 1971.
9. Hamlyn-Harris, 1993, pp.21–23.
10. Quoted from Laffin, 1994, p.81.
11. AWM 52 8/3/31; AWM 52 8/3/33.
12. AWM 52 8/3/25; Draydon, 2000.
13. AWM 52 8/3/39.
14. McCarthy, 1959.
15. AWM 52 8/3/16.
16. AWM 52 8/2/21.
17. AWM 52 8/2/25.
18. AWM 55 Enemy Publications 5/3 No. 33.
19. Quoted from Williams, 2012, p.85.

20. Quotes from Yoshida, 2017, p.210.
21. Collie & Marutani, 2009.
22. Williams, 2008, 2012.
23. Bullard, 2007, p.165.
24. Quoted from Ham, 2004, p.252.
25. AWM 55 Current Translation 3/2 No. 218.
26. AWM MSS0732.
27. Quote from Hull, 1995, p.161.
28. Quotes from Paull, 1958, p.256.
29. Edgar, 2010; Horner, 1978.
30. Bergerud, 1996; Cailey, 2000; Campbell, 2007; Duffy, 2016; Horner, 1998; McCarthy, 1959.
31. Quote from McCarthy, 1959, pp.239–40.
32. Brigg & Brigg, 1967; Blake, 2019; Horner, 1998; McCarthy, 1959; Trigellis-Smith, 1992.
33. Quoted from Trigellis-Smith, 1992, p.70.
34. Quoted from Trigellis-Smith, 1992, p.71.
35. Quoted from Thompson, 2000, p.154.
36. AWM 52 8/3/27.
37. AWM 52 8/3/27.
38. Quote from Brune, 1991, p.174.

Chapter 24
1. Williams, 2008, 2012.
2. AWM 52 8/2/21; AWM 52 8/3/39; Williams, 2008, 2012.
3. AWM 52 8/2/21.
4. Cousens, 2010, pp.6-7.
5. AWM 52 8/3/16.
6. AWM 552 8/2/21; AWM 52 8/3/16; Uren, 1959.
7. AWM 52 8/3/16; Crooks, 1971.
8. AWM 52 8/3/14.
9. AWM 52 8/3/39.
10. Crooks, 1971.
11. Crooks, 1971, p.159.
12. AWM 52 8/3/33; Crooks, 1971.
13. Crooks, 1971, p.159.
14. AWM 52 8/3/33; Crooks, 1971.
15. Crooks, 1971, p.159.
16. AWM 52 8/3/33; Crooks, 1971.
17. AWM 52 8/3/31.
18. AWM 52 8/2/25; AWM 52 8/3/25; Laffin, 1992.
19. Quoted from Johnston, 1944, p.160.
20. Quoted from Hull, 1995, p.162.
21. AWM 52 8/2/25.
22. AWM 52 1/5/14.
23. AWM 55 Enemy Publications 5/3 No. 33; Bullard, 2007; McAulay, 1991; Williams, 2008, 2012.

ENDNOTES

24. AWM 52 8/3/78 39th Battalion Appendix.
25. AWM 55 Current Translation 3/2 No. 266.
26. AWM 55 Current Translation 3/2 No. 218.
27. AWM 52 8/3/27; McCarthy, 1959.
28. AWM 52 8/3/14 Treacy Report.
29. AWM 52 8/3/14 McIlroy Report.
30. AWM 419 14/2 Buckler Report.
31. Quote from Horner, 1978, pp.158–59.

Chapter 25

1. Quoted from Eather, 2003, p.63.
2. AWM 52 8/2/25; AWM 52 8/3/39; Eather, 2003; McCarthy, 1959.
3. AWM 52 8/3/33; Crooks, 1971; Kennedy, 1992.
4. Crook, 1971, p.161.
5. Crooks, 1971, p.161.
6. Quoted from Williams, 2012, pp.142–43.
7. AWM 55 Enemy Publication 5/3 No. 33.
8. AWM 52 8/3/16.
9. AWM 52 8/2/21; AWM 52 8/3/16; Uren, 1959; Williams, 2008, 2012.
10. AWM 52 8/3/14; AWM 52 8/3/16; Russell, 1948; Trigellis-Smith, 1992.
11. Trigellis-Smith, 1992, p.48.
12. AWM 52 8/3/14; Russell, 1948.
13. AWM 52 8/3/14.
14. AWM 52 8/3/16.
15. AWM 52 8/2/21; AWM 52 8/3/25; AWM 52 8/3/39; Draydon, 2000; McCarthy, 1959.
16. Laffin, 1994.
17. Laffin, 1994.
18. Laffin, 1994.
19. AWM 52 8/2/21; McCarthy, 1959; Russell, 1948; Uren, 1959.
20. Hamlyn-Harris, 1993, p.27.
21. AWM 52 8/3/31.
22. AWM 52 8/3/31.
23. Cousens, 2010, p.8.
24. AWM 52 8/3/78.
25. MacAulay, 1991.
26. AWM 55 Current Translation 3/2 No. 266.
27. Bullard, 2007; Williams, 2008, 2012.
28. AWM 55 Current Translation 3/2 No. 218.
29. Quoted from McAulay, 1991, p.224.
30. Quote from Hull, 1995, pp.162–63.
31. McCarthy, 1959; Horner, 1992; Walker, 1957.
32. Horner, 1992; McCarthy, 1959; Trigellis-Smith, 1992.
33. McCarthy, 1959; Brigg & Brigg, 1967.
34. McCarthy, 1959.
35. AWM 52 8/2/21; McCarthy, 1959; Trigellis-Smith, 1992.

36. AWM 52 8/3/14.
37. Quote from Brune, 1991, p.172.
38. AWM 52 8/3/27.
39. AWM 52 8/3/27; Burns, 1960; McCarthy, 1959.
40. Johnston, 1944, pp.157–58.

Chapter 26
1. Williams, 2008, 2021.
2. Bullard, 2007, p.165.
3. AWM 52 8/2/21; AWM 52 8/3/16; Uren, 1959.
4. AWM 52 8/3/16.
5. Grayden, 2015, pp.86–88.
6. Quoted from Williams, 2012, p.145.
7. Cousens, 2010, p.8.
8. AWM 52 8/3/25; Draydon, 2000.
9. Draydon, 2000, pp.110–11.
10. Draydon, 2000.
11. AWM 52 8/3/16; AWM 52 8/3/39; Draydon, 2000.
12. AWM 52 8/3/31.
13. Quoted from McAulay, 1991, p.227.
14. AWM 52 8/3/78 39th Battalion Appendix.
15. McAulay, 1991.
16. AWM 52 8/3/14.
17. AWM 52 8/3/39.
18. Russell, 1948.
19. Quote from Bergerud, 1996, pp.396–97.
20. AWM 52 8/3/14.
21. AWM 52 8/3/25.
22. AWM 52 8/3/14.
23. Russell, 1948, p.171.
24. UNSW AWFA 1223.
25. Quote from Russell, 1948, p.170.
26. Quoted from Trigellis-Smith, 1992, pp.48–49.
27. UNSW AWFA 1319.

Chapter 27
1. Crooks, 1971.
2. Kennedy, 1992, p.50.
3. AWM 52 8/3/39.
4. Crooks, 1971, pp.162–63.
5. Quote from Hogan *et al.*, p,116.
6. Crooks, 1971.
7. AWM 52 8/3/25; AWM 52 8/3/33; Crooks, 1971; Draydon, 2000; Kennedy, 1992.
8. Kennedy, 1992, pp.50–51.
9. Quoted from Kennedy, 1992, p.51.
10. Kennedy, 1992, p.51.

11. AWM 52 8/3/16.
12. Uren, 1959.
13. Cousens, 2010, pp.8-9.
14. AWM 55 Current Translation 3/2 No. 266.
15. Crooks, 1971, pp.164–66.
16. Crooks, 1971.
17. AWM PR00787.
18. Vernon Diary, 1943, p.25.
19. Quoted from Kienzle, 2011, p.156.
20. Quoted from Eather, 2003, p.65.
21. Quote from McCarthy, 1959, p.231.
22. McCarthy, 1959.
23. AWM 52 8/3/33; Crooks, 1971; McCarthy, 1959.
24. Crooks, 1971.
25. Quoted from Eather, 2003, p.75.
26. AWM 55 Current Translation 3/2 No. 218.
27. Burns, 1960.
28. AWM 52 8/3/27.
29. Quote from Brune, 1991, p.176.
30. Quote from Brune, 1991, p.175.
31. AWM 52 8/3/27.

Chapter 28

1. Crooks, 1971.
2. Crooks, 1971.
3. McCarthy, 1959; Draydon, 2000; Eather, 2003; Tanaka, 1980; Williams, 2012.
4. AWM 52 8/3/33; Crooks, 1971.
5. Crooks, 1971, pp.168–69.
6. AWM 52 8/3/16; AWM 52 8/3/25; Draydon, 2000.
7. AWM 52 8/3/14.
8. AWM 52 8/3/16.
9. AWM 52 8/3/25; McCarthy, 1959; Russell, 1948.
10. Cousens, 2010, pp.9-10.
11. McCarthy, 1959.
12. AWM 52 8/3/31.
13. McAulay, 1991.
14. AWM 52 8/3/78 39th Battalion Appendix.
15. Quote from McAulay, 1991, p.223.
16. McAulay, 1991.
17. Quote from McCarthy, 1959, p.232; see also AWM 52 8/2/21.
18. Eather, 2003; McCarthy, 1959.
19. AWM 52 8/2/21; Eather, 2003; McCarthy, 1959.
20. AWM 52 1/10/1 004.
21. AWM PR00787.
22. Vernon Diary, 1943, pp.25–26.
23. Quote from Hull, 1995, p.163.

24. AWM 52 8/2/25.
25. Quote from Hull, 1995, p.163.

Chapter 29
1. Russell, 1948.
2. AWM 52 8/3/14.
3. AWM 52 8/3/39; Kennedy, 1992.
4. Kennedy, 1992, p.58.
5. AWM 52 8/3/31.
6. Quote from Sublet, 2000, p.85.
7. Draydon, 2000.
8. AWM 52 8/3/25.
9. Quoted from Eather, 2003, p.70.
10. AWM 52 8/3/33.
11. McCarthy, 1959.
12. AWM 52 1/5/14; Crooks, 1971.
13. AWM 52 8/3/33.
14. AWM 52 1/5/14.
15. AWM 52 1/5/14.
16. Eather, 2003, p.68.
17. Quotes from Paul, 1958, p.224.
18. Paull, 1958.
19. AWM 55 Current Translation 3/2 No. 266.
20. AWM 55 Current Translation 3/2 No. 218.
21. McAulay, 1991.
22. AWM 52 8/2/21; Burns, 1960; Powell, 2003.
23. AWM 52 8/3/27; Burns, 1960; McCarthy, 1959.
24. AWM 52 8/3/27; Burns, 1960; McCarthy, 1959
25. AWM 52 8/3/27; Burns, 1960; McCarthy, 1959.
26. AWM 52 8/3/14 Treacy Report.
27. AWM 52 8/3/14 Treacy Report
28. AWM 52 8/3/14 McIlroy Report.
29. McCarthy, 1959.

Chapter 30
1. AWM 52 8/2/25.
2. Quote from Hawthorne, 2003, p.212.
3. AWM 52 1/5/14.
4. Quote from Crooks, 1971, p.173.
5. AWM 52 1/5/14.
6. Crooks, 1971, p.174.
7. AWM 52 1/5/14; AWM 52 8/3/33; McCarthy, 1959.
8. AWM 52 1/5/14; AWM 52 8/3/33.
9. Eather, 2003; McCarthy, 1959.
10. Quoted from Johnston, 1944, p.161.
11. AWM 52 8/3/39.

ENDNOTES

12. AWM 52 8/3/25; Crooks, 1971; McCarthy, 1959.
13. AWM 52 8/3/25; Draydon, 2000.
14. AWM 52 8/3/31.
15. McCarthy, 1959.
16. AWM 52 8/3/14.
17. AWM 52 8/3/16.
18. AWM 52 8/3/14 September 1942.
19. AWM 52 8/3/14 September 1942.
20. Pratten, 2009.
21. Quote from Hull, 1995, p.163.
22. PR00787.
23. Vernon Diary, 1943, p.27.
24. McLeod, 2019; Walker, 1957.
25. AWM 419 14/2 Buckler Report.
26. AWM 419 14/2 Buckler Report.

Chapter 31

1. Quoted from Draydon, 2000, p.115.
2. AWM 52 8/2/25; James, 2008; Trigellis-Smith, 1992.
3. Quoted from Ham, 2004, p.253.
4. AWM 52 8/3/78 39th Battalion Appendix.
5. McAulay, 1991.
6. AWM MSS0732.
7. AWM PR00297, Nakahashi.
8. Tanaka, 1980, p.23.
9. McAulay, 1991.
10. AWM 55 Current Translation 3/2 No. 218.
11. Quote from McAulay, 1991, p.239.
12. Quote from McCarthy, 1959, p.249.
13. Quote from Brune, 1991, p.179.
14. McCarthy, 1959.
15. Quote from Bradley, 2012, p.66.
16. Johnston, 1944, p.159.
17. Quote from McCarthy, 1959, p.234.
18. Quote from McCarthy, 1959, p.234.
19. Quote from Paull, 1958, p.247.
20. Bergerud, 1996; Duffy, 2016; Gallaway, 2000; McCarthy, 1959.
21. McCarthy, 1959, p.247.
22. Anderson, 2018; Bergerud, 1996; Day, 2003; Henderson, 1992; McCarthy, 1959; Thompson, 2000; Veitch, 2019.
23. Quote from McCarthy, 1959, p.235.
24. Quote from Robertson, 1981, p.118.
25. Bergerud, 1996; Duffy, 2016; Horner, 1978, 1998; McCarthy, 1959.
26. Quote from McDonald, 2004, pp.342–43.
27. Cameron, 2022.

Chapter 32

1. AWM 52 8/3/33; Crooks, 1971.
2. AWM 52 8/3/33; Crooks, 1971
3. Crooks, 1971, p.175.
4. Crooks, 1971.
5. AWM 52 8/3/25.
6. AWM 52 8/3/25; Draydon, 2000.
7. AWM 52 8/3/31.
8. AWM 52 8/3/31.
9. AWM 52 8/2/21; AWM 52 8/3/39.
10. AWM 52 8/3/14.
11. AWM 52 8/3/14.
12. AWM 52 8/3/16.
13. AWM 52 8/2/25.
14. AWM 52 1/10/1 004; Powell, 2003.
15. Quote from Hull, 1995, p.163.
16. McAulay, 1991.
17. Quotes from Ham, 2004, p.306.
18. AWM 55 Current Translation 3/2 No. 218.
19. Quote from McAulay, 1991, p.243.
20. McCarthy, 1959; Russell, 1948.

Chapter 33

1. AWM 52 8/3/25.
2. McAulay, 1991.
3. AWM 55 Current Translation 3/2 No. 266.
4. AWM 55 Current Translation 3/2 No. 218.
5. Quote from McAulay, 1991, p.246.
6. AWM 52 8/3/39.
7. AWM 52 8/3/33.
8. AWM 52 8/3/31.
9. AWN 52 8/3/14.
10. AWM 52 8/2/21.
11. Russell, 1948; Uren, 1959.
12. AWM 52 8/3/14.
13. AWM 52 8/3/16.
14. Russell, 1948, p.173.
15. Quoted from Ham, 2004, p.250.
16. Blake, 2019; McCarthy, 1959; Trigellis-Smith, 1992.
17. Quoted from Hull, 1995, p.164.
18. Burns, 1960; McCarthy, 1959.
19. AWM 52 8/3/27 Burns Report; also Burns, 1960, pp.124–25.
20. AWM 52 8/3/14 Treacy Report.
21. McCarthy, 1959; Russell, 1948.

ENDNOTES

Chapter 34
1. James, 2008; McCarthy, 1959.
2. Quotes from Bullard, 2007, p.193.
3. Quotes from Milner, 1955, p.99.
4. Quotes from Bullard, 2007, p.193.
5. Bullard, 2007.
6. AWM MSS0732.
7. AWM 52 8/3/25.
8. AWM 52 8/3/39.
9. AWM 52 8/3/31.
10. AWM 52 8/3/33.
11. AWM 55 Current Translation 3/2 No. 218.
12. Quotes from Ham, 2004, p.306.
13. AWM 52 8/3/14.
14. AWM 52 8/3/14.
15. AWM 52 8/3/16.
16. AWM 52 8/2/21.
17. Quote from Hull, 1995, p.164.
18. AWM 419 14/2 Buckler Report.
19. McCarthy, 1959, pp.593–94.
20. Russell, 1948, p.153.
21. McCarthy, 1959, p.594.
22. AWM 419 14/2 Buckler Report.
23. AWM 52 8/3/14 Treacy Report.
24. AWM 52 8/3/27 Burns Report; also Burns, 1960, p.125.
25. Quote from McCarthy, 1959, p.235.

Chapter 35
1. AWM 52 8/3/31.
2. AWM 52 1/5/14.
3. AWM 52 8/3/39.
4. Horner, 1995; James, 2008.
5. Crooks, 1971, p.176.
6. AWM 52 8/3/25; Draydon, 2000.
7. AWM 52 8/3/33.
8. AWM 52 8/2/25.
9. Quote from McAulay, 1991, p.254.
10. McAulay, 1991.
11. AWM 55 Current Translation 3/2 No. 218.
12. Cameron, 2020, 2022.
13. AWM 55 Current Translation 3/2 No. 87.
14. AWM 55 Current Translation 3/2 No. 266.
15. Quoted from McAulay, 1991, p.254.
16. AWM 52 8/3/16.
17. AWM 52 8/3/14.
18. Quote from Hull, 1995, p.164.

19. AWM 652 8/3/31.
20. AWM 52 8/3/14.
21. AWM 52 8/3/14.
22. AWM 52 1/10/1 004; Powell, 2003.
23. AWM PR00787.
24. McCarthy, 1959; Burns, 1960.
25. Burns, 1960; McCarthy, 1959.
26. AWM 52 8/3/27 Burns Report; also Burns, 1960, p.125.
27. AWM 52 8/3/14 Treacy Report; see also Trigellis-Smith, 1992.
28. Quote from Trigellis-Smith, 1992, p.30.
29. Benson, 1957, pp.61-62.

Chapter 36
1. AWM 52 8/3/25; Crooks, 1971; Draydon, 2000.
2. AWM 52 8/3/31.
3. AWM 52 8/3/39; Gillison, 1962.
4. UNSW AWFA 1151.
5. AWM 52 8/3/39.
6. Johnston, 1944, pp.161–63.
7. AWM 52 8/3/14; AWM 52 8/3/16.
8. AWM 52 8/2/21.
9. AWM 52 8/3/14.
10. Quote from McAulay, 1991, p.258.
11. McAulay, 1991.
12. Quote from Hull, 1995, pp.164–65.
13. Blake, 2019; Trigellis-Smith, 1992.
14. AWM 52 8/3/14 Treacy Report.
15. AWM 52 8/3/27 Burns Report; also Burns, 1960, pp.125–26.
16. AWM 52 8/3/144 McIlroy Report.

Chapter 37
1. AWM 52 1/5/14.
2. Crooks, 1971.
3. AWM 52 1/5/14; AWM 52 8/3/31.
4. AWM 52 8/3/25.
5. AWM 52 1/5/14.
6. AWM 52 8/2/25; AWM 52 8/3/39.
7. McAulay, 1991.
8. AWM 55 Current Translation 3/2 No. 266.
9. AWM 55 Current Translation 3/2 No. 218.
10. Brigg & Brigg, 1967; James, 2011.
11. Brigg & Brigg, 1967, p.41.
12. AWM 52 8/3/14; Brigg & Brigg, 1967; Gillison, 1962.
13. AWM 52 8/3/27 Burns Report; also Burns, 1960, p.126.
14. AWM 52 8/3/144 McIlroy Report.
15. Horner, 1998; McCarthy, 1959; Milner, 1955.

ENDNOTES

16. Quote from McDonald, 2004, p.351.
17. Quoted from Rowell, 1974, p.127.
18. Rowell, 1974, p.127.
19. Quote from Mayo, 1975, p.61.
20. Bergerud, 1996; Duffy, 2016; Horner, 1978, 1998; McCarthy, 1959; Mayo, 1975.
21. Horner, 1978, 1992; McCarthy, 1959; Thompson, 2000.
22. Briggs & Briggs, 1967; Horner, 1991; McCarthy, 1959.
23. Bergerud, 1996; Cailey, 2000; Campbell, 2007; Horner, 1978, 1992; McCarthy, 1959; Trigellis-Smith, 1992.
24. Quoted from Trigellis-Smith, 1992, p.72.
25. Quoted from Paull, 1958, p.256.

Chapter 38

1. AWM 52 8/2/25.
2. Quote from Hull, 1995, p.165.
3. AWM 52 8/3/25; Draydon, 2000.
4. AWM 52 8/3/31.
5. Hamlyn-Harris, 1993, p.26.
6. AWM 52 8/3/33.
7. Crooks, 1971.
8. Quoted from Kienzle, 2011, p.160.
9. McAulay, 1991.
10. Quote from Williams, 2012, p.185.
11. McCarthy, 1959; Williams, 2012.
12. McCarthy, 1959; Gillison, 1962; Milner, 1955.
13. Milner, 1955, p.98.
14. Cailey, 2000; Hayashi & Coox, 1959; McCarthy, 1959; Thompson, 2000.
15. Hayashi & Coox, 1959; McCarthy, 1959; Milner, 1955; Sublet, 2000; Tanaka, 1980; Williams, 2008, 2012.
16. Quotes from Yoshida, 2017, p.211.
17. AWM MSS0732.
18. AWM 55 Current Translation 3/2 No. 266.
19. Quoted from Collie & Marutani, 2009, p.141.
20. AWM 55 Current Translation 3/2 No. 218.
21. Quoted from Collie & Marutani, 2009, p.142.
22. Quoted from Collie & Marutani, 2009, pp.142–43.
23. AWM PR00297 Nakahashi.
24. AWM 52 8/3/14.
25. AWM 52 8/3/14.
26. AWM 52 8/3/14; AWM 52 8/3/16; Brigg & Brigg, 1967.
27. Briggs & Briggs, 1967, p.47.
28. AWM 52 8/3/16.
29. AWM 52 8/2/21.
30. AWM 52 8/3/27 Burns Report; also Burns, 1960, p.126.
31. McKernon, 2006.

Chapter 39
1. AWM 52 8/3/25.
2. AWM 52 8/3/25.
3. AWM 52 8/2/25.
4. AWM 52 8/2/25; Draydon, 2000.
5. Quoted from McAulay, 1991, p.267.
6. AWM 52 8/2/25.
7. AWM 52 8/3/25.
8. AWM 52 8/3/33.
9. Crooks, 1971, pp.178–80.
10. AWM 52 8/3/31.
11. AWM 52 8/3/38; Laffin, 1994.
12. Quoted from Yoshida, 2017, pp.211–22.
13. AWM 55 Current Translation 3/2 No. 266.
14. McAulay, 1991.
15. AWM 55 Current Translation 3/2 No. 218.
16. McAulay, 1991.
17. McAulay, 1991.
18. AWM MSS0732.
19. AWM 52 8/3/14.
20. AWM 52 8/3/14.
21. AWM 52 8/3/16.
22. AWM 52 1/10/1 004.
23. AWM 52 1/10/1 004.
24. Trigellis-Smith, 1992, p.72.
25. AWM 52 8/3/27 Burns Report; also Burns, 1960, p.126.
26. AWM 52 8/3/144 McIlroy Report.
27. Quote from McCarthy, 1959, p.237.
28. Quoted from McDonald, 2004, p.352.
29. Quoted from Horner, 1978, pp.153–54.

Chapter 40
1. AWM 52 8/2/25.
2. AWM 52 8/3/33.
3. Crooks, 1971, 1980.
4. AWM 52 8/3/33; Crooks, 1971.
5. Quoted from Eather, 2003, p.78.
6. AWM 52 8/3/31.
7. AWM 52 8/3/25; Draydon, 2000.
8. AWM 52 8/3/39.
9. Quotes from Johnston, 1996, p.159.
10. AWM 52 8/2/25.
11. AWM 52 8/2/25.
12. Quoted from Paull, 1958, p.268.
13. AWM 52 8/3/78 39th Battalion Appendix.
14. McAulay, 1991.

ENDNOTES

15. AWM PR00297, Nakahashi.
16. AWM 55 Current Translation 3/2 No. 218.
17. McAulay, 1991.
18. Quote from Hull, 1995, p.165.
19. Cameron, 2022.
20. AWM 52 8/3/14; Brigg & Brigg, 1967.
21. AWM 52 8/3/16.
22. Briggs & Briggs, 1967, pp.48–49.
23. AWM 419 14/2.
24. AWM 52 8/3/14 Treacy Report.
25. AWM 52 8/3/27 Burns Report; also Burns, 1960, pp.126–27.
26. Quote from McCarthy, 1959, pp.237–38.
27. Quote from McCarthy, 1959, p.238.
28. McDonald, 2004.
29. Quote from McDonald, 2004, p.352.
30. Quote from McDonald, 2004, p.353.
31. Quote from McDonald, 2004, p.391.
32. Archer, 1985, p.83.

Chapter 41
1. AWM 52 8/3/25; Draydon, 2000.
2. AWM 52 8/3/25; McCarthy, 1959.
3. AWM 52 8/3/31.
4. Crooks, 1971, p.181.
5. Crooks, 1971, p.181.
6. Crooks, 1971, p.149–50.
7. AWM 55 Current Translation 3/2 No. 218.
8. Quoted from Collie & Marutani, 2009, pp.144–45.
9. Collie & Marutani, 2009; Happell, 2008.
10. Quote from Ham, 2004, p.316.
11. AWM 52 8/3/39.
12. AWM 52 8/2/25.
13. AWM 52 8/2/25.
14. AWM 52 8/2/25.
15. Quote from Hull, 1995, p.165.
16. Quote from Cox, 2020, p.140.
17. AWM 52 8/3/27 Burns Report.
18. Burns, 1960, p.127.
19. Quoted from Edgar, 2010, pp.184–85.
20. Cameron, 2022.
21. Horner, 1978; McCarthy, 1959.

Chapter 42
1. Crooks, 1971, p.181.
2. AWM 52 8/3/33; Crooks, 1971.
3. Crooks, 1971, p.182.

4. AWM 52 8/3/33.
5. AWM 52 8/3/25.
6. AWM 52 8/3/25; Draydon, 2000.
7. AWM 52 8/3/25; Draydon, 2000.
8. AWWM 52 8/3/31.
9. Hamlyn-Harris, 1993, p.32.
10. AWM 52 8/2/25.
11. AWM 52 8/2/25.
12. Johnston, 1944, pp.164–65.
13. AWM 52 1/10/1 004.
14. Vernon Dairy 1943, p.27.
15. McAulay, 1991.
16. McAulay, 1991.
17. AWM 55 Current Translation 3/2 No. 218.
18. AWM MSS0732.
19. AWM 419 14/2 Buckler Report; Bradley, 2012; Campbell, 2007.
20. AWM 52 8/3/27 Burns Report; also Burns, 1960, p.127.
21. AWM 52 8/3/14; James, 2011.
22. Bergerud, 1996; Horner, 1998; McCarthy, 1959.
23. Quote from McCarthy, 1959, p.237.
24. Quoted from Ham, 2004, p.277.
25. McCarthy, 1959; Milner, 1955.
26. Horner, 1978, p.188.
27. Quoted from Horner, 1978, p.183.
28. Quote from Braga, 2004, p.216.
29. Quoted from Horner, 1992, p.197.
30. Quote from Braga, 2004, p.217.
31. AWM 52 8/3/14 September.

Chapter 43
1. AWM 52 8/2/25; AWM 52 8/3/33.
2. Crooks, 1971, pp.182–83.
3. Quoted from Draydon, 2000, p.124.
4. AWM 52 8/2/25; AWM 52 8/3/25; Draydon, 2000.
5. Quoted from Draydon, 2000, p.124.
6. AWM 52 8/225; AWM 52 8/3/31; Laffin, 1994.
7. AWM 52 8/3/39.
8. AWM 52 8/2/25.
9. Quoted from Eather, 2003, p.79.
10. Quoted from Braga, 2004, p.211.
11. AWM 55 Current Translation 3/2 No. 218.
12. Quote from McAulay, 1991, p.275.
13. Quote from Trigellis-Smith, 1992, pp.30–31.
14. AWM 419 14/2 Buckler Report; Russell, 1948; McCarthy, 1959.
15. AWM 52 8/3/27 Burns Report; also Burns, 1960, p.127.

ENDNOTES

Chapter 44
1. AWM 52 8/3/31.
2. AWM 52 8/2/25.
3. AWM 52 8/3/33.
4. Crooks, 1971, pp.182–84.
5. Draydon, 2000.
6. AWM 52 8/3/25.
7. AWM 52 8/3/25.
8. AWM 52 8/3/25.
9. AWM 52 8/2/21.
10. AWM 52 8/2/21.
11. AWM 52 8/3/31.
12. AWM 52 8/2/25.
13. AWM 52 8/3/31.
14. AWM 52 8/3/39.
15. AWM 55 Current Translation 3/2 No. 218.
16. McAulay, 1991.
17. AWM 52 1/10/1 004.
18. AWM PR00787.
19. Vernon Dairy, 1943, p.27.
20. AWM 52 8/3/27 Burns Report; also Burns, 1960, pp.127–28.
21. AWM 52 8/3/144 McIlroy Report.
22. Benson, 1957, pp.63-64
23. AWM 52 8/3/78.

Epilogue
1. AWM 52 8/3/14 McIlroy Report.
2. AWM 52 8/3/14 McIlroy Report.
3. AWM 52 8/3/14 McIlroy Report.
4. AWM 52 8/3/14 McIlroy Report.
5. McCarthy, 1959; Walker, 1957.
6. McCarthy, 1959.
7. AWM 52 8/3/14 McIlroy Report.
8. AWM 52 8/3/27 Burns Report; also Burns, 1960, p.128.
9. McCarthy, 1959; McLeod, 2019; Vernon Dairy, 1943; Walker, 1957.
10. AWM 52 8/3/27 Burns Report; also Burns, 1960, p.128.
11. AWM 52 8/3/27 Burns Report.
12. AWM 52 8/3/27 Burns Report.
13. (AWM 52 8/3/27).
14. McCarthy, 1959, p.594.
15. Quote from Trigellis-Smith, 1992, p.31.
16. AWM 419 14/2 Buckler Report; Bradley, 2012; Ham, 2004; Russell, 1948; Trigellis-Smith, 1992.

INDEX

2/14th Battalion 3–5, 24, 27, 35, 55, 63, 65, 68, 85, 116, 203, 333–6, 339, 351, 378–80, 389
  Brigade Hill assault 143, 146, 154
  composite battalion with 2/16th 198–9, 203, 294, 311, 319, 359, 368
  Efogi 56, 66, 80
  Ioribaiwa 220, 235, 252–3, 266
  Ioribaiwa Ridge 283–6, 311–12, 319
  Maguli Range 219
  Menari, holding 181
  Menari, retreat towards 158–60, 162, 173–4, 177–8, 181
  Mission Ridge 81, 99, 101, 103, 104, 106, 108, 129, 132–6, 139
  Myola 22–3, 28–9, 39–40, 47–8
  Nauro, retreat to 179, 181, 189, 197
  relief of 395–6, 413–14, 427, 436–7
  soldiers lost in the jungle 4, 36–7, 46, 61, 62, 75, 194–5, 210–11, 259–60, 274, 326–7, 379–80, 482–4
  Uberi 339, 350–1, 359–60, 368, 427
  US forces, briefing 460
2/16th Battalion 3–5, 27, 29, 39, 46–7, 49, 65, 68, 85, 203, 251, 333–6, 339, 351, 378, 389
  Brigade Hill assault 143, 146–50, 154
  composite battalion with 2/14th 198–9, 203, 294, 311, 319, 359, 368
  Efogi 56, 66, 80
  Ioribaiwa 220, 235, 239, 265
  Ioribaiwa Ridge 276–9, 295, 311–12, 319
  Maguli Range 218–19
  Menari, holding 181, 183
  Menari, retreat to 158–60, 162, 173–4, 177–8
  Mission Ridge 81, 99, 105, 108, 123, 128, 132, 134–6, 139–40, 155
  Myola 22–3, 25, 47–8, 52, 63
  Nauro, retreat to 189, 196–7
  reconnaissance 123–6
  relief of 395–6, 414, 415, 427, 436–7
  soldiers lost in the jungle 4, 259
  Uberi 339, 350–1, 359–60, 368, 379, 415, 427

2/25th Battalion 193, 231, 339, 358, 394, 403–4
  Australian offensive 443, 447, 452–3, 467–8, 474–5
  Imita Ridge 331–2, 348–9, 385
  Ioribaiwa 257, 355
  Ioribaiwa Ridge 286, 290, 292–3, 311–12, 319–20, 375, 394, 417, 419–20, 432
  Uberi 238, 239, 255
2/27th Battalion 4, 6, 19, 65, 68, 85, 116, 178, 206, 209
  Efogi 66
  Kagi 30–1, 47–8, 55–6
  Menari, retreat towards 179, 181, 183, 186–8
  Mission Ridge 80, 81, 84, 86, 100–1, 103, 104–6, 108, 130–6, 158, 160–2
  Myola 42, 48
  rejoining Australian forces 343, 353–4, 395–6
  soldiers lost in the jungle 209–10, 274–5, 304–7, 325–6, 342–3, 348, 378–9, 389
  wounded soldiers 231, 246–7, 274–5, 392–3, 438, 477–8, 484–8
2/31st Battalion 193, 231, 256–7, 339, 358
  Australian offensive 432, 443–4, 447, 453–4, 468–9, 475–6
  Imita Ridge 236, 238, 239, 332, 339, 358, 375, 385, 404, 424
  Ioribaiwa 255–7, 265–9
  Ioribaiwa Ridge 281–2, 320, 376, 419, 432, 468
  Nauro offensive 257
2/33rd Battalion 193, 205–7, 256–7, 289, 339, 359, 405, 472
  Australian offensive 431–2, 444, 447, 451–2, 466–7
  Imita Ridge 328, 330, 348–9, 358, 367
  Ioribaiwa Ridge 249, 253–4, 262–3, 282, 290, 311, 319–21, 339, 376, 419–20
  Uberi 228, 231, 239
39th Battalion 1, 12, 93, 175, 360–1, 401, 436
  Efogi 55, 65, 66
  Kagi 32–3, 42, 47–8, 55
  Menari 91–2
  Port Moresby 68, 192, 480–1

INDEX

soldiers lost in the jungle 4
wounded soldiers 37, 207

A

Abau 244
Abuari 25, 34, 36, 51
Advanced New Guinea Force 462
Advisory War Council 344
aerial assaults 108–9, 227, 385, 392, 406
    Japanese 264
    US 101–2, 104, 130, 131, 407
aerial photographs 325
air drops 15, 16, 56, 57, 59, 70, 112, 208, 303, 362, 378, 489
    injuries caused by 56, 60
Airacobra fighters 101–2, 104, 130, 131
Allen, Major General Arthur 'Tubby' 5, 14, 24, 58, 76, 84, 102, 113, 133, 157, 208, 303, 314, 427, 481
    Australian offensive 419, 432, 455
    Blamey, and 465
    Eather, orders to 322–3
    Imita Ridge, visiting 403
    Port Moresby, defence of 273, 344
    Potts' report to 232–3, 243–4
    removal of Potts from command 168–9, 189, 198
    Rowell's letter to 463
Allen, Private Pierce 185
Alola 2, 3, 21, 36, 123, 436
    Japanese occupation 26
Andrews, Captain Colin 467, 474, 477
Applebee, Lieutenant Reginald 48, 55, 65
Araki, Lieutenant 49
Archer, Captain Thomas 307, 451
Archer, Jules 442
Arthur, Corporal 381
Ashby, Private Stanley 449
Ashford, Private Frederick 401
Atkinson, Captain William 367, 386, 395
Attard, Private Joe 184, 185
Audsley, Sergeant John 299–301
Austin, Lieutenant 318, 352
Australian Army
    2nd Australian Imperial Force (AIF) 3, 12
    2/1st Pioneer Battalion 193, 203, 229, 271, 319, 332, 339, 359, 385, 400, 443
    2/4th Field Ambulance 193, 272, 238, 484, 487
    2/6th Field Ambulance 20–1, 108, 272, 238
    2/6th Independent Company 198–9, 203, 221, 235, 245–6, 266, 273, 288, 339, 359, 382, 392, 401, 428, 470, 482, 488
    2/14th Battalion *see* 2/14th Battalion
    2/16th Battalion *see* 2/16th Battalion
    2/25th Battalion *see* 2/25th Battalion
    2/27th Battalion *see* 2/27th Battalion
    2/31st Battalion *see* 2/31st Battalion
    2/33rd Battalion *see* 2/33rd Battalion
    3rd Battalion 94, 113, 166, 189, 196–7, 201, 203, 218, 234, 262–3, 272, 281, 282, 289–90, 294, 296–7, 311, 319, 331, 339, 349, 358, 359, 367, 400, 431, 469
    6th Division 400, 401
    7th Division 5, 14, 400
    14th Brigade 12, 72, 369, 400, 434
    14th Field Ambulance 21, 238
    14th Field Regiment 339, 364, 376, 401, 404
    16th Brigade 13, 400, 401, 460, 481
    17th Brigade 400
    19th Brigade 400
    21st Brigade 3, 11, 18–19, 57, 68, 166, 183, 197, 203, 205–6, 208, 221, 235, 238, 249, 257, 267, 272, 333, 360, 361, 395, 436, 464–5
    25th Brigade 12–13, 24, 164, 192, 203, 205–6, 228, 234–5, 249, 262, 272, 313, 318, 344, 358, 400, 469, 472, 481
    30th Brigade 12, 20, 272, 401
    36th Battalion 414–15, 427, 436–7, 487
    39th Battalion *see* 39th Battalion
    53rd Battalion 3, 29, 31–4, 59, 65, 82, 207, 244, 273
    reconnaissance 123–6
    supplies *see* supplies
    vehicles 73
Australian Army Ordinance Corps (AAOC) 205–6
Australian High Command 345–6
Australian New Guinea Administrative Unit (ANGAU) 15, 34, 39, 52, 182, 245, 302, 315, 351, 427, 477, 484
    Menari aid post 20–1, 108, 112, 175, 183
    Nauro, evacuation of 199–200
    searching for Japanese infiltration parties 100
Australian Wireless Association (AWA) radio 18
Avent, Private William 75, 211, 326, 382, 437

B

B-26 Marauder bombers 101
Baker, Private John 277–8
Baldwin, Hanson 442
Baldwin, Private Raymond 85, 162
Balfour-Ogilvy, Lieutenant John 292
Barclay, Lieutenant Graham 252
Barnett, Lieutenant Archibald 417
Barr, Private Peter 299, 423
Bartlett, Warrant Officer Joseph 311
Beard, Private Tom 150
Beck, Corporal Johnny 291
Beckett, Captain James 289, 294, 295
Beggs, Private 'Bluey' 300
Bell, Lieutenant Frederick 82, 83, 96, 98
Bell, Private Albert 212, 483
Bell, Warrant Officer Dick 201
Bennett, Private George 219
Bennie, Lieutenant John 42
Benson, Father James 76–7, 215, 383–4, 478–80
Beresford, Brigade Major Brian 202
Beveridge, Corporal 122, 129
biscuit bombers 15, 16, 70, 378
Bisiatabu 427, 457
Bisset, Lieutenant Harold 'Butch' 25
Bisset, Lieutenant Stan 25–7, 80–1, 122, 143, 218, 286, 350, 389, 396, 413, 427
   Menari 178, 196
   US forces, briefing 460
Black, Lieutenant Alexander 282, 332
Blamey, Major General Sir Thomas 10–11, 12–13, 195, 346, 465
   Greek campaign 431
   NGF command 346, 397, 429, 439–40, 450, 460–1
   Port Moresby, in 244–5, 246, 374, 397–400
   report on PNG operations 344–5
   Rowell, relationship with 374, 398–400, 402, 416, 429–30, 438–42, 450, 460–2
   'Run Rabbit Run' speech 361
   scapegoats 360–1
Blamey, Tom 441
Blundell, Captain Peter 433
Boag, Captain Thomas 113, 166, 198, 218, 234, 263
booby traps 221, 249–52, 266
Bradford, Corporal Harold 'Brick' 229, 230, 300, 302, 310, 452

Breakwell, Private Alex 291, 472
Bridgefort, Brigadier William 441
Brigade Hill 6, 67, 70, 85, 99, 113, 121, 122–3, 133, 136–7, 139–41
   bayonet assault 140–1
   Brigade Headquarters 105, 122, 126, 127–33, 136–7, 143, 146–51, 155, 156
   'Butcher's Corner' 6, 51, 127, 130, 140–1, 154, 163, 170, 171, 277
   Japanese assault 139–57, 170, 171–2
   Japanese occupation 173
   map *142*
   retreat from 156–7, 158
Bright, Lieutenant Peter 292, 309, 320
Brown, Lieutenant Colonel Arnold 193, 203, 271–2
Brown, Private Douglas 327, 354, 483
Brown, Private William 294
Brown River 14, 272, 482
Brummitt, Major Donald 21, 30
Bryce, Captain James 33, 34, 59, 65, 67, 86
Buckler, Captain Sydney 'Ben' 35, 36, 61–2, 75, 96, 116, 194, 488, 489
   lost in the jungle 210–11, 231–2, 260, 338, 370–1
   rejoining Australian forces 413, 437, 459, 470–1
Bullingham, Sergeant 268
Buna 77, 245, 426
   Allied aerial assaults on 407
   Japanese landing at 1, 2, 4, 5, 11
Burges, Sergeant Douglas 123
Burke, Private Thomas 381, 428, 438, 449, 459–60
Burnett, Les 92
Burns, Corporal John 30, 100, 102, 103, 108, 129, 135
   rescue 484–8
   wounded, care of 362, 373, 381–2, 392, 397, 415, 428, 438, 448–50, 459–60, 471, 477–8, 484–6
Burston, Major General Samuel 416
Butler, Lieutenant Charles 371, 437
Buttrose, Lieutenant Colonel Alfred 205, 236, 254, 292, 298, 307, 309, 311, 348, 394
   Australian offensive 431–2, 443–4, 451, 466
   Ioribaiwa Ridge 262–3, 303, 319, 420
   lost in the jungle 236, 253

INDEX

C
Caddy, Lieutenant Ernest 187
Café, Private Teddy 472
Cahill, Private Wilbur 327, 354, 483
Cairns, Lieutenant Norman 55, 58, 122, 128, 129, 368
    US forces, briefing 460
Cameron, Lieutenant Colonel Allan 1, 33–4, 57, 59, 65, 67, 190, 192, 197, 200, 357
    Australian offensive 446, 469
    Imita Ridge 331, 350
    Ioribaiwa 220, 234, 238, 248, 253
Carlyon, Major 374
Caro, Lieutenant Colonel Albert 5, 25, 27–8, 57, 80, 126, 396, 413
    composite battalion 197, 203, 218, 234, 294, 333–5, 359
    Imita Ridge 350
    Menari 158, 178–9, 181
    Mission Ridge 99, 108, 132, 136, 139, 146
    removal from command 336
    report to headquarters 334–5
Cashmore, Jim 92
casualties
    Australian 4, 48, 49, 51, 61, 100, 121–2, 124–5, 143, 147, 152–3, 156, 161, 162, 169, 171–2, 189, 248, 252, 268–9, 278–9, 286, 302, 313, 360
    friendly fire 252, 450
    Japanese 63, 90, 102, 110, 112, 119, 122, 137, 144, 169, 171–2, 180, 221, 251, 270, 297, 314, 340, 341–2, 378, 388, 445–6, 454, 456
    Papuan carriers 94, 316
Cattley, Lance Corporal Roland 434
Challen, Brigade Major Hugh 24, 27–8, 39, 64, 156, 157, 164, 324
    composite battalion 197
    Ioribaiwa Ridge 249
    relief force 133
    withdrawal from Menari 172, 176, 190, 196, 197
Chigawa, 1st Lieutenant 395
Christie, Corporal 196
Churcher, Private Edward 'Ted' 161
Clark, Corporal George 40
Clements, Lieutenant Jack 86, 181, 413
Clift, Ian 318
Clowes, Captain Trevor 207, 228–9, 236, 263, 290, 298–301, 303, 307, 394, 405, 431
    finding lost men 309–10, 452
    patrol actions 420–4, 466–7, 472, 474
Clowes, Major General Cyril 346, 429, 438
communication lines 18, 72, 100, 130–1, 136, 221, 226
    cutting Japanese 46, 47, 137–8, 226, 245, 361, 386, 395, 425
    Japanese 258, 378
    repairing 286
Condon, Private John 291, 472
Cook, Private Les 310
Cooper, Lieutenant Colonel Geoff 17, 19–20, 29–30, 31, 42, 47, 55, 64–6, 82, 198
    lost in the jungle 209–10, 259, 274–5, 304, 325–6, 343, 373
    Menari, retreat towards 158, 162, 174, 179, 183, 187–8
    Mission Ridge 85, 119, 121, 134
    Potts, assessment of 84–5
    rejoining Australian forces 343, 353–4, 381, 389
    wounded soldiers 231, 246–6, 259, 274–5, 326, 353–4, 362, 373
Cops, Sergeant Major George 32, 68, 91
Corbett, Corporal John 127
Corfield, Private Thomas 449
Cotton, Major Thomas 358, 444
Cousens, Lance Corporal William 52–4, 67, 129, 159–60, 173–4, 178, 183–4
    Ioribaiwa Ridge 249–51, 269–70, 279–81, 296–8, 312–13
Cowan, Captain Ronald 129
Cowey, Sergeant Major Jim 92–3
Cox, Lieutenant Richard 290, 291, 292, 298–9, 301, 421, 422, 424
Crameri, Lieutenant John 368, 396
Crombie, Lieutenant William 433
Crooks, Sergeant William 205, 229–30, 236, 253–4, 263, 309, 330, 348, 375, 405
    Australian offensive 431–2, 444–5, 451, 466–7
    Ioribaiwa, defence of 298–302
    patrol movement 291–2, 420–4, 472–4
    rations, on 472–3
Cruickshank, Private Victor 321
Curtin, John 345–6, 399, 461, 462

D
Daniels, Corporal Donald 93
Darling, Captain Gordon 164, 189

Davies, Warrant Officer John 52, 83, 200, 302, 316, 352
Dawson, Sergeant Major Joseph 91–2
dengue fever 343
Deniki 1, 34–5, 42
   Japanese advance on 2
   Japanese holding 2
diarrhoea 96, 397, 438
   Japanese forces 164, 340
Dingle, Private Leslie 178
Distinguished Conduct Medal (DCM) 51
   McCallum, Charlie 143
   Thornton, Alexander 51
Dodd, Captain Robert 348, 349, 351, 355, 366, 375
Dorobisolo 245, 327, 363, 373, 437, 459, 470, 488
Duffy, Captain (Dr) Don 28, 252
   report to Headquarters 335–6
Dullard, Lieutenant Edward 367, 386–7, 395
Dump No. 44 200, 235, 254, 315, 321, 329, 330, 348–9, 358, 375, 385, 394, 405, 431, 433, 443
   medical aid post (Camp 44) 222–3, 302, 315–18
Dump No. 66 325, 337, 351, 380, 427, 457–8, 477
Dunbar, Lieutenant Colonel Colin 231, 268, 282–3, 303, 319, 424, 404
   Australian offensive 431–2, 443
Dutton, Lance Corporal Hugh 267
Dynon, Private Patrick 56
dysentery 71, 96, 180, 193, 210, 222, 302, 315, 337, 380, 457
   Japanese forces 99, 195, 340, 456

E

Eather, Brigadier Kenneth 24, 193, 208, 221, 228, 262, 266, 398
   Australian offensive 419, 432, 434, 443, 455, 464, 469
   Imita Ridge 330–1, 424
   Ioribaiwa Ridge 234, 249, 253, 276, 290, 295, 303, 307, 309, 314–15
   Maroubra Force command 339, 358–9
   recognition for Papuan carriers 337–8
Ebuchi, 2nd Lieutenant 476
Edgar, Sergeant 187
Edwalla 72, 114, 191

Edwards, Corporal Clive 120, 135, 162, 210, 304, 343
Edwards, George 294
Edwards, Private Leonard 'Blossom' 284–5
Efogi 5, 6, 15, 20, 21, 29, 30, 33, 42, 64–5, 67, 89, 116, 169, 208, 392
   air assault on 101, 103–4, 108–9
   Brigade Headquarters 66
   bypassing 39–40
   Japanese occupation 83–4, 86, 101, 104–5, 458
   medical aid 21, 30
   retreat from 76, 108
   retreat to 55–6, 57, 63–7, 69
Egan, Lieutenant Richard 198
Egerton-Warburton, Lieutenant William 108, 119, 133
Elliott, Private David 286
Eltherington, Private James 403
Emuni River 173
Engineers Ridge 218
Enivilogo 429, 478
Eora Creek 21, 23, 25–6, 35, 39, 44, 98, 426, 481
   Japanese defensive line 469–70, 476–7
Eora Creek Valley 25, 333, 488
Etty, Private Gordon 215
Evans, Lieutenant Ken 368, 380, 396, 436
Ewoki 370

F

Fabian tactics 17
Fargher, Major Lee 34
Faria 370
Fenton, Lieutenant Colonel George 398, 429, 441–2
Fletcher, Private Thomas 371, 372, 437, 459, 489
Forde, Frank 195
Foy, Warrant Officer Tommy 123, 124–5
Franklin, Private Viv 154
Fraser, Lieutenant Burnham 122
Fujisaki, Captain Naoma 170, 458, 47
Furokoa, Sergeant 383–4, 478–9
Fuzzy Wuzzy Angels see Papuan carriers

G

Gallagher, Sergeant Vincent 294
Gardner, Lieutenant Valentine 'Vincent' 64, 167
Gash, Private 387
Geddes, Charles 352, 361, 369

INDEX

Gedye, Corporal Stewart 327, 354, 483
Geraghty, Private Noel 294–5
Gill, Captain Thomas 64, 326, 353–4, 362
Gill, Private 121–2, 129
Gilmore, Captain Joseph 34, 59
Gittos, Sergeant Allen 118
Golden Stairs 201, 229, 256
Goldie River 14, 272, 376, 379, 389
Goldsborough, Corporal Gordon 423
Goldsmith, Captain Douglas 139, 146, 285
Gona 77, 481
   Japanese landing at 1, 11, 377
Goode, Private William 'Nugget' 54, 67
Gorari 211, 459, 481
Gorrie, Lieutenant Peter 342–3
Gosden, Private Les 470
Grahamslaw, Captain Thomas 100, 105, 179–80
Grantham, Major Robert 113, 446
Grassy Plateau/Platform 349, 356, 366, 403, 405, 417, 433
Grayden, Lieutenant William 87, 100, 123–6
   Brigade Hill 140, 144, 146, 149, 151–4, 156–7
   Ioribaiwa Ridge 277–9, 280
   Menari 173
Green, Private Gordon 266, 288
Greenwood, Lieutenant John 181, 333
Greenwood, Private Walter 215
Griffith, Dave 280
Groves, Major 352
Guadalcanal 345–6, 365–6, 407–8
Guest, Sergeant William 68
Gwillim, Sergeant John 488

H

Haddy, Warrant Officer Alan 25, 27 215
Hailo 167–8, 194, 212
Hall, Captain Thomas 215
Hall, Private John 219
Hamlyn-Harris, Private Geoffrey 236, 268, 404, 454
Harcourt, Major Harry 198–9, 427
Harris, Charlie 148
Hartman, Sergeant Thor 418
Harukazu, Lance Corporal Oka 144
Hatanaka, Lieutenant 378
Hawkins, Major Charles 34
Hawkins, Private Oliver 301
Hayashi, Lieutenant 340

Hayden's Lookout 332
Hazard, Lieutenant Robert 453, 468
Hearman, Major John 'Ben' 28, 47, 51, 333
Helia, Lubini 22
Herring, Lieutenant General Edmund 462
Hewitt, Lieutenant Sidney 82, 106, 159
Hicks, Lieutenant George 46, 49
Hill, Private H. 'Snowy' 105–6
Hill, Sergeant Donaldson 238
Hirano, 1st Lieutenant Kogoro 180, 215, 227, 357, 377, 395
   supplies, carrying 242, 259, 271, 304, 324, 342, 353, 368
   withdrawal across the mountains 412, 425, 435, 445, 458, 470, 476
Hobson, Lieutenant Colonel Arthur 338
Hodge, Captain Theodore 148, 150, 164, 174
Hogan, Lance Corporal Patrick 'Nance' 191, 293
Hombron Bluff 73, 272, 401
Honner Force 245–6, 273, 288, 361, 401, 427–8
Honner, Lieutenant Colonel Ralph 3, 4, 33, 42, 47, 55, 65, 93, 273, 428, 481
   39th Battalion parade 91
   Port Moresby 68, 360–1
Hopkins, Ronald 374
Horibe, 1st Lieutenant 70, 89–90, 110, 128, 137, 204
   Ioribaiwa 227, 258, 270, 283, 313, 340–1, 356, 376, 390, 395, 406, 419
   withdrawal from Ioribaiwa 434–5
Horie, Major Tadashi 2, 110, 119, 120, 127, 137, 340–1, 469–70
Horii, Major General Tomitarô 1, 2–3, 4, 5, 44, 99, 102, 132, 170–1, 208–9
   Ioribaiwa Ridge 258, 276, 340, 365
   Nankai Shitai Operational Order A-112 43–4
   Nankai Shitai Operational Order A-113 110–11
   Nankai Shitai Operational Order A-115 224
   Nankai Shitai Operational Order A-116 240
   Nankai Shitai Operational Order A-117 264–5
   orders from Imperial Headquarters 241, 364–5, 406–10
   withdrawal through the mountains 408–9, 424–5, 469
Hosokawa, Corporal 314
House, Corporal Thomas 418
Howes, Lieutenant 418

Howland, Lieutenant Robert 206, 253, 451
Howson, Private Jack 'Laurie' 32, 68, 115
Hozumi, Lieutenant Colonel Shizuo 242, 248
Hubon, Charlie 312
Humphery, Major Ronald 182
Hunter, Arthur 257
Hunter, Private Stanley 370
Hurrell, Captain Leslie 268, 282, 320, 424, 432–3, 443
Hutton, Geoffrey 255, 260, 331
Hyndman, Captain Charles 313

I

Ilolo 72, 192, 203, 323, 352, 361, 369, 400
Imamura, General Hitoshi 365, 406
Imanishi, Sergeant Sadashige 412
Imita Ridge 14, 203, 208, 229, 236, 253, 255–7, 264, 271–2, 472
   aggressive deep patrolling policy 351
   Australian artillery 339–40, 350, 358, 364, 385, 394
   holding 400
   Japanese advance on 328
   map *372*
   withdrawal towards 314–24, 328–30
Imperial Japanese Air Force 102
Imperial Japanese Army 230
   17th Army command 364–5
   41st Regiment 2, 4, 5, 42–5, 46–7, 70, 89, 171, 224, 258, 276, 408, 426
      I/41st 5, 408
      II/41st 4, 22, 49–50, 63
      III/41st 4, 171
   55th Division Medical Unit 70, 342, 353, 378, 426, 435
   55th Mountain Artillery Regiment 89, 242, 341, 435
   144th Regiment 2, 5, 42–4, 69, 89–90, 99, 164, 224, 242, 258, 264, 298, 340–2, 458–9, 481
      I/144th 1, 171, 180, 227, 258–9, 270–1, 324, 342, 353, 378, 395, 408
      II/144th 2, 43, 67, 98, 110, 118–21, 128, 132, 154, 170–1, 227, 248, 265, 270, 276, 313–14, 340, 342, 352–3, 377, 390, 395, 406, 408, 434
      III/144th 2, 42–3, 90, 98–9, 104, 118–19, 130–1, 133, 136, 170–1, 218–19, 265, 276, 279, 283, 289, 290, 298, 309, 328, 408, 419

Eighth Area Army 365
engineers 242
English, calling out in 230, 257, 306
illness and disease 99, 242–3, 456
lantern parade 86–7
No. 5 Special Naval Landing Party 76
prisoners, execution of 76, 207, 215, 225, 371, 489
South Seas Force 1, 2, 224, 241, 258, 276, 342, 365, 413, 435, 458
Stanley Detachment 469–70
supply problems *see* supplies
Takasago Formosan Naval Volunteers 377, 426
Tokoyama Engineer Unit 459
withdrawal 406–13, 458–9, 467–8, 469–70, 475, 476–7
Inglis, Corporal Jock 59, 73, 94, 114, 167, 192, 201, 223, 253, 255, 318, 336, 352, 369, 390, 436
Innes, Lieutenant William 348, 349
Inoue, Lieutenant 121, 144
Ioribaiwa 20, 21, 31, 35, 60, 83, 94, 113–14, 165, 166, 190, 196
   assault on 355–6
   defence of 208, 294, 303
   evacuation of 302
   Japanese advance on 240, 249–52, 264–5, 270–1, 340
   Japanese withdrawal 425, 434–5, 453–5
   retreat towards 182, 186, 199–200, 221
Ioribaiwa Ridge 3, 14, 72, 114, 189, 191, 203
   Australian offensive 292–3, 405, 417–19, 443–5, 451–7, 469
   bombardment 349, 375–6, 380, 394–5, 406, 443
   holding 234–5, 248, 290–2, 307, 309, 314–15
   Japanese advance on 249–52, 258, 276–86, 289–30, 313–14
   Japanese occupation 324–5, 349, 352, 387, 390, 405, 406–13, 433
   Japanese withdrawal from 447, 453–5
   map *287, 308, 372*
   planned offensive operations 228
   reconnaissance 296–7, 405
   withdrawal from 315–24
   withdrawal towards 218–20
Irwin, Sergeant William 27, 379–80, 483
Isaachsen, Lieutenant Colonel Oscar 396, 414
Isurava 3, 98, 304, 324–5, 342, 426, 436

INDEX

Battle of 21, 25, 27, 42
Japanese advance on 2–3
Isurava Rest House 35, 61, 211, 333, 488
Itiki 382, 396, 438, 485
Ito, Sergeant 313

J

Jacobs, Private Clyde 396
Jail Gardens 482
Jardine, Private Harry 'Bluey' 92
Jarrett, Sergeant Hylton 325, 415
Jaure 382, 470–1
Jawarere 245, 381, 389, 485, 487
'Jawforce' 273, 343
Jefferson, Lieutenant Kenneth 286
Jeffrey, Captain John 197
Jeffrey, Corporal James 198
Johns, Sergeant Robert 105–6
Johnson, Captain Ronald 56, 100
Johnston, George 275, 344, 387, 455–7
jungle
  greens 193, 205, 260–1, 302, 316, 448
  nature of 237–8, 437
  warfare 37–8, 230, 233, 260, 281, 331

K

Kagi 15, 17, 21, 32–3, 35, 39, 42, 47–9, 55–6, 62, 97, 445, 458
  Japanese occupation 89, 212, 271, 353, 470
  retreat from 65, 67
Kalikodobu 488–9
Kaloda, Superior Private Teyoki 298
Kamimura, Lieutenant 314
Kamotoa, Leading Private 138
Katayama 411, 426
Katekar, Captain Harry 85, 161, 188, 247, 305, 343
Kemp Welsh River 244, 488
Kennedy, Sergeant Colin 72, 95, 166, 190, 234, 289, 294, 320, 433
Kenny, Private Billy 300
Kenny, Private Samuel 339
Key, Lieutenant Colonel Arthur 225, 363
  capture by Japanese 214–15
  missing 4, 5, 16, 88
Kienzle, Captain Burt 15–16, 39–40, 57, 82, 113, 165–6, 325, 477
  Camp 44 302–3, 315
  Dump No. 66 351–2, 380, 406, 427, 457
  Nauro 182, 200

Kienzle, Carl 15
Kienzle, Katherine 15
Kienzle, Meryl 15, 303, 406
Kila-Kila 258
Kilner, Private Alan 150
King, Private Albert 488
Kingsbury, Private Bruce 450
Kiowa, Major 50
Knights, Private Raymond 370
Knott, Private Horace 'Viv' 161
Koboyahi, Major Asao 364
Koehne, Sergeant Arthur 415, 485
Koitaki 18, 91, 323, 436–7
  Koitaki Cricket Ground 361
Koiwai, Major Mitsuo 4, 28, 44–6, 63, 241, 406, 445–6
Kokoda 42
  first battle of 34–5
Kokoda Front Line 19, 91, 346–7
Kokoda Plateau 1, 175, 271, 426
  battle for 33
  Japanese control 1–2, 476
  plan to retake 208
Kokoda Station 35
Kokoda Track 2, 3, 10, 14, 27, 48, 62, 99, 121, 144, 196, 244, 272, 309, 321, 325, 378, 414
  2nd AIF on 12–13
  Australian advance 385, 387–9, 455–7, 467–8, 481
  conditions 444–5, 468
  fighting withdrawal 3–4, 346, 464–5
  Japanese advance along 199, 208–9, 224–7, 341
  map 7
Kumusi River 245, 304
Kunai Patch 321, 330, 349, 375, 405, 432–3, 443–4
Kusunose, Colonel Masao 44, 70, 89, 98, 110, 137, 242, 248, 258, 264, 377
  withdrawal across the mountains 408–9, 458
Kuwada, Lieutenant Colonel Gen'ichirō 2, 3

L

Laloki–Goldie River 245
Laloki River 14, 272, 401
Lamb, Corporal Roland 252
Lambert, Lieutenant Henry 'Bluey' 50, 149, 151–5
Lang, Corporal Ernest 363

537

Langridge, Captain Brett 'Lefty' 49, 50, 133, 146, 151–5
Large Grass Patch 375
Larkin, Brigade Major George 249, 455
Laula, Havala 21–2, 32
Lawson, Captain Kenneth 290, 292–3, 298
Lawson, Lieutenant Raymond 374
Lee, Captain Arthur 64, 119, 160–1
Lintot, Corporal Charlie 277–8
Little, Private Alec 86
Logan, Lieutenant Ted 299, 301, 310, 452
Lynn, Sergeant Bill 26
Lyon, Captain Geoffrey 133, 157, 164, 178–9

M

McAlister, Sandy 186
MacArthur, General Douglas 10–11, 208–9, 399, 461
  assessment of Australians 97, 116–17, 244, 345–6
  tactics 13, 244, 400
McCabe, Captain John 448
McCallum, Corporal Charles 143
MacCormick, Captain Bruce 59
McCracken, Captain John 197
MacDonald, Private James 370
McDonald, Private Spencer 'Spinner' 403
MacDonnell, Lieutenant Desmond 268
McGavin, Lieutenant Alan 86, 181, 413, 460
McGee, Captain 220
McGee, Lieutenant Grear 49, 50, 139, 181
McGregor, Corporal Roy 478, 485, 486, 487
McIlroy, Lieutenant Robert 36–7, 62
  lost in the jungle 75–6, 96, 116, 167–8, 194, 212, 214, 232, 260, 273, 327, 393, 397, 429, 478
  rejoining Australian forces 482–4
Mackay, Corporal Wally 292, 300, 421
Mackenzie, Captain Forbes 404
McKinnis, Keith 92
McLaren, Captain William 21, 71
McLeod, Private Keith 301, 302, 309–10
MacNider, Brigadier Hanford 244–5
McTaggart, Sergeant Francis 301
Madigan, Captain Bert 51, 128–9, 295–7, 312–13
Magarey, Major Rupert 20, 71, 182
Maguli Range 196–7, 198, 218
mail from home 81–2

Makings, Sergeant Clement 30–1, 108
malaria 71, 93, 337, 380
  Japanese forces 99, 164, 242, 377
Maloney, Private Brian 49
Mannion, Lance Corporal James 'Jim' 186, 348
Manurinumu 436
maps, use of 221
Maroubra Force 15, 24, 65, 132, 244, 246, 321, 328, 357, 460
  Blamey's 'Run Rabbit Run' speech 361
  communication issues 18
  Eather 339
  Porter 168–9, 172, 189, 202–3
  Potts 11, 24, 168–9, 172, 189, 202
Marshall, General George 97
Marson, Captain Richard 262, 475
Martlew, Private Robert 449, 478, 485
Maskiell, Private Clarrie 173–4, 178–9, 202, 208
Mason, Lieutenant Lindsay 311
Matschoss, Private Alexander 327, 354, 483
Matthews, Sergeant John 144
Maxwell, Warrant Officer 302, 457, 477
Mayne, Private Ambrose 181,, 370, 372
medical treatment 20–1, 34–5, 71, 404–5
  2/9th Australian General Hospital 21
  Papuan carriers *see* Papuan carriers
  stretchers, constructing 135
  wounded to fend for themselves 206–7
Menari 19–20, 29, 40, 56, 59–61, 83, 133, 196
  39th Battalion parade 91
  aid post 20–1, 108, 112
  Brigade Headquarters 164
  evacuation of 175–6, 177–8, 183, 199
  Japanese advance on 177–8, 183, 184–6
  Japanese occupation 186–8, 353
  Japanese withdrawal to 435
  relief of 39th Battalion 68
  retreat towards 139, 155, 156–64
  supplies 165, 178, 183, 185, 208
  wounded, evacuation of 93, 112, 175–6
Metson, Corporal John 61, 211, 231, 260, 370, 372, 489
Miles, Private Victor 297–8, 312
Military Cross (MC)
  Honner, Ralph 42
Military Medal (MM)
  Morris, George 51
  Preece, Ronald 485

538

INDEX

Miller, Captain Lindsay 307, 321–2, 328–30, 349, 451
Miller, Lance Corporal Alfred 382
Milne Bay 3, 11, 12, 14, 97, 346, 408, 429
Milne Force 346
Milner, Samuel 407
Misima Ridge 2
Mission Hut 85, 100, 104, 108
Mission Ridge 39, 51, 64–7, 84–7, 89
    Brigade Headquarters *see* Brigade Hill
    casualties 169
    establishing positions 80–1
    Japanese advance on 98–9, 110–12, 127–38, 164–5
    map *107*
Mitchell, Corporal James 417
Moir, Private Jim 145–6
Moloney, Corporal Brian 220
Moore, Major Albert 59–60, 64, 73–5, 94, 114–15, 167, 192, 201–2, 223, 243, 253, 255–7, 318, 336–7, 352, 361–2, 369–70, 379, 403
    messages, delivering 271–2
    Port Moresby 447–8
    scones 390–1, 396, 436
Morris, Corporal Arthur 266
Morris, Major General Basil 427
Morris, Sergeant George 50–1, 132
Mortimore, Lieutenant Harry 192
Mosley, Lieutenant John 299, 300, 302, 309–10, 452
Mules, Lieutenant John 468
Murdoch, Captain Ken 149, 413
Murphy, Private Bernard 'Spud' 18
Musgrave, Private Billy 472
Myola 4, 5, 15, 17, 24, 28, 35, 62, 75, 96, 169, 325
    air drops 16, 208
    Brigade Headquarters 39, 48
    defence of 39–40, 47–9, 56
    evacuation of wounded 20, 30
    Japanese occupation 69–70, 84, 90, 171, 180, 342
    retreat from 24, 55–9, 63–4, 69, 76, 82–3
    retreat to 20–2, 28–9, 40, 42, 52–3
    supplies 33, 40, 244

N

Nakahashi, Lieutenant Kokichi 341, 413
Nakaiama, Captain 76–7
Nakao, Lieutenant 45
Nakao, Private 270
Nanadai 222
Naoro River 190, 198, 326, 392, 484
Naro Ridge 29
native carriers *see* Papuan carriers
Nauro 20, 21, 30, 31, 59–61, 69, 72, 83, 92, 93, 157, 166, 196, 245, 275, 325–6
    air drops 56, 208
    Australian offensives 257, 262–3, 466, 474
    bombing 385
    evacuation of 182, 199–200
    Japanese occupation 200, 204, 209, 224, 247, 304–5, 364, 425
    Japanese withdrawal 434–5, 474
    medical aid post 71, 113, 179
    reconnaissance 196–7
    retreat towards 175–9, 181–2, 188–9, 192
    supply base 198, 208
Nauro Ridge 188
Nelson, Private John 219
New Guinea Air Warning Wireless Company (NGAWWC) 18
New Guinea Force (NGF) 11, 14, 398, 439
    Blamey in command 346, 397, 429, 439–40, 450, 460–1
Newton's Dump 436
Nichols, Lieutenant William 382, 470–1, 488
Nigabaifa 343, 485, 487
Nikaiama, Captain 478
Nishikawa, Sergeant 435
Nishimura, Corporal Kokichi 120–1, 144, 170, 435, 446
Noble, Warrant Officer Wilford 'Wofty' 143, 218–19
Noda, 2nd Lieutenant Hidetaka 29, 42, 70, 90, 111, 119
Norris, Colonel Frank 22, 167, 202, 223
Norris, Corporal Ted 57
Norrish, Corporal Keith 145
Nye, Captain Claude 139, 143

O

O'Brien, Private Frank 300
O'Brien, Private Joseph 321
O'Bryen, Captain Thorlough 404, 417, 420
Ofi Creek 220
Ogawa, Commander 378
O'Grady, Captain Oswald 453

Oidobi Rest House 459
Oivi 1, 481
Okada, Seizo 90, 171, 242, 341, 365, 409–11, 426, 458
Okanoue, Corporal 138
Oldham, Major John 21, 71, 113
Omaramara 361
O'Neill, Captain John 59, 252
Onogawa, 2nd Lieutenant 90, 180, 342, 426, 470
Osborne, Private Allen 'Ossie' 245, 288
O'Sullivan, Private John 215
Owen, Lieutenant Colonel William 1, 18, 33, 34–5, 225
Owen Stanley Range 1, 13, 15, 192, 205, 275, 344–5, 378
    troops lost in 4, 27, 36–7, 53–4, 61, 67, 75–6, 96–7, 146, 181
    weather 30, 90, 229, 233, 236, 405
Owers' Corner 21, 272, 318, 336, 339, 352, 369–70, 376, 381, 414, 436

P

Paine, Sergeant Robert 247
Papuan Army
    1st Infantry Battalion (PIB) 1, 68, 160, 175, 236
Papuan carriers 16, 165, 212–13, 223, 254, 318, 352, 370, 484
    health and wellbeing 34–5, 94, 165, 166, 316, 457
    racial segregation 222
    recognition for 337–8, 487–8
    stretcher bearers 21–2, 30–1, 52, 83, 135, 158, 175, 183, 210, 231, 247, 315, 487–8
Papuan locals
    attacking Australian forces 354, 393, 483
    caring for Australian forces 370–1, 482–4
Papuan police *see* Royal Papuan Constabulary (RPC)
Parer, Damien 18–19, 64, 68–9, 73, 167, 442
    death 347
    Japanese, assessment of 347
    Kokoda Front Line 19, 91, 346–7
Parker, Corporal Leonard 186
Parsons, Lieutenant Alfred 59, 64
Paul, Lieutenant Colonel Albert 72, 94, 113
Paull, Joe 249–50, 312
Peach, Captain Cyrus 358, 405
Pearce, Lieutenant George 27, 64, 379

Peleliu 347
Penney, Corporal Edgar 326
Phelan, Signaller Kenneth 32
Phelps, Lieutenant Reginald 283
Plamer, Lieutenant Alex 95
pneumonia 343, 380
Ponoon 220, 228, 234, 263, 282, 432, 444, 451, 466, 472
Port Moresby
    defence of 72, 257, 272–3, 315, 322–4, 401
    fresh troops, arrival of 193
    Japanese advance on 2–3, 89–90, 203, 224, 241, 243, 276, 324, 340–2, 344, 365–6
    Japanese reconnaissance 1
    Japanese threat, assessment of 12–14
    reinforcements 12–13, 375
    wounded, arrival of 37–8, 93, 192, 381, 437
Porter, Brigadier Selwyn 190, 200, 228, 239, 262, 267, 269, 273, 303, 314, 336, 369, 378
    Imita Ridge 350, 359
    Maroubra Force command 168–9, 172, 189, 198, 202–3
    planned offensive operations 228
    withdrawal to Imita 323, 324
porters *see* Papuan carriers
Potts, Brigadier Arnold 3–6, 11, 24, 39, 48, 64, 84, 113, 121–2, 151, 157, 166, 190–1, 223, 413, 415
    39th Battalion, praise of 91, 464–5
    character 84–5, 137, 186, 202, 208, 233
    defensive retreat 16–17, 29, 346, 464–5
    Efogi 66
    evacuation of wounded 178–9, 181, 182
    letters home 87–8
    Maroubra Force command 24, 168, 169, 172, 189, 198
    Menari 156, 163–4, 172, 174, 178–9, 185
    Mission Ridge 80, 84–6, 98, 102–3, 132, 133, 136, 146, 148, 150
    Myola 55–7, 76
    Nauro 182, 186
    Port Moresby 232–3, 243–4, 323–4, 450
    removal from command 168–9, 172, 189, 198, 202
Potts, Doreen 87, 450
Preece, Warrant Officer Ronald 52, 83, 158, 160, 173, 175, 352, 415, 484–5, 488
    Military Medal 485

INDEX

Price, Private 'Tibby' 472
prisoners of war
    Japanese 76–7, 207, 215, 225, 371, 383–4, 489
Proudfoot, Private Jock 472
psychological warfare 45
Pyramid Hill 309

Q
Quarrel, Private Basil 185

R
Rabaul 365, 406
radio communications 18, 130
Rainey, Lieutenant David 184, 266, 395, 436
Ramsey, Private Clyde 326
Ransom, Private Leslie 'Pappy' 284–5
Redding, Donald 367, 404
Reid, Captain William 59, 64–5, 67
Rhoden, Captain Phillip 5, 16, 27, 40, 80, 218, 221, 252, 333, 391, 396, 413, 427
    mail from home 81–2
    Menari 158, 178, 181, 196
    Uberi 368–9, 389
Rice, Corporal Billy 299, 300
Riches, Corporal Albert 449, 471
Riches, Private Leonard 449
Rigo 244, 488
Roberts, Lance Corporal Benjamin 296
Roberts, Lieutenant Mervyn 'Merv' 293
Roberts, Private Alexander 212–14, 273, 483
Robertson, Sergeant Bruce 320
Robertson, Sergeant Paul 188
Robinson, Major Alfred 273, 343
Robson, Major Ewan 468, 475–6
Rockliffe, Private Frederick 75, 96, 211, 326, 382, 437
Rogers, Brigadier John 246
Ross, Private Jim 288
Ross, Ray 336
Rowell, Lieutenant General Sydney 10–14, 116, 117, 164, 189, 195, 374
    assessment of forces 12–13, 233
    Blamey, relationship with 374, 398–400, 402, 416, 429–30, 438–42, 450, 460–2
    dismissal 460–4
    planned offensive 245
    Port Moresby, defence of 272–3, 344
    Potts' report to 243–4
    removal of Potts from command 168–9
    Withdrawal to Imita, on 322–4
Royal Papuan Constabulary (RPC) 1, 68, 179
    searching for Japanese infiltration parties 100
Royals, Joe 186
Rumble, Lieutenant Noel 367
Russell, Captain William 136, 164, 173, 181–2
    Australian lines, return to 266
    Menari, retreat from 184–6
Ryan, Captain 487
Ryland, Lieutenant Christopher 282, 320, 424, 432

S
Sackley, Sergeant Keith 248
Sakamoto, Lieutenant Atsushi 69, 120, 171, 181, 242, 353, 368, 434, 446
Salvation Army 64, 73–6, 94
    Red Shield Posts 59–60, 114–15, 167, 192, 201–2, 223, 243, 253, 255–7, 318, 361–2, 369–70, 379, 390–1, 403, 447
Sambell, Private Eric 155
Sanananda 76–7, 215, 246, 383, 478
Sanderson, Lieutenant 186
Sandison, Lieutenant Colin 183, 209–10
Sargent, Sergeant Albert 219
Sasaki, Private 283
Sato, Toshio 411, 426
Saubo, Corporal 200
Scammell, Private Geoffrey 29
Schwind, Lieutenant Lindsay 101, 104, 198
Scott, Private Lawrence 363
Scott, Sergeant John 25, 27, 118, 148
scrub typhus 93, 96, 456
Sengai 211, 260, 370, 371, 437, 459, 489
Seven-Mile Airfield 342
Sheppard, Dave 210
Shimmin, Private John 218
Shitamoto, Leading Private 138
Sims, Captain Charles 48, 55, 64, 119, 133, 183, 186–8
    Cooper, looking for 209–10
    rejoining Australian forces 342–3
Skipper, Captain Justin 161, 304
Smith, Brigadier Walter 433
Smith, Captain Peter 274
Smith, Corporal Allan 93, 234
Smith, Corporal Frank 421
Smith, Corporal Lionel 212–14, 273, 483
Smith, Lieutenant Sydney 245

541

SAVING PORT MORESBY

Smith, Staff Captain Peter 173–4, 183
Snelgar, Private Tom 370
Sogabe, Leading Private 138
Solomon Islands 365
Soputa 478–9
Sorimuto
South West Pacific Area (SWPA) 10, 344
Speed, Corporal William 'Bill' 196
Spotter's Hut 265–6, 267, 270, 281, 386, 406, 433, 453, 468, 475
Spring, Major Gerry 34
Spry, Colonel Charles 24, 168, 323
Steel, Lieutenant Stanley 281, 403–4, 417
Steel, Private Alan 300
Steward, Captain Henry 'Blue' 63, 101, 123, 128, 159, 178, 336
Stokes, Private Norman 328
Strachan, Lieutenant Howard 417–19
Subatani 487
Sublet, Captain Frank 28, 46–7, 49–51, 54, 58
   Mission Ridge 81, 139
supplies 207
   air drops 15, 16, 56, 57, 59, 60, 70, 112, 208, 489
   carriers *see* Papuan carriers
   destruction and contamination of 340
   finding lost 305–6
   Imita Ridge, at 340
   Japanese 5, 11, 16, 226–7, 239, 241–2, 304, 316, 324–5, 341, 353, 364–5, 368, 377, 407, 455, 456
   Japanese disruption of Allied 13–14
   logistics, importance of 13
   Menari 165, 178, 183, 185
   pack animals 414
   problems 3–4, 13, 71, 233, 241, 316, 341, 390–1, 414, 455, 456, 472–3
Susuki 383, 384, 478
Sutherland, Major General Richard 13

T
Ta-ai 485, 487
Takamasa, Private Tamotsu 298
Takase, Captain 226
Takeuchi, Private Yoshio 298
Tanaka, Lieutenant Colonel Toyonari 408–11
Tanaka, Staff Officer Kengoro 164, 342
Taylor, Private Keith 488
Tee, Private Cecil 449

Templeton, Captain Sam 1, 93
Templeton's Crossing 4, 5–6, 17, 21, 25, 35, 39, 44, 48, 52, 57–8, 62, 89, 477, 481
   Australian offensive 448
   evacuation of wounded 52
   loss of 22–3
   map *41*
   supply drop zone 15
Teruoka, Private Akira 70, 271, 325, 342, 353, 357, 378, 426, 435
Thompson, Captain Clifford 16–17, 28, 66, 391, 413, 427
Thompson, Sergeant Robert 40, 87
Thompson, WJ 295
Thorburn, Sergeant David 418
Thorne, Sergeant 185
Thornton, Private Alexander 51
Thurgood, Lieutenant John 29
   US forces, briefing 460
Tillyman, Lieutenant Lex 185
Tipton, Regimental Sergeant Major Les 25, 26
Tongs, Sergeant Bede 385
Treacy, Lieutenant Maurice 'Mokka' 35–6, 61–2, 488
   lost in the jungle 75, 96, 115–16, 168, 194–5, 211–12, 232, 259–60, 326–7, 363, 373, 382
   rejoining Australian forces 382, 392, 413, 437–8, 470
Tregarthen, Lieutenant Arthur 266, 288
Trenerry, Lieutenant Aldebaran 105
Tsukamoto, Lieutenant Colonel Hatsuo 1–2, 3, 90, 425, 476
Turnbull, Claude 249–50
Turnbull, Private Clive 18
Turner, Lance Corporal Albert 'Pompey' 161

U
Ua-Ule Creek 114, 253, 255, 271, 388, 405, 431
Uberi 20, 21, 35, 72, 74, 92, 94–5, 185, 207, 302, 318, 319, 333, 350, 400, 436, 457, 477
   medical personnel, withdrawal of 113
   reinforcements 228, 230–1
   supplies 198, 201
ulcers, tropical 343
United States Forces 375, 442
   3rd Bombardment Group 101
   126th Regiment, 32nd Division 244, 459, 460
   128th Regiment, 32nd Division 13
   air power 101–2

INDEX

briefings from Australian troops 460
  Guadalcanal 345–6, 365–6
Upcher, Lieutenant Robin 349, 385
Utsumi, Warrant Officer 49
Uyama, Captain 426, 476

V

Vasey, Major General George 12, 116–17, 401, 462–3
Veale, Private Lawrence 215
Vernon, Captain Geoffrey 'Doc' 34–5, 61, 69, 112, 337, 380, 406, 457–8, 477
  evacuation of wounded 93–4, 112, 165–6, 175–6, 199
  Ioribaiwa 222–3, 315–18
Vial, Private Eric 185
Victoria Cross (VC) 51, 143, 450
  Kingsbury, Bruce 450
Vidler, Private Oscar 267
Viner-Smith, 'Doc' 247

W

Wairopi 245, 407
Walker, Brigadier General 429
Walker, Lieutenant Bruce 403–4, 417
Waller, Corporal Lewis 212, 483
Wallman, Captain Douglas 21, 113
Wanigela 400, 429, 439
Wanigela concept 400
Ward, Corporal Stan 302
Ward, Lieutenant Colonel Kenneth 34
Ward, Private Albert 121
Warhurst, Sergeant Dudley 311
Watanabe, Private Toshi 377–8
Waters, Private Julian 262, 321
Watson, Lance Corporal Robert 158
Watson, Major William 175–6
Watson, Roy 28
Watts, Lieutenant Ross 249–51, 279
Weale, Lieutenant William 253, 263, 451
Westbrook, Corporal Clive 382
White, Major Denis 19–20, 31, 56
White, Osmar 18, 19, 60, 73, 260
Whitfield, Private Harry 277
Wild, Private Raymond 380
Wilkinson, Captain Robert 484, 485, 486, 487
Wilkinson, Private John 'Norm' 104
Wilkinson, Sergeant Jack 35, 166
Williams, Corporal Leonard 381, 397, 415, 449, 460

Williams, Lieutenant Vivian 148
Williams, Sergeant Pel 277–8
Willoughby, Major General Charles 398, 399, 430
Wilmot, Chester 18, 19, 60, 73, 260–1, 431–2
Wilson, Sergeant Harry 148
Withy, Lieutenant Colonel Charles 231, 262, 267, 290, 293, 303, 319, 332, 403, 405, 420, 423, 431, 468, 475
Woodger, Lieutenant Billy 294, 295
Woodland, George 201
Wright, Captain George 105, 139, 144–5, 146

Y

Yamamoto, Superior Private Yasuhide 298
Yazawa, Colonel Kiyoshi 2, 5, 42–4, 46, 258
Yeo, Private Charles 370
Yodda 15
Yokoda, Corporal 313

Z

Zanker, Private Alfred 362, 373, 381–2, 392, 397, 415, 428, 438, 448, 459–60, 471, 477, 484–8

ABOUT THE AUTHOR

David Cameron is a Canberra-based author who has written several books on Australian military and convict history, as well as human and primate evolution, including over 60 internationally peer-reviewed papers for various journals and book chapters. He received 1st Class Honours in Prehistoric Archaeology at the University of Sydney and later went on to complete his PhD in palaeoanthropology at the Australian National University. He is a former Australian Research Council (ARC) Post Doctorial Fellow at the Australian National University (School of Archaeology) and an ARC QEII Fellow at the University of Sydney (Department of Anatomy and Histology). He has participated and led several international fieldwork teams in Australia, the Middle East (Turkey, Jordan, Israel, and the United Arab Emirates), Europe (Hungary) and Asia (Japan, Vietnam and India) and has participated in many conferences and museum studies throughout the world.

More from Big Sky Publishing

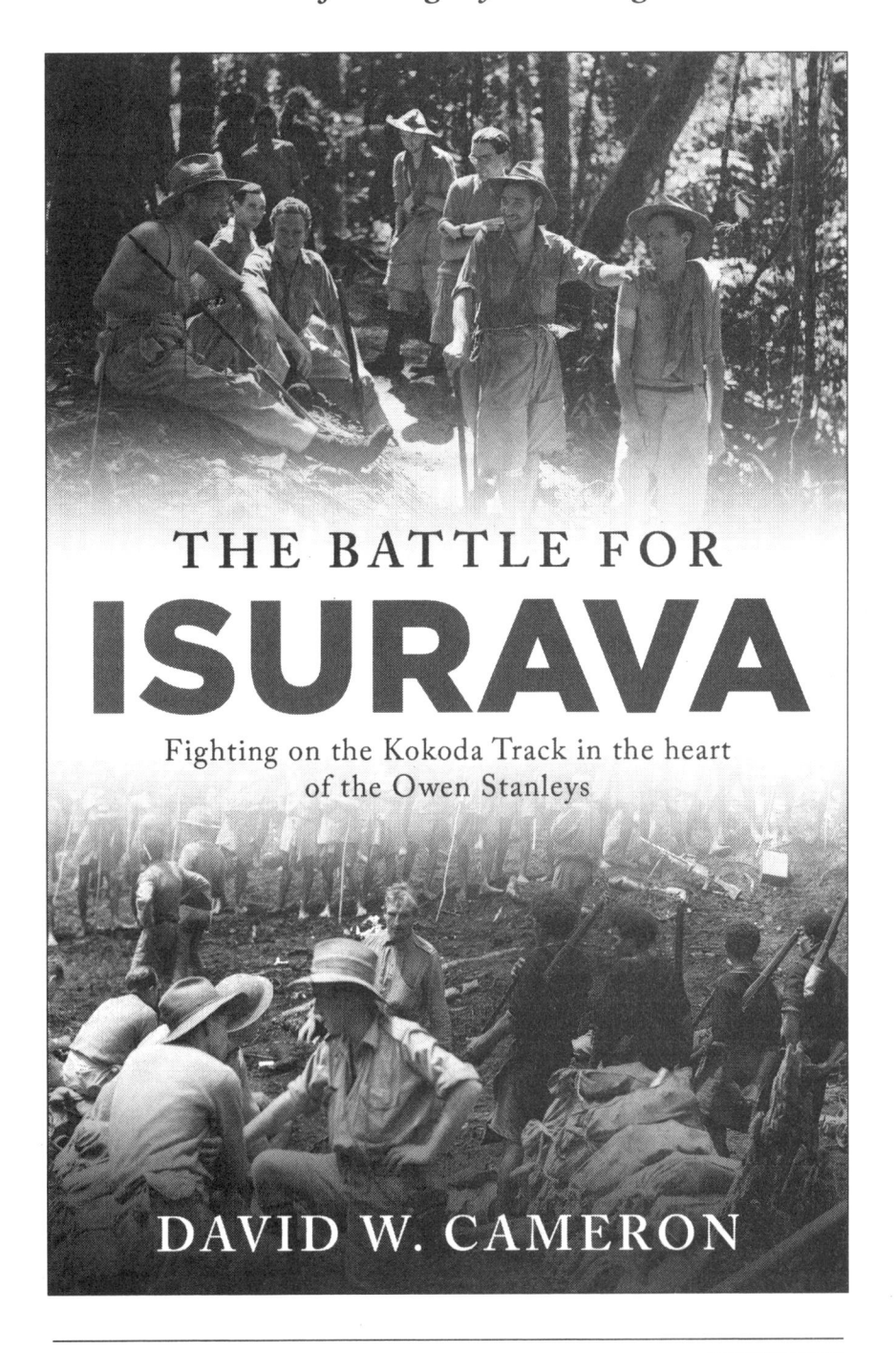

THE BATTLE FOR
ISURAVA

Fighting on the Kokoda Track in the heart of the Owen Stanleys

DAVID W. CAMERON

View sample pages, reviews and information on this book and other titles at
www.bigskypublishing.com.au

For more great titles visit
www.bigskypublishing.com.au